\#15-10 BK Bud   Mar. 2010

Local
Environmental
Movements

日米環境問題の比較

Local Environmental Movements

A Comparative Study of the United States and Japan

Edited by
Pradyumna P. Karan
and Unryu Suganuma

Cartography by Dick Gilbreath

THE UNIVERSITY PRESS OF KENTUCKY

Scholarly publisher for the Commonwealth,
serving Bellarmine University, Berea College, Centre College of Kentucky,
Eastern Kentucky University, The Filson Historical Society, Georgetown College,
Kentucky Historical Society, Kentucky State University, Morehead State University,
Murray State University, Northern Kentucky University, Transylvania University,
University of Kentucky, University of Louisville, and Western Kentucky University.
All rights reserved.

Editorial and Sales Offices: The University Press of Kentucky
663 South Limestone Street, Lexington, Kentucky 40508-4008
www.kentuckypress.com

12 11 10 09 08   5 4 3 2 1

Library of Congress Cataloging-in-Publication Data

Local environmental movements : a comparative study of the United States and Japan /
edited by Pradyumna P. Karan and Unryu Suganuma.
    p. cm.
 Includes bibliographical references and index.
 ISBN 978-0-8131-2488-9 (hardcover : alk. paper)
 1. Environmentalism—United States. 2. Environmentalism—Japan. 3. Environmental
protection—United States. 4. Environmental protection—Japan. I. Karan, Pradyumna P.
(Pradyumna Prasad) II. Suganuma, Unryu, 1964-
 GE197.L63 2008
 333.720973—dc22   2008016856

This book is printed on acid-free recycled paper meeting the requirements of the American
National Standard for Permanence in Paper for Printed Library Materials.

♾ ⊛

Manufactured in the United States of America.

 Member of the Association of
American University Presses

Contents

List of Illustrations and Tables   vii
Preface   xi

Part I: Perspectives on Local Environmental Movements

1. Local Environmental Movements: An Innovative Paradigm to Reclaim the Environment   3
   Pradyumna P. Karan and Unryu Suganuma

2. A Comparative History of U.S. and Japanese Environmental Movements   13
   Richard Forrest, Miranda Schreurs, and Rachel Penrod

3. Virtual Grassroots Movements: The Role of the National Geographic Society as a Sustained Promoter of Environmental Awareness   39
   Stanley D. Brunn

4. Going Global: The Use of International Politics and Norms in Local Environmental Protest Movements in Japan   45
   Kim Reimann

Part II: Protesting the Effects of Nuclear Radiation and Chemical Weapons

5. Citizen Activism and the Nuclear Industry in Japan: After the Tokai Village Disaster   65
   Nathalie Cavasin

6. Citizen Advisory Boards and the Cleanup of the U.S. Nuclear Weapons Complex: Public Participation or Public Relations Ploy?   75
   John J. Metz

7. Grassroots Environmental Opposition to Chemical Weapons Incineration in Central Kentucky: A Success Story   111
   David Zurick

Part III: Seeking to Preserve Rural and Urban Landscapes

8. The Role of Local Groups in the Protection of Urban Farming and Farmland in Tokyo   131
   Noritaka Yagasaki and Yasuko Nakamura

9. From Horse Farms to Wal-Mart: The Citizens' Movement to Protect Farmland in the Central Bluegrass Region of Kentucky   145
   Dan Carey and Pradyumna P. Karan

10. Farmers' Efforts toward an Environmentally Friendly Society in Ogata, Japan   165
    Shinji Kawai, Satoru Sato, and Yoshimitsu Taniguchi

11. The Administrative Process of Environmental Conservation and Limits to Grassroots Activities: The Case of Kyoto   177
    Masao Tao

12. The Grassroots Movement to Save the Sanbanze Tidelands, Tokyo Bay   187
    Kenji Yamazaki and Tomoko Yamazaki

Part IV: Seeking to Preserve the Natural Environment

13. Citizens for Saving the Kawabe: An Interplay among Farmers, Fishermen, Environmentalists, and the Ministry of Land, Infrastructure, and Transport   207
    Todd Stradford

14. The Efforts of Japan's Citizens and Nongovernment Organizations to Maintain People-Wildlife Relations in Rural Japan: A Case Study of Monkeys in Mie Prefecture   219
    Kenichi Nonaka

15. The Grassroots Movement to Preserve Tidal Flats in Urban Coastal Regions in Japan: The Case of the Fujimae Tidal Flats, Nagoya   229
    Akiko Ikeguchi and Kohei Okamoto

16. The Protection of the Shiraho Sea at Ishigaki Island: The Grassroots Anti–Ishigaki Airport Construction Movement   245
    Unryu Suganuma

17. The Management of Mountain Natural Parks by Local Communities in Japan   259
    Teiji Watanabe

Part V: Protesting the Effect of Military Activity

18. Antimilitary and Environmental Movements in Okinawa   271
    Jonathan Taylor

19. Grassroots Participation in Hawaiian Biodiversity Protection and Alien-Species Control   281
    Christopher Jasparro

List of Contributors   295
Index   297

Illustrations and Tables

Photographs

5.1.   JCO Corp 69
5.2.   Location of the accident site at the JCO plant 69
5.3.   Bullet train station, Tokai Village 69
5.4.   The Tokai antinuclear activist Yatabe Yuko 69
5.5.   The Tokai antinuclear activist Fujii Gakusho 70
5.6.   The Mito antinuclear activist Nemoto Gan 71
6.1.   Meeting of the advisory panel 94
6.2.   Fernald nuclear facility, 1988 99
6.3.   Fernald nuclear facility after cleanup, 2006 100
7.1.   Underground igloos store chemical weapons awaiting safe disposal 112
7.2.   Kentucky citizens' demonstration against incineration 114
7.3.   Ammunition inspector among M-55 rockets with nerve agent payloads 115
7.4.   The active involvement of local environment groups 116
8.1.   A part of Nerima Ward in 1948 132
8.2.   A part of Nerima Ward in 1992 133
8.3.   Private School for Green and Agriculture 136
8.4.   A vegetable stand using coin-operated lockers 139
8.5.   An urban farm surrounded by homes 140
8.6.   Farmers' shop operated by the agricultural cooperative 141
8.7.   Photographs of farmers supplying produce 141
8.8.   A ward resident's farm 141
8.9.   A citizen's farm 142
8.10.   Agriculture experiment farm 142
9.1.   Commercial and residential development 145
9.2.   Kentucky Horse Park 149
9.3.   A former horse training track 150
9.4.   Scattered rural estate development 156
9.5.   The design of new four-lane Paris Pike 158
9.6.   Protected farmland east of Lexington 160
11.1.   Old Kyoto 178
11.2.   Traditional houses, Kyoto 178
11.3.   Kyoto Tower and Station Building 183
11.4.   Kyoto Station Building 183
11.5.   The huge Isetan Department Store 183

12.1.   An aerial view of the Sanbanze Tidelands 188
13.1.   Flooding along the Kuma River in Sakamoto Village 207
13.2.   Artist's rendition of the Kawabe Dam 208
13.3.   Farmers protesting water fees from the new dam 210
13.4.   Information dispersal on the streets of Fukuoka 212
13.5.   Gathering signatures in Kumamoto City 212
13.6.   Lecturing during a break while planting trees 213
13.7.   Junior/senior high school being built 217
14.1.   Japanese monkey (*Macaca fuscata*) 220
14.2.   Rocket-type fireworks driving monkeys away 221
14.3.   Examples of crop damage caused by monkeys 222
14.4.   An antimonkey cage 223
14.5.   Information displayed on a cellular phone 225
14.6.   Web site information displayed on a computer 225
15.1.   The Fujimae Tidal Flat within the industrial complex 234
15.2.   One of the life forms inhabiting the Fujimae Tidal Flat 235
16.1.   Shiraho coral reefs 247
16.2.   Shiraho Sea floor with rich marine life 247
16.3.   An anti–airport construction meeting 247
16.4.   The APYSCR appeals to the public to save the Shiraho Sea 249
16.5.   An anti–Karadake Plan meeting 254
17.1.   Boardwalk constructed as a big-budget public works project 261
17.2.   Boardwalk constructed by local volunteers 262
18.1.   The effects of artillery bombardment 274
18.2.   Village of Henoko near the proposed site of the heliport 278
19.1.   Cattle damage to native koa trees 282
19.2.   Volunteer field crew at Hakalau National Forest Reserve 285
19.3.   Lines of koa trees planted by volunteers 285

Figures

1.1.   Location of case studies in Japan and the United States 6
4.1.   Nagara River Estuary Dam 51
4.2.   Isahaya Bay Land Reclamation Project 55
5.1.   Tokai Village 66
6.1.   The U.S. nuclear weapons complex 76
7.1.   Chemical weapons storage sites in the United States 112
7.2.   Bluegrass Army Depot 113
8.1.   Farmland in the central twenty-three wards of Tokyo 134
8.2.   Three levels for sustaining urban farming and farmland 137
8.3.   Organization of urban farming and farmland in Nerima Ward 143
9.1.   Percentage of nonfederal land developed in Kentucky 146
9.2.   Farming on the edge 147
9.3.   Fayette County, Kentucky. Growth of the urban area 148
9.4.   Cost of community services 149

9.5.   Core agricultural and rural land 157
9.6.   Determining the value of development rights 158
9.7.   Clustered and scattered development 159
9.8.   PDR-protected farms 160
10.1.   Akita Prefecture 166
10.2.   Ogata Village 167
10.3.   Birth of Ogata Village 167
11.1.   Kyoto 180
12.1.   A woodblock print of what is now the Tsukuda district 189
12.2.   Land reclamation: the Tokyo scheme 190
12.3.   Land reclamation: the Chiba scheme 190
12.4.   Mitsui Real Estate Company's investment in reclamation 191
12.5.   Stages in the reclamation of Tokyo Bay 193
12.6.   The reclamation of Tokyo Bay: number of fishermen 194
12.7.   Meiji era map of the Sanbanze Tidelands 195
12.8.   Reclamation area and present coast of Funabashi 197
12.9.   Value of fish catch at Funabashi 198
12.10.   Structure of the grassroots movement to preserve Sanbanze 203
13.1.   Kawabe River Dam site and Kumamoto Prefecture 208
13.2.   Itsuki, households and population 209
13.3.   Articles on the Kawabe River 214
14.1.   Location of Mie Prefecture 221
14.2.   Distribution of monkey groups in 2002 shown by SIS 224
14.3.   The tripartite collaboration 226
14.4.   Relations among monkeys and people 227
15.1.   Fujimae Tidal Flat 230
15.2.   Change tidal-flat area 231
15.3.   Change tidal-flat area 233
15.4.   Articles about Fujimae Tidal Flat reclamation 238
16.1.   Potential new airport sites 246
16.2.   Analysis of the APYSCR's activities 248
17.1.   Mountain national parks mentioned in the paper 260
17.2.   A cross section of a deteriorated trail 260
17.3.   Present and proposed future management structures 265
18.1.   Military bases in Okinawa 272

Tables

6.1.   Major facilities of the U.S. nuclear weapons production complex 78
6.2.   Operational arrangements of SSABs 81
6.3.   Nuclear testing impacts 86
6.4.   Human experiments 87
6.5.   Health impacts on workers at nuclear facilities 88
6.6.   Processes leading to the establishment of citizen advisory boards 91
6.7.   SSAB guidelines 92

6.8.   Conflicts between the DOE and SSABs 97
7.1.   Chemical stockpiles and citizens' movements 121
10.1.   Environmentally friendly agriculture practices 165
12.1.   Reclaimed land 192
14.1.   Information on discovered monkeys 225
16.1.   Economic effects in Yaeyama County 253
16.2.   Ishigaki airport status 255
16.3.   Cargo freight at Ishigaki Harbor 255
16.4.   Passenger and cargo flights from different routes 256
19.1.   OISC volunteer support 286
19.2.   OISC volunteer administrative support 287
19.3.   Volunteer support of plant-eradication efforts 290
19.4.   Volunteer/administrative support 290

Preface

Japanese and American citizens are taking threats to the environment more seriously than ever and looking for ways to make a difference. In the face of an onslaught of pollution and depletion of natural resources, it may seem hopeless to mitigate the damage to the environment. But local groups in both Japan and America are reluctant to surrender the environmental fates of their communities and resources. Citizens in the two largest economies in the world are attempting to preserve the environmental futures of their specific locales. Environmentalism has waxed and waned in importance in the political ecologies of Japan and America, but it appears to be on the upswing now. The number of local environmental groups in both countries is up. These groups have confronted the sociopolitical structures in both countries in an effort to preserve local environments and ways of life. They include in their numbers old conservationists, young deep ecologists, former socialists, local activists, and concerned citizens.

These local environmental movements may be viewed as a new kind of politics, replacing the old politics of the industrial society with a new postmaterial, value-oriented politics. These movements emphasize a new, postmodern environmental consciousness and call for new relations between society and nature, emphasizing the combination of thought and action. These movements are slowly progressing toward defining a model of development to replace the current one, which has created severe ecological instability in various parts of the world.

The extent of the development of local environmental groups not only is an important intellectual question but also is important for understanding public policy that might be adopted for the environment. Recognizing their common intellectual and policy problems, a team of scholars from Japan and the United States held an international conference at the University of Kentucky in the spring of 2003. The essays in this book, based on papers presented at the conference, focus on the complex dynamics of local environmental groups' influence in Japan and the United States.

Part I of the book includes chapters on local environmental movements (Karan and Suganuma), the history of environmentalism in Japan and the United States (Forrest, Schreurs, and Penrod), the role of the National Geographic Society in promoting environmental awareness (Brunn), and links between local and international movements (Reimann). Part II of the book consists of case studies of local movements that were formed to save environments from nuclear radiation and chemical weapons disposal (Cavasin; Metz; Zurick). Part III deals with movements aimed at protecting rural and threatened urban landscapes (Yagasaki and Nakamura; Carey and Karan; Kawai; Sato and Taniguchi; Tao; Yamazaki and Yamazaki). Part IV discusses movements aimed at preserving features of the natural environment, such as streams, wildlife habitats, tidal flats, coral reefs, and national parks (Stradford; Nonaka; Ikeguchi and Okamoto; Suganuma; Watanabe). Part V provides case studies of movements involving the military and environmental protection (Taylor; Jasparro).

The essays in this book explore the challenges facing environmental movements in Japan and the United States and tackle questions that are crucial to understanding the influence of citizen groups on local policy. For instance, what are the opportunities and constraints facing Japanese and American environmental groups? What distinguishes cases in which citizens were successful from cases in which they did not achieve their goals?

A conference of this intellectual scope and vigor could not have happened without the assistance of many individuals and the support of a number of key institutions. Our appreciation goes to the Vice President for Research at the University of Kentucky and to the College of Arts and Sciences for generous financial assistance. The Department of Geography at the University of Kentucky and the Asia Center also deserve recognition for their help in organizing the event that produced the essays in this book. Finally, we would like to extend our heartfelt thanks to participants of the conference, who so graciously accepted our invitation to share their knowledge of this fascinating aspect of grassroots democracy to save threatened landscapes and environments.

Part I

Perspectives on Local
Environmental Movements

Local Environmental Movements

An Innovative Paradigm to Reclaim the Environment

Pradyumna P. Karan and Unryu Suganuma

The environment is under siege nearly everywhere. Residential, commercial, and industrial development threatens national parks, streams, wildlife habitats, tidal flats and coral reefs, urban landscapes, and farmland in both Japan and the United States. No single private or public entity can counter this trend; the answer lies in a creative partnership involving local people, citizens' groups, and governments at all levels. Strengthening local environmental groups and local governments can help protect and preserve the environment as a whole.

There are many local movements organized to pursue through collective action a variety of goals. The focus of this volume, however, is on those movements in Japan and the United States organized in response to environmental problems. These local environmental movements have geographic attributes. They originate in particular places. Their planning processes and subsequent campaigns occur in specific locales. The very existence of a local organization is rooted in notions of space and place—particularly of place. People develop strong emotional attachments to their local environments (Tuan 1974; Buttimer and Seamon 1980). Their sense of place is often especially acute when beloved environments—rivers, tidal flats, traditional urban landscapes, natural landscapes, or farmlands—are threatened or under attack. The chapters in this volume argue that local activism (1) creates a new focus on environmental concerns and ecosystem protection, (2) integrates a complex set of local issues that contextualize local movements, making it easier for such movements to coalesce with and strengthen national and regional environmental organizations, and (3) starts a process of local mobilization that propels future activism and appears to be effective in redirecting development policies toward both protecting the environment and enhancing social opportunities.

Local communities in Japan and the United States are facing a number of similar environmental issues, including the loss of farmland, the disposal of chemical waste, the problems associated with the use of nuclear power, and the preservation of historical-cultural landscapes and wildlife habitats. All these issues are being tackled by local groups, as evidenced, for example, by myriad local newspaper reports and, more specifically, a recent guidebook of citizens' movements in Japan's Miyagi Prefecture that includes contact details for seventy-five environmental groups (Sendado Kurabu 1991). The environmental movements discussed in this volume represent very diverse networks of people. Some of them are relatively loosely organized and some very highly organized. Most of them promote a holistic view emphasizing the preservation of nature and the restraint of unsustainable development. Many had their origins in the 1970s and 1980s.

Environmental issues are no longer the preserve of a few visionaries or intellectuals, having begun to engage the consciousness of the millions in the United States and Japan who fear the excessive cost of uncontrolled development, which non-Western countries other than Japan began to see in the 1980s as whole ecosystems came under threat. Transnational companies, forbidden from polluting their own backyards, used developing countries as dumping grounds for their industries' most poisonous by-products and rode roughshod over attempts to regulate their destructive practices. As a result, local environmental movements in Asia and Africa have allied themselves with anti-imperialist movements.

Grassroots movements with local priorities should be distinguished from such worldwide movements with global priorities as the Green Movement. Local, grassroots movements engage thousands of people by offering them the opportunity to *participate,* whether in consensus decisionmaking, media events, focus groups, or nonviolent protests. The issues raised by the grassroots movements are, and will remain, of genuine concern for local communities. For example, when it turned out that neither market forces nor government bureaucracies were willing or able to step up and do anything about environmental degradation—first noticed in the early 1970s and the result of the postwar boom in both the United States and Japan—citizens' groups in both countries emerged to take up the fight. In general, such groups have shown that the problem of environmental degradation can be tackled in the same manner as the problems of sustainable development and human rights.

Despite the fact that, in this first decade of the twenty-first century, national and international groups have dominated the headlines, local environmental movements are among the most vibrant, diverse, and powerful social movements on the planet. Their global distribution allows us a window on the multitude of ways in which different cultures mobilize. This volume is based on the premise that there is no one, single environmental movement and that the differences between the many environmental movements far outweigh their similarities.

In contrast to large nonprofit organizations, which represent national constituencies, grassroots support and advocacy groups represent local constituencies. And the accountability relations within these grassroots organizations flow from the bottom up rather than from the top down. The balance of power between the leadership and the rank and file often ebbs and flows over time, the two not necessarily seeing eye to eye at all times. Still, in order to maintain its control, the leadership must make sure that, by and large, it represents group members' wishes.

There are, of course, top-down influences on local movements as well. Often, such movements address problems that are global in scope and attempt to implement solutions that were first proposed in the international activist arena. And they are also affected by their own national political and economic cultures. The United States in particular has proved to be fertile ground for environmental protest, particularly with regard to uncontrolled development. But this should come as no surprise to anyone since the time of Alexis de Tocqueville, the French political thinker and historian who more than two hundred years ago interpreted protest movements as the products of the politics of hope. That is, there must be, as there is in the United States, the assumption in place that certain rights do, in fact, exist before people will protest attempts to curtail those rights (Pakulski 1991, xxi).

Social movements, by their very definition, are amorphous political forms, living entities often undergoing rapid change and, therefore, often unable to be characterized by the methods of traditional institutional history. Social movements focusing on the environment are especially difficult to characterize because they so often function extra-institutionally, beyond the realm of government and nongovernment organizations, churches, and corporations. It is this latter factor that makes researching environmental social movements fascinating. What is revealed is politics in the raw, the nature of power relations and decisionmaking processes at the beginnings of issue cycles, before formalization. It was within environmental movements that a modern environmental sensibility was born in the 1960s in both Japan and the United States.

Studying local networks, informal groups, and developing movements demands somewhat different methodological techniques than does studying established institutions and corporate bodies. That is, while both primary and secondary sources on grassroots activism have been compiled and collected, much of the relevant research material has not, necessitating more than the usual amount of ethnographic fieldwork and ferreting out of supporting documentation. All the case studies in this collection are, thus, premised on the need to find better ways to engage the reality of environmental politics and, thus, it is hoped, better ways to resolve, even if only partially, environmental issues/problems.

While discussing specific movements, the essays deal with four key features identified by Offe (1985): (1) the actors in those movements (e.g., farmers, fishermen, women, the general public); (2) the goals of the movements (e.g., the protection of environmental interests); (3) the values of the movements (e.g., an ecological view of place); and (4) the mode of action undertaken (e.g., the advocation of local self-determination). Often, the movements are in conflict with technocrats, private entrepreneurs, and governments. Most of the essays reveal that integrating multiple objectives without losing track of the initial ground-level protests is vital to a movement's success.

The movements discussed in this volume (see figure 1.1) fall into four categories: those protesting the effects of nuclear radiation and chemical weapons; those seeking to preserve rural and urban landscapes; those seeking to preserve the natural environment; and those protesting the effect of military activity. The strength and popularity of these movements can be attributed to their long history, objectives, committed leaders, and stature at the local and regional levels. It is especially to their credit that, in the face of the serious challenges presented by the popularity of the backers of development among the media and international agencies, they have kept concern for the environment and awareness of environmental issues alive at the grassroots level.

Perspectives on Local Environmental Movements

As background and context for case studies of local environmental movements in Japan and the United States, Richard Forrest, Miranda Schreurs, and Rachel Penrod, the authors of chapter 2, analyze the context and history of local environmental movements in the two countries, contrasting the movements themselves as well as the cultural, religious, political, legislative, and technological influences that have made environmentalism distinctive in both countries. The environmental movement in Japan is in some ways different from those in the United States and other Western nations. Japanese movements are victim oriented, concerned more with the negative impact on people than with the impact on the natural environment (McKean 1981). The goals of the movement are, therefore, very focused, the solution of specific local problems. American environmentalism is, by contrast, more broadly concerned with the preservation of the environment in its natural state, even when specific local issues are involved. Japanese environmental groups use a variety of methods—for example, direct action, petitioning, and filing lawsuits—in their effort to save the local environment. Unlike American groups, these groups are not hierarchically structured, which makes them

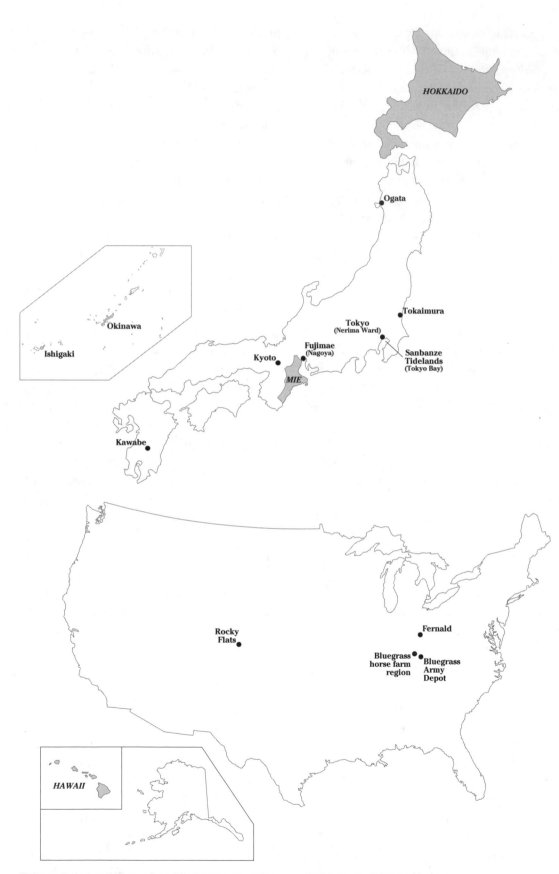

Figure 1.1. Location of case studies in Japan and the United States.

more difficult to integrate into existing power structures. As a result, the Japanese environmental movement is far less unified than is that of the United States. Yet some Japanese movements have had a striking and unique impact on policymaking in Japan.

The last quarter of the twentieth century was a time of fertile intellectual activity, and many of the leading environmental thinkers developed ideas and skills during this period that shaped the modern environmental movement. The National Geographic Society is one example. It developed a powerful theory—fully articulated in many articles in *National Geographic,* the society's monthly magazine—that humans and nature are interdependent but that humans have the power to cause irreversible damage to nature. The images of landscape published in the magazine helped make millions of people aware of the need for environmental protection. *National Geographic* photographs played a powerful role in the modern environmental movement by introducing readers worldwide to the beauty of nature and, subsequently, the need to preserve it. The photographs became concrete manifestations of the objects of environmental protection. In chapter 3, Stanley D. Brunn analyzes the role of the National Geographic Society in promoting environmental awareness among the readers of *National Geographic.*

By the beginning of the 1990s, environmental NGOs had become increasingly important actors in national and global environmental politics. They began routinely attending and influencing conferences dealing with environmental issues. They have found points of leverage that they can often bring to bear to help resolve problems. Flexible groupings of NGOs, local, national, and international, have proved crucial for progress. The complementary roles played by different types of NGOs can add more to an environmental movement than mere members of participating groups. In chapter 4, Kim Reimann explores the emergence and growth of local environmental movements in Japan and the connection between local and global social movements, showing how grassroots activists and NGOs in Japan were influenced by global trends and seized new international support to advance their cause.

The Case Studies

Protesting the Effects of Nuclear Radiation and Chemical Weapons

Nuclear power plants and their reactors dot the American and Japanese landscapes. In the United States, there are also several chemical weapons storage facilities. In fact, the United States and Japan are the first- and third-ranked countries in the world in terms of number of nuclear reactors (including those producing weapons in the United States). There were 118 reactors in the United States in 2000 and 55 in Japan in 2006. The 55 reactors in Japan—which generate a little more than 30 percent of that country's electricity—are located in an area the size of California, many within 150 kilometers of each other, and almost all built along the coast (where seawater is available to cool them). Many of the reactors in Japan and a few of those in the United States have been sited on active seismic faults. Those so situated in Japan are located in the subduction zone along the coast, where major earthquakes (magnitude 7 or greater on the Richter scale) occur frequently. There is almost no geologic setting in the world more dangerous for nuclear power than Japan.

The performance of Japan's nuclear power plants has been exemplary during earthquakes; either they have sailed through unruffled, or reactors have shut down automatically. Nonetheless, human error and cover-ups have generated unease. Japan's Monju reactor, located in Tsuruga, Fukui Prefecture, has been shut down since it caught fire following a sodium coolant leak on December 8, 1995. An accident on September 30, 1999, at a nuclear processing plant in Tokai Village, Ibaraki Prefecture—caused by employee carelessness—killed two workers and exposed more than 660 nearby residents to radiation. In 2002, details of falsified inspection and repair reports forced the temporary closure of seventeen plants. And, in 2004, steam from a broken pipe that had not been inspected for two decades killed five workers in western Japan. In April 2006, the governor of Aomori Prefecture approved the location of Japan's first commercial nuclear fuel reprocessing plant in the prefecture. In the town of Rokkasho at the northeastern tip of Honshu Island, Japan is working to get around the limits of the uranium supply. Inside a new $20 billion complex, workers service cylindrical centrifuges for enriching uranium and a pool partly filled with rods of spent fuel. Spent fuel is rich in plutonium and leftover uranium—valuable nuclear material that the plant is designed to salvage and reprocess into a mixture of enriched uranium and plutonium that can be burned in some modern reactors and could stretch Japan's fuel supply for decades or more. Due to start in 2007, the Rokkasho complex includes housing for the inspectors from international agencies whose job it is to ensure that the plant is put to entirely peaceful purposes. Such a safeguard does not satisfy the opponents of nuclear energy. Deep suspicion of nuclear energy and its regulation remains in the country. Local opposition has, since 2003, forced three utilities to shelve plans for new nuclear power plants. In chapter 5, Nathalie Cavasin deals with citizen activism since the Tokai accident.

In chapter 6, John J. Metz examines the involvement of the public in dealing with safety and severe environmental problems at major nuclear weapons–production sites in the United States. During the early 1990s, the U.S. Department of Energy (DOE) radically altered its relationship to local citizens at these sites by establishing a variety of public participation programs, including citizen advisory boards. Many of the nuclear weapons–production sites had been closed in the late 1980s because of severe environmental and safety problems. The DOE's move to incorporate local citizens in its decisionmaking process was a complete reversal of the policy prevailing for the previous five decades, when no information whatsoever was shared with the public.

Metz indicates that the DOE's history of neglect and deception with regard to environmental and safety issues led many local groups to question the sincerity of its commitment to citizen involvement. He traces the conditions that led to the establishment of the local advisory boards and determines that the program has met with mixed results: boards at sites that were being closed were more successful than those at sites with ongoing programs. But even successful boards had no real power, the DOE being under no obligation to follow their advice. And the Bush administration is curtailing public participation as it begins producing new weapons and building the facilities required to manufacture them. Some local boards are, however, complaining. And the pool of informed and concerned citizens is larger than it has been in the past and ready to challenge the government.

As the chapters by Cavasin and Metz indicate, in both Japan and the United States citizen involvement helped fuel the public debate about nuclear energy by dramatizing the catastrophic risks involved. It has also helped speed up the symbolic greening of private and public life. Even industry realizes the importance of maintaining environment-friendly operations. Environmental protection has been firmly established in the policies of large Japanese and American companies. Citizen involvement has helped make environmental management central to life in the twenty-first century.

Over the past year, the high price of oil

Pradyumna P. Karan and Unryu Suganuma

and natural gas and a growing concern with global warming have lodged the debate over nuclear power in Japan and America even more firmly in the environmental arena. Some environmental activists have taken the position that global warming requires an embrace of new nuclear plants because, unlike gas- or coal-powered plants, they produce electricity without emitting greenhouse gasses. And of course the nuclear industry has seized the opportunity to promote itself as the green energy option. Most environmental groups, however, have not softened their opposition to the nuclear industry.

Despite the hazards posed by nuclear power plants, their abandonment would pose a tremendous challenge for both the United States and Japan. Overall, the 103 nuclear reactors active in the United States today supply about 20 percent of the nation's electricity. Some states of course are more dependent than others: New York gets 29 percent of its power from nuclear energy, New Jersey 52 percent. A substantial amount of Japan's electricity also comes from nuclear energy. The challenge facing both countries is finding a clean power source to substitute for nuclear energy—before what some activists see as the inevitable catastrophic accident occurs.

Thirty miles to the south of Lexington, Kentucky, some of the world's most lethal chemical weapons are stored in earth-covered bunkers at the Bluegrass Army Depot, protected from terrorists and monitored for leaks by the army. An international treaty obligates the United States to destroy such stockpiles by 2012. To do so, the army built incinerators in Anniston, Alabama, and Tooele, Utah, and on remote Johnston Atoll in the South Pacific. But the Kentucky chemicals are too dangerous to be transported in a safe and environmentally friendly manner. After widespread opposition to both movement of the materials and incineration on-site, the Pentagon agreed to build a $2 billion plant to chemically neutralize them.

In chapter 7, David Zurick discusses this successful grassroots movement.

Seeking to Preserve Rural and Urban Landscapes

Farmland and open space are essential natural resources but ones that are both limited and fragile. The threats to open space in the United States have been well documented. The forests of the Pacific Northwest, the rangelands of the Rocky Mountain West, the wetlands of Florida, the farmlands of Virginia, New Jersey, and Kentucky, all are under heavy pressure from residential and commercial development. Both in Japan and in the United States, considerable areas of farmland are lost to nonfarm use. This fact has been the source of innumerable headline stories since 1975, when the farmland issue first began to come into public view. No single private or public entity can counter the trend; the answer lies in innovative partnerships involving local groups, landowners, private conservation groups, and governments at all levels.

The social, economic, and cultural pressures that shape Japanese and American society are different, and these differences affect land use and the response to land loss in the two countries. In chapter 8, Noritaka Yagasaki and Yasuko Nakamura discuss the role of local groups in protecting urban farming and farmland in the Tokyo region, while, in chapter 9, Dan Carey and Pradyumna P. Karan describe preservation tools that local groups in the Bluegrass region of Kentucky have employed to save the horse farms and farmland of central Kentucky. In chapter 10, Shinji Kawai, Satoru Sato, and Yoshimitsu Taniguchi examine two distinctive grassroots movements in Ogata (Akita Prefecture) to grow organic rice and create an environmentally friendly farming society. Farms in Ogata are shifting toward organic production because consumers willingly pay much more for organic food and local groups have raised the land-stewardship ethic.

The high cultural value of the traditional-style buildings in Kyoto, Japan's old capital city, has long been recognized, representing as it does the nation's architectural heritage. They attract visitors from all over the world. But the cityscape is changing rapidly with the construction of large glass and concrete buildings. In chapter 11, Masao Tao discusses the urban conservation movement in Kyoto, highlighting conflicts between development and preservation interests. Unfortunately, the development interests seem to be winning. This is partly because the policymaking process suffers from structural problems that result in a disregard for both particular circumstances and government authority.

Tidal lands along the shores of Tokyo Bay are easy to reclaim and have long been favored for urban development. A good example is the post–World War II reclamation of a seashore area by the Chiba prefectural government to create the Keiyo industrial complex. Nevertheless, the prefectural government has called a halt to its reclamation program and decided, instead, to revive the Sanbanze Tidal Flats. Sanbanze covers an area of about sixteen hundred hectares in the innermost part of Tokyo Bay, fronting the cities of Funabashi, Ichikawa, and Urayasu. Many species of marine creatures live there. The tidal flats are also a wintering site for ducks and other migratory birds. In chapter 12, Kenji Yamazaki and Tomoko Yamazaki discuss the grassroots effort to save Sanbanze.

Seeking to Preserve the Natural Environment

River-management policy in Japan since the Meiji era has mainly relied on dam construction to serve irrigation and flood-control needs (Takahasi 2004). The 1970s saw the rise of anti–dam construction movements that intensified in the 1980s and 1990s. There is, in fact, considerable evidence of the damage caused by dam construction: the deterioration of water quality, sedimentation in dammed lakes, and a decline in river fishing. Meanwhile the economic slowdown of the 1990s and the decline in population growth reduced the demand for water to a level much lower than previously projected. In chapter 13, Todd Stradford describes the role of the local environmental movement in protecting the Kawabe River from unnecessary dam construction.

Monkey populations are spreading across Japan, a tidy, compact nation more often associated with urban landscapes than with wildlife. From a scant postwar 15,000, the monkey population has increased tenfold in half a century, to more than 150,000 (Brooke 2002). Increasingly, farmers deal with the populations as they would an infestation. Monkeys cause at least $7 million in crop damage each year. The number of monkeys killed by humans has risen during the last twenty-five years to about 10,000 a year. Yet, like coyotes in North America, the monkeys continue to increase. And, with an improved diet owing to a move from the mountains to farmland, they are becoming larger and more aggressive. In chapter 14, Kenichi Nonaka discusses the latest technological attempts to control the monkeys, using Mie Prefecture as a case study.

In chapter 15, Akiko Ikeguchi and Kohei Okamoto discuss the environmental movement to preserve the tidal flats along the coastal areas of Japan using as case studies Fujimae in Nagoya. Tidal flats in Japan have for hundreds of years been the targets of reclamation projects. The success of the grassroots movement to save the Fujimae Tidal Flats, the first case in Japan in which a reclamation project was contested on the grounds of biodiversity, has had enormous significance for the environmental movement in Japan. But more needs to be done if the lessons learned there are to be put into practice systematically and nationwide.

Pradyumna P. Karan and Unryu Suganuma

In chapter 16, Unryu Suganuma discusses the local movement opposing the construction of a new airport on land reclaimed from or in the vicinity of the Shiraho Sea. This new airport, first proposed in the early 1970s, has generated local as well as international opposition because of the likely damage that would be caused to what may be the best preserved coral reef in the archipelago. Better than building a new airport would be the expansion of the existing airport. But expansion doesn't serve the interests of the island's business community and, thus, the local government. At the time of writing, the outcome was unclear.

The national parks in Japan protect the last remaining wilderness in the country. They preserve multiple marvels—rocky peaks, active volcanoes, grassy meadows with alpine flora, rapid rivers, waterfalls, tranquil lakes, beautiful forests, and various species of animals. In chapter 17, Teiji Watanabe discusses the rapid environmental deterioration of national parks, a result of poor management, road construction, and development, and the role of local groups in protecting national parks and developing an improved management plan.

Protesting the Effect of Military Activity

In modern times, the military, through its bases, has often driven economies in that it is the single largest utilizer of resources. Military bases have taken their toll on the entire spectrum of natural resources, often highly visibly so. In chapter 18, Jonathan Taylor examines the antimilitary attitude and the environmental movement in Okinawa, where the plan to move the Marine Corps Futenma Air Station remains mired by opposition from local environmental movements and the prefectural governor. In chapter 19, Christopher Jasparro discusses the participation of local groups in biodiversity protection efforts in Hawaii.

The Future of Local Environmental Movements

Local environmental movements in the twenty-first century, buoyed by their achievements during the last decades of the twentieth century, are attempting to instill a strong current of environmental responsibility into American and Japanese political systems dominated by economic fundamentalism (the idea that an unregulated market is the best and most efficient way to meet human needs). The challenges that they face in this new century remain essentially the same: publicizing their ideas, fund-raising, lobbying decisionmakers, and mobilizing their limited resources as efficiently as possible. Above all, they must continue to avoid adopting an environmental fundamentalism that, in the current political climate, would be considered a direct attack on the system and emphasize instead the local nature of the problems they address.

Despite the creation of new environmental agencies in the United States and Japan, the political will to implement the spirit of environmental protection can, as we have seen, be lacking at the local level. Local governments often give priority to short-term economic interests rather than long-term ecological considerations. Nevertheless, grassroots movements have had some successes. The case studies presented in this volume offer us the opportunity to analyze what made for success in some cases and failure in others. In general success is influenced by a set of opportunities and constraints that can be grouped in two broad categories: (1) the character of the local environmental movement and (2) the political and social context in which that movement operates.

Crucial among the elements of a successful movement's character is effective leadership, that is, leadership that is flexible, determined, and able to read people accurately, strategize and frame issues appropriately, project the right image, and maintain unity. Also important is a movement's ability to demonstrate its technical command of the issue in question and use the media to promote its message. Networking too can be important; often, an active association with professional groups and academics can have a positive influence. Such elements are most important in cases where the political and economic stakes are the highest. And, in such cases, policymakers are often willing to risk environmental damage.

Contextual factors can also contribute significantly to a movement's success. The nature of an issue (e.g., the threat to personal safety posed) is a key element that can jumpstart a movement and power it to victory. Key too can be the support of powerful political figures. Timing can also be crucial, especially when it comes to mobilization efforts, as can international support.

The general trend in both Japan and the United States has been that government is less resistant to a cooperative attitude than a confrontational one. Efforts to change the status quo can easily be constrained by government response. Other factors that influence decisionmakers—and, thus, affect environmental movements—are the power of other interest groups; the close link between approval of development projects, money, and votes; the desire to promote economic growth; and the fear of lawsuits being brought by developers denied permits.

In sum, the case studies in the volume offer several lessons for those wishing to maximize the influence of local environmental movements. First, it is paramount that they look at the entire political and social structure with which they are confronted, obtain expert advice and assistance, and develop a support network. Second, they need to cultivate opportunities assiduously and strategize carefully. Effective leadership and clear, consistent goals can harness public opinion and allow grassroots environmental movements to achieve their goals.

References

Brooke, James. 2002. "Nikko Journal; By Leaps and Bounds, Monkeys Overrun Japan." *New York Times*, April 12.

Buttimer, A., and D. Seamon, eds. 1980. *The Human Experience of Space and Place*. London: Croom Helm.

McKean, Margaret A. 1981. *Environmental Protest and Citizen Politics in Japan*. Berkeley and Los Angeles: University of California Press.

Offe, Claus. 1985. "New Social Movements: Challenging the Boundaries of Institutional Politics." *Social Research* 52, no. 4:817–68.

Pakulski, Jan. 1991. *Social Movements: The Politics of Moral Protest*. Melbourne: Longmans Cheshire.

Sendado Kurabu [Sendado Club], ed. 1991. *Kaettekita Sendado mappu* [The map of Sendai citizen's activities]. Rev. ed. Tokyo: Katatsumurisha.

Takahasi, Y. 2004. "Evolution of River Management in Japan: From a Focus on Economic Benefits to a Comprehensive View." *Global Environmental Change* 14:63–70.

Tuan, Y.-F. 1974. *Topophilia: A Study of Environmental Perception, Attitudes and Values*. Englewood Cliffs, NJ: Prentice-Hall.

Pradyumna P. Karan and Unryu Suganuma

Chapter 2

A Comparative History of U.S. and Japanese Environmental Movements

Richard Forrest, Miranda Schreurs, and Rachel Penrod

The environmental movements of Japan and the United States have quite different histories. Nevertheless, the story of environmentalism in each country can be seen as a centuries-long struggle to define proper relationships among individuals, nature, and society.

The Japanese and American environmental movements have been shaped by the very different geographies, cultures, and political histories of their countries. But there are some similarities between their early stages. In both countries, groups of people, often local residents, responded to the need to preserve forests and other landscapes and to prevent the unchecked spread of hazardous pollution of various kinds that accompanied modernization and industrialization.

Some similarities resulted from a conscious borrowing and cross-pollination of influences. While the American environmental movement has not yet been influenced significantly by Japan's environmental history, the Japanese environmental movement has frequently adopted or followed responses first seen in the United States and other industrialized nations. This is to a large extent due to the influence of efforts begun in Japan during the Meiji Era (1868–1912) to learn from the West. Japan was forcefully pulled out of two centuries of self-imposed isolation in the middle of the nineteenth century by the United States. Responding to the humiliation this caused the country, Japan's Meiji Era leaders set out to modernize the country, an effort encapsulated in the rallying cry, "Fukoku kyohei" [Rich country, strong military].

The government brought in Western advisers and teachers and sent hundreds of Japanese abroad to study everything Western, including science and technology, political systems and constitutions, entire bodies of law, and cultural and social traditions. The success of these efforts to learn from the West made Japan more open to influence from abroad when drafting early environmental laws, including the formation of regulations to address urban air pollution and the development of a national park system.

At the same time, there are many important differences between the early environmental movements and laws of Japan and the United States. Early movements in Japan were almost exclusively locally focused and emerged either to preserve the immediate surroundings or to challenge industrialization that threatened human health and livelihoods. While efforts in the United States also addressed pollution problems, this was accompanied by a concern with the conservation and preservation of nature and wildlife, often in far-removed locations.

While there were numerous cases in which citizens organized to protest pollution in Japan, especially in the face of pollution accompanying rapid economic development in the 1950s and 1960s, these efforts usually focused on negotiating on behalf of affected residents; very few organizations were formed to pursue more abstract environmental goals per se. In the United States, in contrast, local efforts more easily evolved into national movements and organizations that worked to pursue environmental principles and goals; many of these organizations still exist today.

In this chapter, we explore some of the salient historical elements, especially sociocultural and political factors, that have influenced environmental movements in Japan and the United States over time. We provide a brief overview of the key issues, cases, and trends that gave birth to the very dissimilar situations we see today.

The Rise of Environmentalism in the United States

The early histories of the United States and Japan as nations could not be more different. The United States was formed from a group of colonies that rebelled against English kings and the idea of hereditary leadership in a war for independence waged between 1776 and 1783. A formal constitution was established in 1787, one based on the ideas of democracy, open political debate, and the election of political leaders by, and for, the people (albeit conceived, initially, as only white men).

In the century thereafter, a key focus was on efforts to gain control over and settle vast areas of land through development programs and the subjugation of indigenous peoples. In 1803, the Louisiana Purchase greatly expanded the nation's territory, providing an impetus for expansion, settlement, the clearing of land, and the damming of rivers to irrigate farmland. By the early years of the nineteenth century, New England had already been largely deforested (Shutkin 2000, 51–52), and immigration from Europe led to pressures to expand settlements elsewhere. With the passage in 1862 of the Homestead Act, which provided free land to frontier settlers, the government instituted a policy essentially requiring that land be developed for farming or other purposes. Especially after the Civil War, many people moved west to create homesteads, in keeping with the notion of Manifest Destiny,

the belief that the United States was destined to control the continental landmass from the Atlantic to the Pacific, an area some twenty-five times as large as Japan.

There were many factors contributing to the rise of an American environmental movement. A culture of civic activism emerged out of the many self-help societies that formed in newly established towns. The need to work together to survive contributed to the development of the strong civil society famously remarked on by Alexis de Tocqueville in his 1835 *Democracy in America*.

Just as Americans believed that their fate lay in expansion into the West, they were attracted to the bucolic beauty of seemingly untrammeled nature. Writers and painters who helped spread ideas about conservation and preservation found a receptive audience. George Catlin, a Romantic landscape painter, traveled the frontier in 1832 and popularized wilderness areas that would later become national parks. He was an important forerunner of preservationism and originated the national park concept. Ralph Waldo Emerson, whose writings, such as the 1836 essay "Nature," greatly influenced Thoreau and other Transcendentalists, advocated the preservation of natural serenity and simplicity. The naturalist John James Audubon contributed to early conservationism by writing on forest depletion and popularized the appreciation of wildlife through his impressive paintings of birds (Benton and Short 1999, 18, 62–63; Kline 1997; Nash 1989, 36–37).

Others influenced environmentalism for a long time to come, if more indirectly. Henry David Thoreau not only wrote extensively about appreciating and living in harmony with nature (such as in the 1854 *Walden*), but, in the 1848 essay "On the Duty of Civil Disobedience," he also laid the foundation for nonviolent social movements resisting unjust policies—a strain of thought that led, through Gandhi and Martin Luther King, to the later "direct action" tactics of environ-

Richard Forrest, Miranda Schreurs, and Rachel Penrod

mental groups such as Greenpeace. Direct action, bearing witness, and conscientious objection (such as to military service by religious groups such as the Quakers) became important threads in American thinking and the later environmental movements challenging authority.

The advance of the natural sciences in the United States led to increasingly sophisticated arguments for the conservation of soils, ecosystems, and species. George Perkins Marsh's 1864 *Man and Nature* noted the historical effects of unsustainable resource management and advocated the conservation of natural resources in keeping with ecological views. Marsh defined basic principles of conservation, describing scientifically the relationships between soil, water, and vegetation. His work contributed to the establishment of the National Forestry Commission in 1873 (Benton and Short 1999, 70–71).

The key schools of thought in American environmentalism emerged as conservationism and preservationism. Conservationism, heavily influenced by Marsh, was a movement initiated in the 1890s by the forester Gifford Pinchot and later led by Aldo Leopold. The movement embodied Enlightenment ideals of rationality, science, and utilitarianism. Stressing the efficient utilization of resources, particularly sustained-yield forest management, it led to the early creation of government bureaucracies to manage resources, such as the U.S. Forest Service (established in 1905) to manage the vast federally owned forest areas and the Reclamation Service (established 1902, later the Bureau of Reclamation) to undertake the irrigation of arid Western regions that would allow settlement and farming. This was an era that saw a flowering of scientific societies and organizations aimed at the rational study and management of natural resources. These included private scientific societies, such as the American Fisheries Society (1870), the American Forestry Association (1875), and the American Ornithologists

Union (1883) (Clepper 1966, 8–15). Elite outdoor and sportsmen's clubs, such as the Boone and Crockett Club (founded 1885), advocated for game conservation legislation, and their sportsmanship codes acted as a type of conservation policy (Clepper 1966, 13; Dorman 1998, 35). The 1890 U.S. Census Report ignited public opinion in favor of conservationism, calling attention to the quickly disappearing supplies of timber and arable land in the nation (Gottlieb 1993, 21).

In response, the competing philosophy of preservationism, embodied in the views of the naturalist and mystic John Muir (who co-founded the Sierra Club in 1892), was based on the Romantic ideologies of William Wordsworth and Jean-Jacques Rousseau, which held that wild, primitive, and ideal forms of nature deserve preservation in pristine form. Preservationism thus held that nature reserves and parks should be off-limits to resource use, in contrast to the efficient utilization of resources for human benefit championed by the conservationists. American history was to see various seesawing battles between these different attitudes toward the environment and natural resources.

Protecting the Land

The first efforts to set aside natural areas for conservation occurred in cities. In 1853, New York City authorized the creation of Central Park. Not insignificantly, the designer of that park, Frederick Law Olmsted Jr., first proposed (in 1865) that Yosemite should be a protected area.

John Muir was the first nature writer to form his ideas into a political ideology, and he was successful in advocating for the establishment of parks such as Yellowstone and Yosemite, as well as a national park system, and for the preservation of wilderness (Benton and Short 1999, 67–69). Preservationism grew with the popularity of outdoor recreation clubs, and, while influential, it was, like conservationism, essentially the concern of the upper class (Shutkin 2000; Dowie 1995; Hays 1959; Benton and Short 1999; Gottlieb 1993).

The conservation and preservation movements thus both contributed in different ways to the establishment of a national parks system, which began with the creation of Yellowstone National Park, the world's first national park, in 1872. In the United States, the view evolved that the national parks were inviolable: those living in or from the land in the areas designated as parks were relocated when possible (as in the case of Yellowstone, where the indigenous inhabitants were forcibly removed).

By the mid- to late nineteenth century, there were a growing number of environmental groups in the country. The Sierra Club, a California hiking society founded in 1892 by John Muir and Robert Underwood Johnson, lobbied for the protection of wild areas. Even more staid groups such as the National Federation of Women's Clubs mobilized women around conservation and preservation, in 1904 petitioning Congress to save California's Calaveras Grove of Big Trees, and in 1913 winning a long campaign conducted in conjunction with the Audubon Society to ban the use of the feathers of exotic birds in hats (Dorman 1998, 155–57).

The extermination of wildlife led to new policies. In 1900, the Lacey Act, prohibiting interstate commerce in wildlife killed in violation of state law, was passed in response to both public outcry about the extinction of the passenger pigeon through excessive hunting and the lobbying of scientific societies for game management (Anderson 2000).

Political support in high places was also crucial to the early successes of the American environmental movement. The American environmental movement benefited greatly from a president who was an outspoken proponent of conservation. Theodore Roosevelt was a hunter and outdoorsman and a great supporter of Gifford Pinchot's conservationism.

A pioneer of national parks and wildlife refuges, he believed that conservation was necessary to protect the long-range prosperity of the country and that forest and water problems in particular were vital internal security issues. In 1903, Roosevelt joined John Muir in a hike in the Yosemite Valley and became more open to preservationist lobbying, later introducing a bill to prevent exploitation and damage in the park. In 1908, he convened the first Governors' Conference on Conservation and established the National Conservation Commission (Shutkin 2000, 94; Dorman 1998, 164–65).

The environmental movement in the United States was by no means always united in its efforts to set aside land, however. Tensions between preservationists and conservationists, for example, escalated in 1908 when the mayor of San Francisco planned a dam and reservoir in the Hetch Hetchy Valley in central California in order to supply water to the San Francisco Bay Area metropolis. In this case, Gifford Pinchot supported utilization of the resource, while John Muir condemned it as undermining the significant aesthetic value of the valley. The dam was signed into law by President Wilson in 1913, but the controversy was important to the establishment of the National Park Service under Stephen Mather in 1916 (Shutkin 2000, 93–97; Benton and Short 1999, 67–69; Kline 1997, 54–63).

These competing views—of use versus preservation—have colored many local debates. The establishment of the Everglades National Park in 1934 was a victory for preservationists as the area was marked for protection, not so much because it was monumental or awe-inspiring as because it was considered important both ecologically and as a wildlife habitat (Benton and Short 1999, 169–71).

Two major environmental groups to form at this time were the Wilderness Society (1935), a strong advocate of preservationism, and the National Wildlife Federation (1936), which boasted a broad membership, including many hunting groups, and was, thus, a strong advocate of conservationism. Interestingly, the Wilderness Society (founded by the preservationist Robert Marshall and the conservationist Aldo Leopold) was based on the philosophy that social equality was tied to wilderness protection (Clepper 1966, 155–56).

Early Antipollution and Sanitation Movements

In large part, the environmental movement in the United States was focused on the protection of nature and wildlife. This was mostly a reflection of the historical development of a largely agrarian society. By the turn of the twentieth century, however, the United States, like England before it, was grappling with problems associated with urbanization and industrialization. Social reformers active on a wide variety of issues, such as labor and women's rights, began to press for concerted action to address pollution, using scientific evidence (e.g., epidemiological data demonstrating the harm caused by pollution) in the process. Their activities led to the creation of organizations that could influence policy on a national level.

Once again, writers played an important role in shaping public attitudes. Like Charles Dickens in England, Upton Sinclair exposed the appalling conditions faced by industrial workers in his novel *The Jungle* (1906) (Gottlieb 1993, 194–96). Women also played an active role in furthering environmental protection as it pertained to human health and worker safety. In 1889 in Chicago, Jane Addams established Hull House, which became a hotbed of reform activity addressing urban issues, including sanitation, public health, and labor conditions. Another Hull House participant, Alice Hamilton, a physician and professor of industrial medicine, was a pioneer urban grassroots environmental organizer who worked on the problems of industrial diseases and occupational hazards and was

famous for her studies of "phossy jaw" (necrosis of the jaw caused by the phosphorous used in match manufacturing) and widespread lead poisoning (Gottlieb 1993, 47–55).

Grassroots activism often had a political impact. The U.S. Public Health Service was established in 1912 as a response to efforts by urban reformers. The service developed bacteria-based water quality standards and institutionalized professionalism in public health advocacy. Similarly, the International Association for the Prevention of Smoke, the first nationwide association to address pollution control, successfully lobbied for national air pollution controls, resulting, eventually, in the 1955 Air Pollution Control Act (Gottlieb 1993, 55–59).

The Rise of Antipollution and Conservation Movements in Japan

In contrast to the United States, a nation-state created in recent centuries by European colonization, Japan has a culture stretching back millennia. The country's long written history also allows for clearer insight into early environmental activities there, unlike in the United States, where the indigenous populations left no written records before and during the colonization period.

Confined to the Japanese archipelago, the Japanese people successfully exploited the diverse biological riches of their forests, wetlands, and seas to develop a culturally and organizationally advanced society. Pressures on the land, however, increasingly began to be felt.

The main influence of the Japanese on their environment before the advent of modern technology was on the forests that originally blanketed nine-tenths of the archipelago and nurtured Japan's culture, providing building materials, firewood, charcoal, mushrooms, and edible plants. Japan underwent several historical periods when forests were seriously affected. In the first, in the seventh century, during what Conrad Totman has called the "Ancient predation" (Totman 1989, 3), forests in the Kinkai Basin near the early capitals of Nara and Heian (later Kyoto) were logged to provide building material for monumental government and religious buildings. Another predation occurred with the unification in 1600 of the country under the Tokugawa *bakufu,* or military government, when timber was requisitioned from throughout Japan's three main islands (the fourth and northernmost main island, Hokkaido, was yet to be developed by the Japanese and was the home to the indigenous Ainu). Growing populations, moreover, led to increased use of wood for fuel and building materials (including rebuilding after the frequent devastating fires in the capital city that came to be known as the "flowers of Edo"). Tanaka Yuko has noted: "In the century from 1540 to 1640 Japan . . . cut down countless trees to construct castle towns, crisscrossed the country with canals, reclaimed land, and created large tracts of new rice fields" (Tanaka 1988, 12–13).

There was some consciousness of the effect of deforestation and the need for village-level forest management. According to Totman: "Surviving documents from the village of Imabori in Omi province provide clues to the operation of communal forest control. Later in the year 1448 . . . the villages met, discussed the problem of damage that wood gatherers had inflicted on forests in the past, and agreed on a policy to prevent it in future. They posted regulations specifying the punishment for anyone who cut trees without proper authorization, whether on village land or their own" (Totman 1989, 42).

Villagers were, thus, taking action to protect and sustainably manage their forests as early as the fifteenth century. Indeed, Japan's rural regions developed sophisticated traditional communal management systems for irrigation, farming, the use of forests (known as *iriaichi*), and fisheries management. These "common property" resources and areas were not owned by any individual in the community, but different groups were recognized as having rights to use them (for a variety of purposes), and such uses were managed by the communities themselves (Totman 1989, 84; McKean 1981). Various folk beliefs, customs, taboos, and sanctions governed the management of resources (Oyadomari 1989, 32).

Much of the effort to conserve forest resources, however, was advanced in a top-down fashion by the elites. For example, regulations were promulgated to respond to overuse and to preserve stands of trees reserved for use by the feudal rulers (Totman 1989, 84). The leaders of the *han,* or the feudal domains, also developed systems to forbid the harvesting of certain trees, arrangements that later evolved into the establishment of forest preserves, such as in Matsumoto Han in 1642 (Totman 1989, 71). Tanaka records that, in 1665, the Owari and Hirosaki domains declared certain areas to be *tomeyama* [closed mountains], meaning that "it was forbidden to tamper with standing trees in any way." He also notes that, in 1666, in order to control flooding, the shogunate issued its first edict forbidding tree felling; this order also mandated reforestation and restricted the development of rice fields in erosion-prone areas. Other areas, known as *suyama* [nesting mountains], were preserved to raise birds used in falconry. "In addition," Tanaka states, "tree felling by individuals was restricted in common forests, and in some *akeyama* (open mountains), forests open to free access, it was forbidden to fell certain types of trees" (1988, 13). Also, from the middle of the seventeenth century, the central government created forestry offices and designated protected forests, preventing unregulated cutting and other activities leading to soil erosion, downstream silting, and flooding. The planting of vegetation was also promoted to prevent flooding, especially of the Yodo River, the flooding of which was linked to the deforestation of its watershed (Tanaka 1988, 13–14).

Richard Forrest, Miranda Schreurs, and Rachel Penrod

While much of the populace was illiterate, conservation efforts were undoubtedly influenced by the writings of key authors. One who influenced resource management concepts in this early period was the Confucian scholar Kumazawa Banzan, whose 1687 *Daigaku Wakumon* [Questions on the Great Learning] noted that "eight out of ten mountains of the realm have been denuded" and recommended "the cessation of lumbering, reforestation, and planned lumbering." Other recommendations were "refraining from building new Buddhist temples and Shinto shrines and . . . recycling lumber" (Tanaka 1988, 14–15). Yamaga Soko, another important seventeenth-century scholar, also developed the idea of conservationist forest management, advocating that wood be cut only in the appropriate season, that overcutting be avoided, and that harvested areas be replanted (Totman 1989, 117).

In response to the problems of deforestation and erosion, a movement to rationalize silviculture arose in the mid-seventeenth century, leading to the development of official regulations for forest protection and enhanced production methods that allowed Japan to achieve sustained-yield forestry and avoid ecological devastation and economic impoverishment (Totman 1989, 70). While manuals for farming and forest management also existed and technical innovations were adopted from Dutch traders and others in industry and medicine, Western-style science and management of resources did not spread widely. Nevertheless, the conscious management of forest resources allowed Japan to escape the devastating national impoverishment that would likely have occurred if these resources had not been managed as well as they were (Totman 1989).

Antipollution and Sanitation Movements

During the Edo Period (1603–1867), Japan was a largely self-reliant and environmentally sustainable nation. During its two and a half centuries of near total isolation, it developed a highly structured and sophisticated society. While the majority of Japanese lived in rural hamlets, by 1720 the capital city, Edo, was the world's largest, with a population of more than 1 million.

Japan managed high-density urban development relatively well, utilizing human waste as fertilizer in surrounding agricultural areas. Edo had a remarkably successful water and sewer system. According to Tanaka: "When the American zoologist Edward Morse arrived in *Tokyo* (as Edo had been renamed) in 1877, he was amazed to learn that its mortality rate was lower than Boston's. Investigating the reasons, he noticed that there was no dysentery, cholera, or malaria. In short, Japan had no diseases spread by poor sanitation" (1988, 15). Tanaka further elaborates: "In his book *Japan Day by Day* (1917), Morse attributed the lack of such diseases to the fact that all human waste was carried away from the city and used as fertilizer. He observed that in America, by contrast, untreated sewage was allowed to run into bays and inlets, contaminating water and killing marine life, while people were continually assailed by the noxious fumes of putrefaction and sewage" (1988, 15).

Japanese authorities thus appeared to understand the importance of controlling some harmful effluents. In fact, according to Tanaka, in 1649 "toilets debouching directly into rivers and moats were torn down and it was forbidden to haul away sewage by boat and dump it into the water, and around 1656 farmers began carting off sewage to use as fertilizer. . . . Thanks to this cycle, there was no need for sewerage systems even in the city. Instead the water-supply system was developed. . . . Since there was no sewerage system and the waterways were not contaminated by waste, drinking water did not transmit disease" (Tanaka 1988, 15).

In America and the West, it was only later, after John Snow's 1854 demonstration that

London's Broad Street pump was linked to the incidence of cholera, that rational sanitation and urban pollution control were pursued. Interestingly, authorities in Edo, unaware of this specific finding, nevertheless were able to control the spread of such infectious diseases.

Antipollution Movements

During the Edo period, the citizenry in Japan was highly regulated (in part due to Confucianism, which promoted a strict hierarchical social order advantageous to the military rulers), and challenges to authority were uncommon. Because Japanese society had been ruled for centuries by emperors and, later, by a military government (the shogunate), deference to authority was expected. However, increased contact with the West led to changes that would build the foundation for later environmental movements.

After the opening of Japan in 1853 by Commodore Matthew Perry of the U.S. Navy, who sought trade and port services for whaling ships, Japan sought to "catch up" with the West, adopting foreign technology and industrialization. New ideas of popular rights also spread through the land in the 1870s and 1880s, promoting new forms of representative government (Irokawa 1985). But, when, in 1889, Japan's Meiji leaders were considering what kind of constitution to adopt, they rejected the American model as being too liberal. They instead used as a model the authoritarian parliamentary system of Prussia. Thus, while some democratic elements were included in the constitution, such as an elected house of representatives, the idea of multiple political parties, and male suffrage (for landowners), power resided largely with the emperor and his unelected advisers and, later, the military. Japan thus developed a hybrid of authoritarian and democratic traditions. The vitality of the country's early environmental protest movement reflects the larger trends in the politics of the time.

The best-known early antipollution movement in Japan occurred in response to pollution caused by the Ashio copper mine in Tochigi Prefecture. The mine began operation in 1610, and, in 1871, it was sold to the Furukawa Corporation (Suzuki and Oiwa 1996, 214). Acidic deposition from the mine's smelting operations and toxic metals from slag heaps and wastewater resulted in the destruction of surrounding forests and the pollution of the Watarase River. Deforestation associated with the mine also contributed to flooding (Hane 1982, 263). Pollutants discolored the water, killed fish, and made those who ate the fish ill. In 1878, local citizens complained that the mine was discharging toxins that harmed their fields when the river flooded (as it naturally and regularly did). Tanaka Shozo, a member of parliament, fought as leader and spokesman for the protest movement that arose in 1891 when evidence linked the Ashio mine to the pollution problems. Tanaka worked for decades to assist the locally affected people and the natural environment. Minor compensation payments—in the form of a *mimaikin* [a solatium or sympathy payment], rather than compensation based on admission of responsibility for damages caused—were received, "calculated more to bribe village elders than in proportion to damages suffered" (Tsuru 1999, 34), along with the misleading promise that an imported "filtering device" would alleviate pollution. Affected farmers and residents continued to seek relief for years. Between 1897 and 1900, they participated in five protest marches (McNeil 2000, 134), which they called *oshidashi*. The last march, known as the Kawamata Incident (Hane 1982, 262), became a major protest in which four thousand marchers clashed with police on their way to Tokyo and more than seventy were arrested.

Since the Ashio mine was important for military and foreign exchange purposes, the government resisted the people's demands for changes in its operations. Nonetheless:

Richard Forrest, Miranda Schreurs, and Rachel Penrod

"Tanaka decided to risk his life and submit a petition directly to the emperor. When he did so in the fall of 1901, the authorities sought to dismiss his action as that of a mad man and succeeded in curbing the protest movement" (Hane 1982, 263). In 1902, an investigative committee was set up by the government, but it consisted of elite bureaucrats who sided with the company and recommended flood countermeasures rather than pollution control. In 1907, the government destroyed the village where protest was centered (Tsurumi 1977, 9), banishing some 450 households (McNeil 2000, 135). Thus, ironically, the local protest movement resulted, not in a restoration of the integrity of the community and the productivity of farmers' fields, but, rather, the forced relocation of the village owing to the construction of a dam to create a flood catchment basin to prevent toxins from entering the river (Brecher 2000; Tsuru 1999; Notehelfer 1975, 361–69). There would be many more such forced removals of villages to make way for dams and other river-engineering works in the century to come. Finally, in 1974, after the Ashio mine had closed (as late as 1972), local farmers "won millions of dollars in compensation for a century of air and water pollution" (McNeil 2000, 135). As would be seen again later, even partial restitution in cases of pollution damage in Japan could often take many decades. Japan's tragic and drawn-out involvement with industrial pollution, later to be known as *kogai*, had begun.

There were other major movements against pollution during the time of Ashio. One was against the Besshi copper mine in Ehime Prefecture, which in 1883 installed a smelting plant that emitted sulphurous gas. In 1893, local farmers complained that emissions from the mine were harming their crops and rose up in protest. The corporation and the police first suppressed the citizens' movement, but, when in 1894 the wheat crop failed, sparking more violent protests, the

plant's operator, the Sumitomo Corporation, bought the affected land as a buffer zone and eventually relocated the pollution-producing facility, at an unprecedented cost, to an uninhabited island twenty kilometers from shore (Tsuru 1999, 38). While this may have seemed like a triumph, the source of pollution was simply relocated, and the emissions continued, affecting the productivity of different farmlands and sparking new protests. In 1909, a special government investigative team conclusively linked the pollution to the facility, leading to national attention and mediation in 1910 by the minister of agriculture and commerce (Tsuru 1999, 39). The company finally agreed to offer farmers decent annual compensation for damages and to restrain their activities by scaling back production during key grain-development periods. This was one of the few instances in Japanese environmental history of a successful citizens' movement and a company willing to admit wrongdoing, and it was significant in that the Besshi farmers sought to eliminate the source of the *kogai* itself (Tsuru 1999, 40).

In another case, in 1906 smoke pollution from the Hitachi copper mine in Ibaraki Prefecture was reported to have damaged farm and forest areas. From 1911, Seki Tenshu led protests by the victims of the pollution, interrupting a diplomatic career to assist the local community, and developing means of quantifying the impact of smoke damage. A plant administrator, Kaburagi Tokuji, led advancements in pollution-control technology, researching meteorologic effects, the use of chimneys for smoke diffusion, and pollution-resistant products for farmers. Citizen pressure caused the government to appoint a commission that, eventually, forced the Hitachi Company to install pollution controls. By 1909, conditions were improving, but the company continued to make minor payments to the farmers. This was the first *kogai* case where pollution-control technologies were used (Tsuru 1999, 40–44).

While these movements were important, they did not result in significant changes in understanding pollution and finding means to systematically analyze the relationship between the environment and technology. Moreover, as Tsuru notes, the Ashio struggle was fought as a political rather than a legal battle (Tsuru 1999, 35) and did not lead to new laws tempering the "economic growth at all costs" goal of the government and industry (Tsuru 1999, 35–36). Similarly, later, during the prewar and wartime periods, industrialization was emphasized and citizen organizing restricted by the central government, especially the Interior Ministry.

While media attention was also useful for these causes, it tended to come only after difficulties: "There is a certain consistency in the manner in which a local kogai problem was reflected in a nation-wide public discussion. For one thing, the mass media would not publicize an issue unless it becomes an 'incident' involving violence of some sort. Secondly, in a chauvinistic social atmosphere of war time, which prevailed frequently enough in Japan in those decades after the Meiji Restoration, protesting victims of kogai would feel constrained to keep quiet for at least a while" (Tsuru 1999, 46).

These cases reveal that strong leaders play a key role in protest movements. Some, like Tanaka Shozo, would demonstrate a dedication to their causes for many decades.

The twentieth century witnessed early movements addressing severe localized pollution and the struggle between development and the environment. In the pre–World War II period, there were not many new major antipollution movements, with the exception of the 1926 Bamboo Spear Affair, in which farmers staged a protest at the smelter of the Kosaka copper mine (McNeil 2000, 27). This is likely partly due to the fact that, while there was a brief flowering of citizen activism and public freedoms during the period known as the Taisho Democracy (during the Taisho

Era [1912–1926]), after 1925 government restrictions on public assembly began to be strengthened in reaction to the growth of left-wing Communist and socialist movements in the country. In 1925, although universal male suffrage had recently been established, conservatives pushed through the Peace Preservation Law, which made organizations and movements that had as their goal changing the political system illegal. A 1928 amendment by emergency imperial decree strengthened the law and made left-wing activism essentially punishable by death. A 1941 amendment went one step further and allowed for preventive arrest. The Peace Preservation Law marked the beginning of the empowerment of ultrarightists and the elimination of open political debate in Japan (Duus 1976; Mitchell 1976). It also effectively silenced environmental protest movements.

Protecting the Land: Parks and Wildlife

The religious foundations of Japanese society are based in Shinto ("the way of the *kami*," *kami* being deities), an animistic religion that emphasizes the worship of the spirits in nature, including its wrathful aspects, such as erupting volcanoes, earthquakes, tsunamis, and heavy rains. Buddhism, which later arrived from China, also inculcated beliefs promoting harmony with other living beings. Not inconsequently, some of the best-preserved natural areas are the precincts of Shinto shrines and Buddhist temples.

Influenced by Western ideas, the first public parks were created in 1873 from the former estates of *daimyo* [feudal lords] in Tokyo (Tanaka 1988; Sutherland and Britton 1995, 6). Japan's Imperial Game Law, passed in 1892, an early conservation policy with the aim of maintaining populations of popular game at a sustainable level, was primarily use oriented rather than preservation oriented (Brecher 2000). In 1906, a national policy of rationalizing local administrative jurisdictions led to the amalgamation of shrines (Tsurumi 1977, 10)—a move made possible by state Shinto, the government control of religion—leading to the destruction of local shrines and surrounding forested areas. A scientist, Kumagusu Minakata, protested the resulting loss of forests and their benefits, using arguments from microbiology to demonstrate the resulting destructive effects on crops and fisheries.

The first effort to preserve public land appears to have been a 1911 petition brought by concerned citizens calling for the protection of Nikko, a beautiful forested area where the shogun Tokugawa Ieyasu had a splendid, albeit somewhat ostentatious, temple erected for his memorial. Two years later the National Parks Association [Kokuritsu koen kyokai] was created.

Foreign examples and visitors played instrumental roles in advancing the national park movement in Japan. In 1918, the Historical Spot, Scenic Beauty, and Natural Monument Preservation Law was enacted. Interestingly, Motoko Oyadomari notes, it was "formulated not as the result of genuine grass-roots nature conservation movements, but as the result of conceptual imports by the elite from overseas" (1989, 25). Similarly: "Paradoxically, it was an Englishman, Walter Weston, who started the craze of mountaineering at the turn of the [twentieth] century" (Sutherland and Britton 1995, 7). This led to an appreciation of natural areas for recreation.

Following the American example, in 1931 Japan passed the National Parks Law. Between 1934 and 1938, twelve areas were designated as national parks (to be administered by the Ministry of the Environment). According to Oyadomari, the National Parks Law "was also promulgated for obtaining foreign currency by promoting international tourism" (1989, 26). It should be noted that, unlike in the United States, in Japan only about half the land within the national park system was owned by the state. Given differ-ent historical developments and geographic realities, there was much less open land in Japan than there was in the United States. Thus, in Japan, the national parks remained fragmented, a mix of state-owned, commercial, and residential land (Sutherland and Britton 1995, 8).

Paralleling these efforts to set aside land for protection, there was some interest in wildlife protection in Japan as well. One of the oldest existing environmental groups in Japan today, the Wild Bird Society of Japan [Nihon yacho no kai], was created in 1934, following the example of numerous bird-protection societies, especially those in Great Britain.

Environmentalism in the Post–World War II Period

The United States

The United States escaped World War II largely undamaged. When the war ended, conservation and preservation movements took up where they had left off before the war started.

Environmental groups began to work together to lobby Congress for the preservation of more wilderness areas and the protection of wildlife. Howard Zahniser, a staffer at the U.S. Biological Survey, became the director of the Wilderness Society in the mid-1940s. He believed that environmental organizations could more effectively further their cause by forming coalitions lobbying for similar goals. After David Brower assumed the directorship of the Sierra Club, he and Zahniser lobbied together to have Congress designate 50 million acres in the West as wilderness (Dowie 1995; Gottlieb 1993, 41–43). In 1956, these environmental movements also defeated government plans to dam the Grand Canyon as well as Echo Park in Dinosaur National Monument (Dowie 1995; Dorman 1998,

217). They also successfully lobbied for the passage of the Wilderness Act of 1964, setting aside pristine forests for preservation.

Environmental groups also successfully lobbied for the passage of the Land and Water Conservation Fund Act of 1964 to fund federal purchases of land in order that they be protected, the Endangered Species Act of 1966, and the Wild and Scenic Rivers Act of 1968 to protect rivers of special character (Dowie 1995; Kline 1997, 94–98).

David Brower was also a pioneer in efforts to turn U.S. environmental groups into mass-membership organizations. As the first salaried executive director of the Sierra Club, he helped expand the membership of the organization greatly. He also helped transform the environmental movement into a much more activist force (Switzer 1997, 6). A variety of new environmental groups formed at this time, including the Soil Conservation Society of America (1944), the Nature Conservancy (1951), Resources for the Future (1952), and the World Wildlife Fund–USA (1961).

The Rise of the Modern Environmental Movement Public concern began to grow in the postwar period about the impact that rapid industrialization and modernization were having on the earth and living species. A series of publications began to change public attitudes toward environmental protection. In 1948, Fairfield Osborne published *Our Plundered Planet,* one of the first books to sound the wake-up call about agricultural practices. Rachel Carson's groundbreaking 1962 *Silent Spring* became a national and international best seller. The book asks its readers to imagine a world in which there are no birds, a threat Carson considered to be all too real because of the devastating impact that DDT and other pesticides were having on the bird population. Her book is a chilling critique of the chemical and agricultural industries as well as of a society that was so enamored with modern inventions that it was disregarding what it was doing to the environment. Ralph Nader, a founder of the postwar U.S. consumer-protection movement, brought attention to the resistance of the automobile industry to safety standards with his 1965 *Unsafe at Any Speed,* opening the door to citizen and consumer activism in many fields, including the environment. Paul Ehrlich's 1968 *Population Bomb* raised concerns about global population growth rates and the carrying capacity of the planet.

A series of accidents also helped shock the public. In 1969, an explosion on an oil platform off the coast of Santa Barbara created a huge oil slick that affected thirty-five miles of California coastline, killing marine life and birds. The same year, the Cuyahoga River in Ohio, which was contaminated with chemicals, caught fire. In 1970, Senator Gaylord Nelson of Wisconsin, a Democrat, called for a national teach-in focusing on the environment, an idea that took off and led to an estimated 20 million people participating in the nation's first Earth Day celebration (Lewis 1990).

As a result of these changing understandings of the relation between human activities and the environment, new environmental groups began to form, and existing environmental groups began to expand their range of activities. The environmental movement became part of a larger countercultural movement that included the anti–Vietnam War movement, the civil rights movement, and the women's rights movement. James Gustave "Gus" Speth, the creator of the Natural Resources Defense Council (1970), one of the most powerful groups regularly involved in environmental litigation in the country, stated that he was inspired to create an organization lobbying and conducting lawsuits on behalf of the environment by the example of the National Association for the Advancement of Colored People (Shabecoff 1999, 117).

Other new groups to form at this time included the Environmental Defense Fund

Richard Forrest, Miranda Schreurs, and Rachel Penrod

(1967), the Friends of the Earth (1969), and Greenpeace-USA (1971). The focus of the environmental movement broadened to include not only nature conservation and wildlife preservation but also air and water pollution, nuclear energy, chemical use, incinerators, toxic landfills, and many other issues. It is important to note that, while these groups are national, they are often major players in local environmental cases with national ramifications.

These environmental groups became powerful political forces, making use of the federal government system to lobby policymakers at multiple levels. Environmental activists were also empowered in the 1970s by Congress and the courts. The Administrative Procedures Act of 1946 had provided citizens with some avenues to learn about and be involved in decisionmaking related to development, the environment, and other issues (Buck 1996, 67). Environmental groups' ability to influence regulatory direction expanded substantially, first with the passage in 1969 of the National Environmental Policy Act (NEPA), then with the Supreme Court's ruling in the 1972 Mineral King case brought by the Sierra Club. NEPA is one of the most important pieces of environmental legislation in the United States. It requires that environmental impact assessments be conducted for all major federal projects. It also includes provisions for citizen suits and transparency in decisionmaking on environmental matters, provisions that were included in many subsequent pieces of environmental legislation as well. The Sierra Club lost the Mineral King case, but, in its ruling, the Supreme Court expanded its interpretation of the concept of "standing to sue," indicating that if an organization is able to show that its members would be affected by an action, even if not directly injured by the action, it could have standing to sue (Buck 1996, 67).

On the basis of this, environmental organizations undertook suits to compel admin-

istrative and corporate improvements. Key U.S. environmental organizations, such as the National Resources Defense Council and the Environmental Defense Fund (now Environmental Defense), were created to undertake NEPA cases in order to preserve local areas. In the process, they became strong national-level players that were able to transfer their concerns to other locales facing destructive development. Environmental groups in the United States turned to the courts to force regulatory change when faced by intransigent political administrations. U.S. environmental groups used these provisions to sue violators of environmental laws and government agencies that were not performing duties required of them by the law. The federal courts thus became very active in reviewing executive actions on environmental policy and not infrequently requiring federal agencies and states to adopt new regulations.

Antipollution Movements: The Case of Love Canal The 1970s were a period of dramatic environmental policy change in the United States. Changes in federal environmental law were often driven by the impact of local events. Love Canal is among the best-known pollution cases in the United States. For ten years beginning in 1942, Hooker Chemicals and Plastics used the abandoned Love Canal, near Niagara Falls, New York, as a dump for chemical wastes, including halogenated organics, pesticides, and benzene. In 1953, the site was covered and then deeded to the Niagara Falls Board of Education, and, in an area near the site, a school and homes were built. By the 1960s, residents of the area were complaining about bad odors and residues, problems that grew worse in the 1970s as the groundwater table rose and runoff began contaminating the nearby residential areas and also the Niagara River (Environmental Protection Agency 2007).

Emblematic of the postwar U.S. antipollution movement is the work of Lois Gibbs, a recipient of the prestigious Goldman Environmental Prize.[1] In 1978, Gibbs became worried about reports she had read about the contamination of the Love Canal site, where she lived, and she started to wonder about the possible connection between the site and her children's and neighbor's health problems. Many in the community were suffering from chronic headaches, respiratory problems, and skin ailments. There were also high incidence rates of cancer and deafness. Gibbs decided to form the Love Canal Homeowners Association and began to battle local, state, and federal government agencies to address the problem. That same year, President Carter declared a state of emergency at Love Canal, and 239 families were permanently evacuated from the area. When studies showed that there was chromosome damage in residents of the region and that this could increase the risk of cancer and reproductive problems, a second state of emergency was announced, and the entire community of 900 families was evacuated. Gibbs became nationally famous because of the media attention her work attracted.

While Love Canal was the case that caught the nation's attention, there were actually thousands of hazardous waste sites across the country that were not properly being handled, and many of these were ticking time bombs. A series of hazardous waste accidents helped convince Congress of the need to act. In 1977, in Bridgeport, New Jersey, six people died and thirty-five were injured when sparks from a worker's torch ignited chemicals that had accumulated in a waste-storage facility. In 1978, in Riverside, California, the erosion of a retaining dam for the Stringfellow Waste Pits threatened to release huge quantities of toxic pollutants into the Santa Ana River. The state released some of the contaminated water to ease the strain on the dam, exposing children who were playing in nearby water to dangerous chemicals. In the same year, in Toone, Tennessee, residents filed a lawsuit against a chemical company that dumped pesticides into the area's landfill, causing contamination of drinking water (Environmental Protection Agency 2000).

In December 1980, Congress established the Superfund program (a financial program established under the Comprehensive Environmental Response, Compensation, and Liability Act) to deal with locating and cleaning up uncontrolled and abandoned hazardous waste sites. The success of Lois Gibbs's work attracted many requests from around the country for her help. In response, Gibbs created the Citizens Clearinghouse on Hazardous Waste to fight pollution caused nationwide by the dumping of toxic waste. According to Philip Shabecoff: "By the late 1980s the clearing house was working with over 5,000 local organizations around the country, some with as few as fifty members, some with over a thousand. It provides its members with scientific expertise and other information and helps them organize" (1999, 237).

Restoring the Environment   In the period since the 1970s, many local and regional programs were initiated by environmental groups and local communities to restore the quality of their environments. Examples include efforts to restore the ecology of the Chesapeake and San Francisco bays; reclamation of abandoned mines in Appalachia and throughout the Midwest; urban waterfront revitalization projects in many older cities, including New York City, Baltimore, Buffalo, Toledo, Detroit, and Seattle; the cleanup of the Great Lakes and the prevention of large-scale water diversion initiatives; restoration of grasslands; and the reintroduction of wildlife species.

The Chesapeake Bay Foundation (CBF), for example, was created in 1967 by a group of Baltimore businessmen who shared a common interest in sailing, fishing, and waterfowl hunting. In 1976, they convinced Maryland senator Charles "Mac" Mathias Jr. to introduce into Congress a bill to have a major study done on the quality of the Chesapeake. The seven-year study found a bay that was threatened by pressures from human activities. Through the work of the CBF and other environmental organizations, the first interstate meeting of the governors of Maryland, Virginia, and Pennsylvania and the mayor of Washington, DC, was held and led to the passage of the first Chesapeake Bay Agreement. Since then, the organization has been involved in numerous other campaigns and educational efforts to address threats to the bay, including nonpoint source runoff and the restoration of wetlands and oyster beds. The CBF had a membership of over 146,000 members in 2006.[2]

The Conservative Backlash The 1970s are often referred to as the *environment decade*, as hundreds of new environmental laws and programs were established and new environmental groups formed throughout the country. As a result of the new legislation, air and water quality improved significantly. Numerous chemicals were banned, and new measures controlling the production and use of hazardous chemicals were introduced.

Environmental legislation, however, began to be challenged by corporate and local government interests, which complained of the costs of complying with unfunded federal mandates and the inflexibility of command and control regulation. Opposition to the establishment of new environmental regulations has grown, and new groups intent on protecting property rights, eliminating environmental restrictions, or allowing enterprise greater freedom to "wisely" exploit the land (the wise use movement) have emerged (Helvarg 1994; Switzer 1997, 191–226, 247–80).

Using many of the same tactics that environmental groups have used, the property rights movement has worked to stop what it perceives as infringements on constitutional rights. There are now numerous groups across the country that have as their goal defending property rights and promoting freedom from what they see as environmental regulations that prevent them from using their property as they desire and without proper compensation. According to Professor Steven J. Eagle, a conservative supporter of the property rights movement: "Endangered species laws aimed at protecting large mammals have led to regulations protecting habitat for kangaroo rats at the expense of homes that burned because the habitat could not be disturbed and to a major dam that goes unused in order to protect a small fish called the snail darter" (Eagle 2001, 13).

The property rights movement worked during the 1990s to weaken the Endangered Species Act and was successful in restricting congressional funding for the enforcement of the act. Although environmental groups would like to see the act broadened to assure greater ecosystem preservation, they are spending their time fighting to ensure that it survives the attacks by the property rights movement. Thus, there are now multiple groups working for different purposes related to many conservation and preservation issues. The Endangered Species Coalition, a coalition of 360 environmental organizations and individuals, seeks to protect endangered species (and the Endangered Species Act) through education, information dissemination, and lobbying. On the other side of the fence, the National Endangered Species Act Reform Coalition, representing 150 groups, including farmers, electric utilities, home builders, and others, would like to see the Endangered Species Act revised so that compensation is provided to property owners when restrictions are placed on them because of endangered species found on their land.[3] The property rights movement was one of the voices opposed to U.S. involvement in the Convention on Biological Diversity.

Local Environmental Initiatives for Global Environmental Protection Another aspect of what some have termed a *conservative back-*

lash has been the strong opposition that has grown, especially within the Republican Party, to the introduction of new environmental regulations. This has made it difficult for environmentalists to gain sufficient support within the Senate to ratify a number of international environmental agreements. As a result, the United States has not ratified the Kyoto Protocol to the Framework Convention on Climate Change, the Biodiversity Convention and its Cartagena Protocol on Biosafety, or the Basel Convention on the Control of Hazardous Wastes and Their Disposal.

While environmental groups in the United States have not always succeeded at the federal level—and, by extension, the international level—in achieving their policy goals, they have made inroads at the local level (Rabe 2004). For example, as of November 2004, there were 148 cities in the United States that were members of the Cities for Climate Protection campaign of the International Council for Local Environmental Initiatives. These cities have agreed to work to reduce their emissions of carbon dioxide, methane, and other greenhouse gases through energy conservation initiatives as well as public transportation, waste-reduction, recycling, and renewable-energies projects.[4] Another initiative is the New England Governors/Eastern Canadian Premiers 2001 Climate Action Plan, in which six New England states and five Canadian provinces agreed to take joint action to reduce greenhouse gas emissions. The New England Climate Action Coalition, a group of 160 state and local environmental groups, was established to ensure that the action plan's goal of reducing the region's emissions to 1990 levels by 2010 is met.[5] Many other initiatives have been established throughout the country, partly at the urging of local environmental groups.

Japan

During World War II, significant areas of Japan were deforested for wartime produc-tion, and cities were devastated by bombing. Bombing also occurred in national parks, which were used for military practice. When the war finally ended, Japan's industrial might had been broken, millions of Japanese soldiers and civilians had died or been wounded, and the future of the country remained uncertain. During seven years of Allied occupation, numerous changes to Japanese government structures and laws were made. These included the establishment of a new constitution, which took sovereign power away from the emperor and gave it to the people. A functioning parliamentary democracy was reestablished, and freedom of speech and assembly were guaranteed.

In the ensuing years, Japan's new democracy took on many unique characteristics. The Liberal Democratic Party (LDP), formed in 1955, became the dominant party and has ruled the country almost continuously ever since (since the 1990s in coalition governments as the overwhelmingly dominant party). Until the 1990s, its main rival was the Social Democratic Party (SDP), a party with Marxist leanings. Environmental concerns often received a more sympathetic hearing from the SDP and the Communist Party of Japan, which meant that the LDP and businesses allied with it could dismiss activist concerns as overly extremist. The LDP chose to align its foreign policy with that of the United States, to tie its security to that of the United States, and to focus its efforts on economic recovery and development.

Environmental Conservation in Japan: The 1950s Amid Japan's staggering postwar changes, some strides were made with environmental conservation. In 1948, in one of the first postwar cases of environmental activism, the Oze Marsh Conservation Union [Oze shotaku hogo renyokai] formed to protest construction of a hydroelectric dam in the Oze wetland in Nikko National Park. This group became the basis for the Nature

Richard Forrest, Miranda Schreurs, and Rachel Penrod

Conservation Society of Japan [Nihon shizen hogo kyokai], founded in 1951. The Nature Conservation Society was successful in saving the Oze marsh (Schreurs 2002, 36). The preservation of nature at Nikko was further advanced in 1964 when the "Takosugi" lawsuit to preserve trees on shrine land was successful (Oyadomari 1989, 28).

One of the government's first environmental actions in the postwar period was the expansion of the national park system. Between 1946 and 1955, five more national parks were created, and, in 1957, a more comprehensive National Parks Law replaced its 1931 predecessor (Sutherland and Britton 1995, 7). These developments suggest at least some concern for nature conservation in Japan. These success stories, however, were soon to be overshadowed by the issue of industrial pollution, which had gone largely ignored.

Industrial Pollution and the Rise of Antipollution Movements: 1950s–1970s Development and urbanization brought new forms of *kogai*. The pollution problems eventually were met with new opposition movements, mainly on the part of those directly affected. To take one early example, to illustrate the degree of pollution in Kitakyushu, an industrial city in northern Kyushu, a women's group hung out white sheets and shirts; then, with the blackened sheets and shirts as evidence, in 1951 it successfully lobbied the local government for the installation of antipollution equipment (City of Kitakyushu Environment Bureau 1999, 7).

Progress on citizen campaigns was slow, however. Companies fought claims made against them with their own (often distorted) science or hid from the public information linking pollutants to disease. They bullied, intimidated, and socially ostracized their critics in an effort to silence them. They portrayed the members of the citizen campaigns as radicals trying to prevent industrial progress. In the disputes that became increasingly

common throughout the country, both the national and the local governments initially sided implicitly or explicitly with the corporations and did little to protect residents from pollution.

One of the earliest antipollution movements to have a national political effect dealt with water pollution. In 1958, after refusing the small reparation payments offered to them by the Edogawa-area Honshu Paper Manufacturing factory, whose effluents were killing fish populations in Tokyo Bay, fishermen in Urayasu, Chiba Prefecture, mounted violent protests at the plant. These protests, which some characterized as riots (Watanuki 1984, 2), directed national attention to the problem of water pollution. The fact that the United States had recently passed a water quality act was also influential. In 1958, the Japanese government passed the Law for the Regulation of Factory Effluents and the Water Quality Conservation Act (Tsuru 1999, 60). These laws designated as "special water areas" bodies of water serving as sources of public water supply and limited discharge from factories, mines, and sewers into them (Tsuru 1999, 60). These measures proved insufficient to protect the population from many other serious pollution problems, however.

Four communities—Minamata, a small fishing village in Kumamoto Prefecture, on the island of Kyushu; Niigata, a much larger city; rice-farming villages along the Jinzu River in Toyama Prefecture; and the city of Yokkaichi—played a particularly momentous role in heightening the national awareness of the threats posed by industrial pollution to human health and in pressuring the government to take pollution more seriously (Gresser, Fujikura, and Morishima 1981; Upham 1987; McKean 1981; Huddle and Reich 1987). In these four communities, the dumping of toxic waste into bodies of water or the release of toxic emissions into the air (in the case of Yokkaichi) over long periods of time resulted in severe communitywide health problems. Victims complained to the industries responsible for the pollution, requested compensation, and demanded the end of polluting activities. While victims did manage to get some small payments, nothing was done to stop the pollution, and new pollution-caused disease cases emerged. After years of frustration, and with the aid of outside lawyers and journalists, these four communities took their cases to court, an unusual step in Japan's relatively nonlitigious society. The national attention that these cases attracted, along with hundreds of other environmental movements that formed across the country, resulted in major grassroots pressures for political change and the eventual adoption of new pollution-control legislation, often first at the local level, and then at the national level.

Minamata Disease   Minamata disease is the best known of Japan's "pollution illnesses" (George 2001). It is a form of mercury toxicosis affecting the central nervous system and is caused by exposure to methyl mercury, a catalyst used in the production of acetaldehyde. The victims of Minamata disease experience severe physical disabilities, including trembling and problems with coordination, leading ultimately to an inability to walk or engage in routine activity, and disfigurement. The disease is often fatal, especially among infants, babies born to victims suffering from debilitating birth defects.

Minamata disease first developed in people eating seafood (mostly fish) caught in Minamata Bay. The initial cases were the result of the Chisso Corporation dumping waste containing methyl mercury into the Shiranui Sea in Kumamoto Prefecture, a practice the company began in 1932. The first indication of the disease was in 1949 when dead fish were found floating in Minamata Bay and shellfish began emitting a noxious odor when opened. Catches started to decline, and, by 1953, there were reports of cats "dancing" in circles be-

fore collapsing and seabirds spiraling into the sea. The disease was first recorded in humans in 1953, but it was not officially recognized until 1956. Because of its debilitating effects, victims were dependent on their families for care, and there was a devastating economic impact on the communities where the disease was prevalent. Fear that Minamata disease was communicable led to the ostracization of victims and their families, with the result that traditional networks of community aid were unavailable to them (Maruyama 1996, 41–43). Such ostracization of the victims of pollution was to become commonplace, victim movements often being linked to political movements disapproved of by the central government, including leftist and Communist movements.

In 1956, the Minamata Disease Patients' Families' Mutual Aid Society was founded. In 1957, the Minamata Fishermen's Cooperative called for a cessation of dumping and the installation of adequate wastewater treatment facilities, but Chisso rejected the request. In 1959, mercury was identified as the disease's cause by Kumamoto University researchers, and the fishermen petitioned for aid again. Only a token payment (*mimaikin*)—specifically not to be considered reparation—was negotiated by the mayor's office. Fishermen stormed the plant, but the governor simply brokered another token payment deal. Chisso installed a cyclator, supposedly making its waste safe. In 1965, Minamata disease was discovered in Niigata (caused by effluent from another company). In 1968, with public concern growing, the government officially recognized that Chisso's waste was the cause of the disease in Minamata, and the next year the victims filed suit in court, finally winning significant reparations in 1973 (Tsuru 1999, 79–101; George 2001; Notehelfer 1975, 351). At one point, there were over 12,000 certified victims of Minamata disease, with perhaps as many as 200,000 affected (Maruyama 1996, 43).

Photographs of the effects of Minamata

disease, including those by Kuwabara Shisei and the American Eugene Smith were extremely significant in gaining support for its victims. The Open Forum for Citizens on Environmental Disruption, founded by arguably the best-known environmentalist in Japan, the Tokyo University researcher Ui Jun, brought together a group of university students that organized lectures on pollution incidents and diseases (Brecher 2000).

Itai-Itai Disease Itai-Itai [lit. "it hurts, it hurts"] disease is a form of cadmium poisoning that makes bones extremely brittle. It was first reported along the Jinzu River in Toyama Prefecture in 1955. However, it took until 1967 for doctors to prove that it resulted from discharges from the Kamioka mine's zinc-refining plant. Dr. Hagino Noboru testified before the House of Councilors in 1967 on these findings, making the case of Itai-Itai disease the first major pollution event to come to trial in the court system. Almost all the victims of this disease were women, and most died. The Anti–Itai-Itai Disease Council of Toyama Prefecture was the most active citizens' organization. Its representatives visited the Kamioka mining plant of the Mitsui Metal-Mining Company to demand reparations in the same year that Hagino took them on in court. Mitsui lost the suit in 1971, and the precedent was set that epidemiological research indicating a relationship between pollution and disease, if not the precise mechanism, could hold up in court; Mitsui was forced to sign a pollution-prevention agreement with four different victim organizations (Tsuru 1999, 104–5).

Yokkaichi Asthma Air pollution was also a major focus of citizen action. An industrial complex, or *kombinato*, including a thermoelectric power plant, a petroleum refinery, petrochemical plants, and other heavy industries, began operations in Shiohama, a suburb of Yokkaichi, in 1959. The next year, citizens

and fishermen appealed to the municipal office for relief from noise and pollution. That same year, fish caught in the Yokkaichi area were refused at Tokyo markets, and, in 1961, the foul-smelling fish were officially linked to industrial waste. The petrochemical complex also emitted large quantities of sulfur dioxide into the air, causing serious respiratory problems for many individuals in the region, and people began to talk about Yokkaichi *zensoku* [asthma]. While the government did initiate studies investigating the problem and made some policy recommendations regarding improving air quality in the region, pollution there remained serious. In 1967, a small group of victims filed suit against several of the plants in the *kombinato* (Wilkening 2004, 124–26).

Antipollution Movements and Voluntary Pollution-Control Agreements The antipollution movements of this period were usually locally oriented (Lam 1999, 80). They emerged in response to specific pollution and quality-of-life concerns within their communities and had as their main goal finding ways to resolve those concerns by negotiating with or pressuring local industries and governments. They succeeded in forcing local and national governments to address noise pollution, vibration, and pollution in urban canals; protect homeowners' "right to sunshine" when new construction threatens a blockage of sunlight (McKean 1981); assure food safety (Broadbent 1998); and require the installation of pollution-control devices.

The groups rarely networked with each other, most were temporary assemblages, and most kept their attention focused locally. Nevertheless, they led the way toward the striking of tens of thousands of "voluntary" local pollution-control agreements. These agreements were made between local governments and companies and generally required facilities to adopt pollution-control measures that were more stringent than those required by national law. Despite the fact that they were voluntary, companies often had no choice but to accept them or risk the wrath of local citizens or the denial of government permits (Matsuno 2005).

The citizens' movements did not win all their battles. Efforts to prevent the building of the New Tokyo International Airport at Narita, Chiba Prefecture, succeeding only in delaying construction by years, and later efforts to halt the airport's expansion succeeded only in curtailing the extent of that expansion. Citizens' movements also failed to prevent the building of many dams, nuclear power plants, roads, apartment complexes, and the like. The most difficult cases for citizens' movements were those in which they found themselves pitted against the powerful economic and construction and agriculture ministries.

Consumer-Environmental Movements: The 1970s to Today   In Japan, the antipollution movement has been linked to consumer concerns. Throughout the country, many housewives started to promote the adoption of natural soaps (made from used cooking oil) instead of synthetic detergents and to recycle milk cartons in an effort to contribute to the cleaning of waterways and the reduction of household waste. Such recycling activities are labor intensive and tedious; still, they are viewed as important to the long-term fight to change consumption practices detrimental to human health and the environment. Japan's historically high incidence of pollution-related diseases and the very real problem of solid waste disposal have contributed to the questioning of modern lifestyles and a growing interest in recycling, organic food cooperatives, and direct producer-consumer distribution channels.

A remarkably large and successful consumer cooperative association, the Seikatsu Club Consumers' Cooperative Union [Nihon seikatsu kyodo kumiai rengokai], formed in

the 1970s. It is as of 2007 an association of twenty-six consumer cooperatives (Seikatsu clubs). Found in many communities throughout the country, the cooperatives buy organic food in bulk and then sell it to member households. The Seikatsu clubs succeeded in growing into a national association by focusing on issues of concern to housewives, for example, selling nonpolluting soap after pollution from detergents caused eutrophication, damaging waterways. They were inspired by the success of a citizens' movement whose complaints about the pollution of Japan's largest freshwater body, Lake Biwa, resulted in the Shiga Prefectural Assembly prohibiting the use of synthetic soap in 1979 (Lam 1999, 106). Other activities promoted by these movements include the promotion of organic food through an alternative distribution method, known as *teikei* (face-to-face), that links consumers directly with farmers. One cooperative association (Seikatsu club) grew to have a membership of at least a million people participating directly or indirectly as of 1997 (Lam 1999, 80). The Seikatsusha Network [Seikatsusha nettowaku, or Netto (a Japanese transliteration of *network* or *net*)] was established as the political arm of the Seikatsu Club Consumers' Cooperative. Recognizing that efforts to restrict the use of agricultural chemicals, get schools to buy organic foods, and reduce packaging require political change if they are to succeed, the Seikatsusha Network was formed in an effort to support the election of sympathetic representatives to local government assemblies. Over the years, it has been somewhat successful in its efforts to win local elections, in the process also increasing the number of women active in local politics.

Healing the Land: Efforts to Save Rivers and Wetlands   Land is at a premium in Japan, which has a population of 127 million people (roughly 43 percent of the U.S. population) in an area the size of Montana. The fact that

Richard Forrest, Miranda Schreurs, and Rachel Penrod

most of Japan's population and economic resources are concentrated on what are natural floodplains has led the government to undertake intensive management of rivers. Rivers have also been extensively dammed for the hydropower they can generate.

In the past few decades, environmental movements have become increasingly active in opposing large infrastructure projects that threaten Japan's wetlands, bays, and rivers, such as the Island City artificial island project at Wajiro Tidal Flat in Fukuoka City, the reclamation of Isahaya Bay in Nagasaki Prefecture, the Nagara River Estuary Dam, and the building of an airport over the Shiraho Coral Reef at Ishigaki Island, Okinawa Prefecture. As these projects were often agreed on decades earlier and supported by powerful government and industrial interests, they have been difficult to alter or halt. Activists have had a mixed success record in their effort to prevent such ecologically damaging construction projects (Karan 2005, 359–75).

A very visible example of their failure is the case of the Nagara River Estuary Dam, one of the biggest domestic environmental issues in the early 1990s in Japan. The dam was ostensibly built to protect sixteen hundred hectares (roughly six square miles) of farmland from salinization. Environmentalists argued that the project was unnecessary and would result in the damming of one of Japan's last remaining free-flowing rivers. According to Maggie Suzuki, a longtime environmental activist in Japan, in this case, as with previous pollution-related controversies, national and local government authorities withheld crucial information from the public (Suzuki 1998). Credible arguments that the projects were not needed and were, in fact, being pursued owing to bureaucratic inertia and a desire to maintain departmental budgets were ignored. Despite nationwide protests, the Nagara dam was completed in 1995. For citizens' groups, it remains difficult to alter the course of major development projects.

Another example is the Isahaya Bay Land Reclamation Project, which was promoted by the Ministry of Agriculture, Forestry and Fisheries and Nagasaki Prefecture at a cost of ¥237 billion. The stated purpose of the massive dike project was disaster prevention, but environmentalists argued that it would threaten an important habitat for migratory birds and spawning marine life and damage water quality. The benefits of the project, they argued, were far outweighed by its costs. Their efforts were unsuccessful. There have been massive die-offs of marine organisms and algal blooms in the cut-off bay and in the areas offshore of the dike (Suzuki 1998).

On the other hand, with backing from some of the world's leading coral reef experts, Japanese environmental groups were successful in preventing the building of Ishigaki airport directly on top of the Shiraho Coral Reef, which is composed of colonies of rare blue coral; the airport location was altered. Activists also succeeded in stopping Nagoya City's Fujimae Tidal Flat Garbage Landfill Project in January 1999.

While their protest activities often involve attracting media attention, environmental groups have also at times tried to alter policy by getting politicians favorable to their cause elected to office. An interesting example is related to the opposition to the Hosogochi Dam in Kitoson Village, Tokushima Prefecture, which successfully supported antidam mayors (Suzuki 1998). Another example was with the election of Domoto Akiko, a former World Conservation Union council member and member of the House of Councilors as the governor of Chiba Prefecture. In 1998, environmentalists succeeded in collecting 120,000 signatures on a petition opposing the Sanbanze Land Reclamation project in Chiba; on becoming governor, Domoto canceled the project and replaced it with a wetland-restoration project modeled on restoration activities in California's San Francisco Bay.

Linking Local, National, and International Environmental NGO Activities: The 1990s On Compared with the environmental movement in the United States, that in Japan has remained mainly local in focus. This has started to change since the 1990s, with growing concern about the deteriorating state of the global environment. New environmental problems on an international scale have forced Japan's environmental movement, the government, and industry to reconsider the organization and goals of environmental groups and the role they play in the formulation and implementation of policy.

Many of the larger environmental nongovernment organizations (NGOs) in Japan are national branches of international organizations, such as Greenpeace, the Friends of the Earth, the World Wildlife Fund, and the Climate Action Network. In addition, some locally grown national environmental groups have formed, such as the Japan Tropical Action Network and A Seed Japan, a youth group. In their early years, many of these groups had to rely on foreign financial support and information for their activities, and for some this situation continues to this day. Japanese environmental NGOs at the national level have been handicapped by small budgets, making it difficult for them to hire professional staff; small numbers of members (many groups have only a few dozen to a few hundred members, and even the largest have memberships only in the tens of thousands), which weakens their political clout and keeps their finances constrained in the absence of significant charitable foundation giving; and a political establishment that closed its doors to them, making it very difficult for them to obtain legal status as nonprofit organizations and exert any direct influence on environmental policy at the national level.

This situation is starting to change. The importance of national environmental NGOs has increasingly been recognized by the government and industry in Japan. One reason for this is the spread of international environmental norms. The international community has worked to spread a normative understanding that NGO participation in environmental decisionmaking and policy implementation is important. In international environmental negotiations, NGOs are now expected to participate as observers. Agenda 21, an international action plan for sustainable development formulated at the 1992 UN Conference on Environment and Development and signed by the government of Japan, calls on nations to foster citizen participation in sustainable development initiatives. Another reason is that, as the Japanese government becomes increasingly involved in international environmental protection efforts, it has come to recognize that it cannot implement its programs without the help of other groups, including environmental groups. A third is the example of local voluntary initiatives. After the devastating 1995 Hanshin-Awaji Earthquake in Kobe, it was the massive efforts of thousands of volunteers that helped prevent the number of fatalities from soaring even higher than the more than five thousand killed when buildings and bridges collapsed. Soon after the earthquake, and motivated by a recognition of the efforts of NGOs assisting in Kobe, efforts began to pass the Nonprofit Organization Law, which was unanimously approved by the National Diet of Japan in 1998. That law removed many, if not yet all, of the obstacles that existed in the past to getting approved as a nonprofit group. The next year, the Public Information Disclosure Law, similar to the U.S. Freedom of Information Act, was passed (Pekkanen 2006; Schreurs 2002, 225).

The government has also tried to ease the financial situation of Japan's NGOs, which do not have the benefit of as many independent philanthropic groups as exist in the United States. In 1993, for example, the Japan Fund for the Global Environment was set up to make it possible for NGOs to apply for funds

Richard Forrest, Miranda Schreurs, and Rachel Penrod

for environmental protection and sustainable development projects. The Ministry of Foreign Affairs also set up a special-assistance fund so that NGOs could undertake projects in developing countries.

With these funds, Japanese NGOs are increasingly networking with NGOs elsewhere in Asia and worldwide to work cooperatively on developing countries' environmental problems, many of which have regional or even global effects (Wong 1998). Thus, the character of Japan's environmental movement is slowly changing.

Conclusion: Confronting Challenges to Sustainability in a Globalizing World

Japan's environmental movement has until recently been locally focused, in terms of both its organization and its goals. Local environmental movements have pursued a wide variety of initiatives focused on such issues as preventing environmentally destructive development plans, forest preservation, recycling, and the development of renewable energy. In many ways, local environmental initiatives in Japan are not so dissimilar from those in the United States. They share similar goals and often must engage in long-term efforts to win public and political support for their causes. They are also often up against powerful industrial and corporate interests.

Where the movements differ is that local environmental groups in the United States also have the backing of many large, professionalized national environmental NGOs. In contrast, in Japan, until quite recently, there were very few national environmental groups that could lend support to local environmental initiatives. In addition, U.S. environmental groups have been empowered by the courts, which have granted them fairly broad rights of standing even in instances where they or their members are not the injured parties.

Japanese NGOs do not enjoy such a broad interpretation of the right of standing in the courts, and many court victories affected only the victims of pollution. In general, U.S. environmental groups are also better financially endowed and have larger membership bases. They are also more likely to have paid professional staff. Thus, the Japanese environmental NGO movement has had to struggle against the powerful development and construction industries and their ministerial backers with little more than their own strong will and ambition and whatever media support they can generate. Since the 1980s, however, the U.S. environmental movement has faced increasingly strong industrial and congressional opposition to its efforts to establish new and stronger environmental regulations. In many cases, it has had to go on the defensive to protect existing environmental laws and programs. In addition, new movements, such as the property rights movement, are working against the goals of the environmentalists. Japanese environmental groups in general have yet to face such assaults.

As the many case studies in this volume suggest, local environmental groups in both Japan and the United States have done much to protect local environments and to push for local and national policy change. What makes the comparison of local environmental case studies in the chapters that follow so interesting is that we can observe how those concerned with the environment in both countries have had to create innovative responses to quite different sets of political, economic, and cultural constraints. In both countries, however, the goals have been similar—finding ways to protect local environments and to promote sustainable development. As the United States, Japan, and other countries confront further environmental challenges, more creativity—and creative borrowing from one another's experiences—will be needed.

Notes

1. For more on the Goldman Environmental Prize, see http://www.goldmanprize.org.

2. For more on the CBF, including updated membership figures, see http://www.cbf.org.

3. For more on the National Endangered Species Act Reform Coalition, see http://www.nesarc.org.

4. For more on the International Council for Local Environmental Initiatives, see http://www.iclei.org.

5. For more on the New England Climate Coalition, see http://www.newenglandclimate.org.

References

Anderson, Terry L., ed. 2000. *Political Environmentalism: Going behind the Green Curtain.* Stanford, CA: Hoover Institution Press.

Benton, Lisa M., and John Rennie Short. 1999. *Environmental Discourse and Practice.* Oxford: Blackwell.

Brecher, W. Puck. 2000. *An Investigation of Japan's Relationship to Nature and Environment.* Lewiston, NY: Edwin Mellen.

Broadbent, Jeffrey. 1998. *Environmental Politics in Japan: Networks of Power and Protest.* New York: Cambridge University Press.

Buck, Susan J. 1996. *Understanding Environmental Administration and Law.* 2nd ed. Washington, DC: Island.

Carson, Rachel. 1962. *Silent Spring.* New York: Fawcett Crest.

City of Kitakyushu Environment Bureau. 1999. *Pollution Countermeasures of the City of Kitakyushu, Japan.* Kitakyushu: City of Kitakyushu, March.

Clepper, Henry, ed. 1966. *Origins of American Conservation.* New York: Ronald.

Dorman, Robert L. 1998. *A Word for Nature: Four Pioneering Environmental Advocates, 1845–1913.* Chapel Hill: University of North Carolina Press.

Dowie, Mark. 1995. *Losing Ground: American Environmentalists at the Close of the Twentieth Century.* Cambridge, MA: MIT Press.

Duus, Peter. 1976. *The Rise of Modern Japan.* Boston: Houghton Mifflin.

Eagle, Steven J. 2001. "The Birth of the Property Rights Movement." *Policy Analysis* (Cato Institute, Washington, DC), no. 404 (June 26): 1–39.

Ehrlich, Paul. 1968. *The Population Bomb.* New York: Ballantine.

Environmental Protection Agency. 2000. "Superfund: Twenty Years of Protecting Human Health and the Environment." EPA 540-R-00-007, OSWER Directive 9200.5-16. http://www.epa.gov/superfund/20years/20yrpt1.pdf (accessed March 1, 2008).

———. 2007. "Love Canal, New York, EPA ID#: NYD000606947." April 3. http://www.epa.gov/region2/superfund/npl/0201290c.pdf (accessed November 8, 2007).

George, Timothy S. 2001. *Minamata: Pollution and the Struggle for Democracy in Postwar Japan.* Cambridge, MA: Harvard University Press.

Gottlieb, Robert. 1993. *Forcing the Spring: The Transformation of the American Environmental Movement.* Washington, DC: Island.

Gresser, Julian, Koichiro Fujikura, and Akio Morishima. 1981. *Environmental Law in Japan.* Cambridge, MA: MIT Press.

Hane, Mikiso. 1982. *Peasants, Rebels, Women, and Outcastes: The Underside of Modern Japan.* New York: Pantheon.

Hays, Samuel P. 1959. *Conservation and the Gospel of Efficiency: The Progressive Conservation Movement, 1890–1920.* Cambridge, MA: Harvard University Press.

Helvarg, David. 1994. *The War against the Greens: The "Wise Use" Movement, the New Right, and Anti-Environmental Violence.* San Francisco: Sierra Club Books.

Hesse, Steve. 1998. "100 Years of Japan's Environment: Where to Go from Here?" *Japan Environment Monitor* 96, no. 4 (April–May), e-mail ed.

Huddle, N., and M. Reich. 1987. *Island of Dreams: Environmental Crisis in Japan.* Rev. ed. Cambridge, MA: Schenkman.

Irokawa, Daikichi. 1985. *The Culture of the Meiji Period.* Edited by M. B. Jansen. Princeton, NJ: Princeton University Press.

Karan, Pradyumna P. 2005. *Japan in the Twenty-first Century: Environment, Economy, and Society.* Lexington: University Press of Kentucky.

Kline, Benjamin. 1997. *First along the River: A Brief History of the U.S. Environmental Movement.* San Francisco: Acada.

Lam, Peng-Er. 1999. *Green Politics in Japan.* New York: Routledge.

Lewis, Jack. 1990. "The Spirit of the First Earth Day." *EPA Journal*, January/February. http://www.epa.gov/history/topics/earthday/01.htm.

Maruyama, Sadami. 1996. "Responses to Minamata Disease." In *The Long Road to Recovery: Community Responses to Industrial Disaster*, ed. James K. Mitchell. Tokyo: United Nations University Press.

Matsuno, Yu. 2005. "Local Government, Industry and Pollution Control Agreements." In *Environmental Policy in Japan*, ed. Hidefumi Imura and Miranda A. Schreurs. Cheltenham: Edward Elgar.

McKean, Margaret A. 1981. *Environmental Protest and Citizen Politics in Japan*. Berkeley and Los Angeles: University of California Press.

McNeil, J. R. 2000. *Something New under the Sun: An Environmental History of the Twentieth-Century World*. New York: Norton.

Mitchell, Richard H. 1976. *Thought Control in Prewar Japan*. Ithaca, NY: Cornell University Press.

Morse, Edward S. 1917. *Japan Day by Day, 1877, 1878–79, 1882–83*. Boston: Houghton Mifflin.

Nader, Ralph. 1965. *Unsafe at Any Speed: The Designed-in Dangers of the American Automobile*. New York: Grossman.

Nash, Roderick Frazier. 1989. *The Rights of Nature: A History of Environmental Ethics*. Madison: University of Wisconsin Press.

Notehelfer, F. G. 1975. "Japan's First Pollution Incident." *Journal of Japanese Studies* 1, no. 2 (Spring): 351–83.

Osborne, Fairfield. 1948. *Our Plundered Planet*. Boston: Little, Brown.

Oyadomari, Motoko. 1989. "The Rise and Fall of the Nature Conservation Movement in Japan in Relation to Some Cultural Values." *Environmental Management* 13, no. 1:23–33.

Pekkanen, Robert. 2006. *Japan's Dual Civil Society: Members without Advocates*. Stanford, CA: Stanford University Press.

Rabe, Barry. 2004. *Statehouse and Greenhouse: The Emerging Politics of American Climate Change Policy*. Washington, DC: Brookings Institute.

Schreurs, Miranda A. 2002. *Environmental Politics in Japan, Germany, and the United States*. Cambridge: Cambridge University Press.

Shabecoff, Philip. 1999. *A Fierce Green Fire: The American Environmental Movement*. New York: Hill & Wang.

Shutkin, William A. 2000. *The Land That Could Be: Environmentalism and Democracy in the Twenty-first Century*. Cambridge, MA: MIT Press.

Sutherland, Mary, and Dorothy Britton. 1995. *National Parks of Japan*. Rev. ed. Tokyo: Kodansha International.

Suzuki, David, and Keibo Oiwa. 1996. *The Japan We Never Knew*. Toronto: Stoddart.

Suzuki, Maggie. 1998. "21st Century Committee Questionnaire Reveals 'Worst 100 Wasteful Public Works Projects.'" *Japan Environment Monitor* 15:8–15.

Switzer, Jacqueline V. 1997. *Green Backlash: The History and Politics of Environmental Opposition in the U.S.* Boulder, CO: Lynne Riener.

Tanaka, Yuko. 1988. "The Cyclical Sensibility of Edo-Period Japan." *Japan Echo* 25, no. 2:12–16. Translation of "Edo shomin no chie ni manabu risaikuru," *This Is Yomiuri*, January 1998, 8190.

Totman, Conrad. 1989. *The Green Archipelago: Forestry in Pre-Industrial Japan*. Berkeley and Los Angeles: University of California Press.

Tsuru, Shigeto. 1999. *The Political Economy of the Environment: The Case of Japan*. London: Athlone.

Tsurumi, Kazuko. 1977. *Social Price of Pollution in Japan and the Role of Folk Beliefs*. Institute of International Relations for Advanced Studies on Peace and Development in Asia, Research Papers, ser. A.30. Tokyo: Sofia University.

Upham, Frank. 1987. *Law and Social Change in Postwar Japan*. Cambridge, MA: Harvard University Press.

Watanuki, Joji. 1984. *Politics and Ecology in Japan*. Tokyo: Sophia University, Institute of International Relations.

Wilkening, Ken. 2004. *Acid Rain Science and Politics in Japan: A History of Knowledge and Action towards Sustainability*. Cambridge, MA: MIT Press.

Wong, Anny. 1998. "The Anti-Tropical Timber Campaign in Japan." In *Environmental Movements in Asia*, ed. Arne Kalland and Gerard Persoon, 131–50. Copenhagen: Nordic Institute of Asian Studies.

Chapter 3

Virtual Grassroots Movements

*The Role of the National Geographic Society as a
Sustained Promoter of Environmental Awareness*

Stanley D. Brunn

The National Geographic Society (NGS) is widely recognized internationally as one of the world's premier organizations committed to informing children and adults about the places, regions, and landscapes in which we reside. It was founded in 1888 with the specific objective "to increase the diffusion of geographic knowledge." During the ensuing years, the NGS has been among the premier environmental groups supporting research on various subjects, including alpine environments, polar and circumpolar areas, undersea discoveries, nature and natural history (especially in biology), and ethnology and archaeological expeditions and explorations. The society has in the past century consistently supported the exploration of the frontiers of such fields as biology, geology, archaeology, anthropology, and alpine and marine environments. Research, as well as exploration, into the areas listed above has also been a hallmark of the society as it pursues its objective of informing both the scientific community and the general public about human, physical, and human/physical environmental processes and problems.

Since its inception, the NGS has served in various ways as a gatekeeper of environmental information. It has also played a crucial role in presenting information about the earth's plants, animals, and humans to both the scholarly community, especially earth and natural scientists, and the wider general public. Much of that information has been presented in visual form. The society's maps are perennially popular. And of course its most well-known magazine, *National Geographic,* is replete with outstanding photographs of places, landscapes, cultures, economies, and societies. During the past several decades, the society has moved more toward advocacy and activism, informing its supporters, scientists and laypeople alike, about crucial environmental issues and the state of the world's environments. It conveys its message most vividly in its several monthly magazines, especially *National Geographic,* but it also utilizes such other media as Discovery (its popular television channel),[1] its Web site,[2] and the wide variety of instructional materials it provides elementary, middle, and high school as well as college and university teachers.

My objective here is to present the NGS as a major and long-standing professional organization with strong commitments to environmental awareness and education, especially to conveying information visually, initially via photographs, and more recently through television and the Internet. I discuss the society's early interest in environmental subject matter, specifically exploration and discovery projects related to the natural sciences and earth history. I also examine *National Geographic*—specifically the change in content over the years—and the society's funding of geographic research. Finally, I examine the NGS's recent commitment to the promotion of environmental issues.

Funding for Research and Exploration

In its early years, the NGS clearly focused on exploration and discovery, especially in polar, circumpolar, and alpine regions (Raven 2000). Its first sponsored expedition in 1890 was a landmark survey of Alaska's St. Elias Mountains that mapped this region along the U.S.-Canadian border. It was on this expedition that Canada's second highest peak, later named Mt. Logan, was discovered. Later expeditions included that led by Robert T. Hill in 1902 to the site of the Mt. Pelée eruption in Martinique and that of William Beebe and Otis Barton in 1934 to the ocean floor in a bathysphere.

As of early 2008, nearly nine thousand research and exploration grants have been awarded. Funding has historically been supplied, not only for surveying and data collection, but also for the taking of photographs. The peak period for such geographic research was 1910–1940. Some important physical environment expeditions funded during those years were those of W. E. Clyde Todd to Hudson Bay and Labrador in 1912–1917, Frank Chapman to the Urubamba Valley of Peru in 1916–1917, Frederick Wulsin to Inner Mongolia and China in 1923–1924, and Lincoln Ellsworth to Death Valley in 1931.

The range of fields funded expanded over the years. The earliest funding went, not surprisingly, to geography. Vulcanology and glaciology were first funded during the first decade of the twentieth century, biology during the second, and mammalogy and ornithology during the 1920s. Funding of botany and oceanography began in the 1930s and of zoology in the 1940s. Special subfields began to be supported in later years, for example, ichthyology and entomology during the 1950s and herpetology and malacology during the 1960s.

During the last half of the twentieth century a number of scientists received multiple NGS grants; some of their names are familiar to those in the general public who follow National Geographic mentions in the media. These include Jacques-Yves Cousteau for his multiple Calypso expeditions from 1952 to 1967, Dian Fossey for her work on mountain gorillas in East Africa from 1967 to 1983, Gordon Frankie for his work on bees from 1979 to 1985, and Archie Carr for his work on green turtles and Harold Rehder for his on marine mollusks in the 1960s and 1970s.

More detailed insights into NGS research funding can be obtained by looking at the major categories and subcategories of research and exploration from 1888 to 1988. During this hundred-year period, there were 2,450 research projects funded (see NGS 1989). As already noted, some individuals received multiple grants, often in the same research area; also, some grant categories are overlapping. Still, the breakdown of funding in major categories as outlined in NGS (1989) is informative: 490 in anthropology, 253 in mammalogy, 234 in paleontology, 227 in ornithology, 207 in botany, 151 in vertebrate zoology, 120 in ethnology, 119 in entomology, and 100 in geology. The total for geography (conceived as encompassing research in ecology, physical anthropology, astronomy, and geophysics) would be 125 if the subcategories physical geography, human geography, and cartography were included. Other, minor categories include astronomy, ichthyology, and planetary science.

Further insights can be gained by examining the specific types of research conducted. Forest, plant collecting, and lichens were the major subcategories of botany. The breakdown of vertebrate zoology is more complicated. More of the mammalogy grants were awarded for the study of mammals generally, monkeys, whales, and bats than for that of bears, gorillas, seals, dolphins, elephants, lemurs, and sheep. More of the herpetology grants were awarded to the study of reptiles (lizards and turtles) than to the study of frogs and other amphibians. The ornithology grants covered finches, seabirds, falcons, and flamingos. And ichthyology grants covered various fishes as well as sharks. As for invertebrate zoology, more of the entomology grants were

Stanley D. Brunn

awarded for the study of butterflies, ants, spiders, and bees than to that of wasps, flies, and beetles.

The major funding categories from 1888 to 1988 were in three broadly defined environmental fields: biology (48 percent), anthropology (28 percent), and paleontology (11.5 percent). Geology and geography each garnered only 5 percent of the grants. Fewer went to astronomy and astrophysics (2.5 percent combined) and even fewer to history and sociology projects. Minuscule amounts were allocated for history and sociology. Specific funding for geographic research picked up again during the 1980s, when both physical and human geography proposals received support. Physical geography research was supported in twenty-four different locations and human geography in forty-one different locations.

Publications and Other Programs

The exploration of the natural world lends itself easily to visual learning. One can, indeed, learn about a physical environment, a village, a vegetation type, or a wildlife habitat simply by observing the photographs in *National Geographic* or, in a current technological context, on the Internet.

The NGS's publications for scholars, teachers and students, and a wider reading public have always been paramount in three areas. First, not surprisingly, exploration and discovery were important components for any initiative promoted by the society; these two foci are integrated into the organization's most fundamental goal, which is to advance the public's awareness of and thinking about matters geographic *and* environmental (McCarry 1997). Second, the society has been active since its inception in promoting studies that are wide-ranging in nature: biological, geologic, anthropological, and archaeological. These investigations may be carried out

in the rain forests and on the savannas of East Africa, in river basins and alpine areas of Central Europe, or in villages in coastal and highland Central America. The third focus is on what is commonly referred to in contemporary geography parlance as the *visual*. A focus on visual learning involves more than simply including a location map indicating where a field investigation was carried out or a photograph of a volcano erupting in Southeast Asia or a village market in South Asia (Carter 1999; Newman 2002; Poole 2004). A true visual understanding encompasses the messages behind the images. Sometimes—especially when complex and controversial materials are presented—those messages can be hidden and multiple (Lutz and Collins 1993). Questions about context must be addressed: Who took the photograph? For what purposes? When and why was the photograph taken? etc. Similar questions could be raised about maps, which are also constructed.

The visual learning component continues to be an important feature of any NGS initiative. One is consciously aware of the importance of visual learning when accessing the society's Web site or one of the many hyperlinks on it to spectacular photographs of wildlife, rivers and coasts, landforms and the atmosphere, human livelihoods, and the human condition of the planet. The society's Web site is discussed below.

There are several ways one can "read" NGS publications to learn about places, environments, people, and landscapes (NGS 1989). One can read them for the narratives they contain, that is, the written words provided about a subject. One can also read them only for the photographs and maps (NGS 1999; Bendavid-Val 2001; Newman 2001). For example, some readers might search out photographs of exotic landscapes or unusual features and use them to trigger geographic fantasies (Rothenberg 1994; Steet 2000), fantasies that can range from the educational and insightful to the voyeuristic. Other readers might search out photographic evidence of Western (i.e., developed-world) imperialism (Rothenberg 1994). Still other readers, considering the society's publications as gateways to information about environmental issues, might seek out information about threatened animal and plant populations.

As far as specific titles are concerned, *National Geographic,* the long-familiar bright yellow monthly magazine, is the society's signature publication. It is translated into more than twenty different languages, has more than 10 million subscribers worldwide, and is one of the most widely recognized magazines in the United States, found in homes and in the offices of physicians, corporate executives, dentists, and travel agents throughout the country. It is taken by more than six thousand libraries worldwide—not only public libraries in small towns and large cities, but also libraries in public and private colleges and universities.

Through the 1960s, *National Geographic* confined itself to presenting balanced views of physical and human geography. But, in the early 1970s, the magazine began to take a more activist stance, particularly where the environment was concerned. This change can be traced through some representative article titles. Offerings for 1969, for example, included "Taiwan: The Watchful Dragon" in the January issue, "Wild Elephant Roundup in India" in March, "Macao Clings to the Bamboo Curtain" in April, "Switzerland and Europe's High-Rise Republic" in July, "Sailing Iceland's Rugged Coasts" in August, and "Florida's Manatees: Mermaids in Peril" in September. The January 1970 issue, however, saw the article "Berlin: Both Sides of the Wall," and in the December issue there were three articles and a map related to the theme "Our Ecological Crisis." The following year, the importance of environmental awareness was again raised by the articles "Alaska's Tundra: Its Uncertain Future" in the March issue, "Fragile Nurseries of the Sea: Can We

Save Our Salt Marshes" in June, and "Will Oil and Tundra Mix? Alaska's North Slope Hangs in Balance" in October.

The NGS published two academic journals to promote the findings of those scientists whose projects it funded. *National Geographic Research* appeared from 1985 to 1990 and its sequel, *Research and Exploration,* from 1991 to 1994. These journals are now available in more than five hundred libraries worldwide. Over their combined life spans these journals published 268 articles, 165 of which (a little more than 60 percent) had an environmental focus. (It should be noted that I don't consider articles on archaeology, paleontology, and ethnography as having an environmental focus.) Sometimes entire issues would be devoted to an environmental topic. Examples from *Research and Exploration* include the Spring 1993 issue, which has seven articles on global warming, and the November 1993 issue, which has seven articles on water.

Of the many other NGS magazines currently being published, some are oriented toward teachers and young people, others toward a wider public interested in geography and the environment, and still others, including newsletters, toward more specialized audiences. *National Geographic Kids* is for children, and *National Geographic Explorer* is for elementary and middle school classrooms. *National Geographic Traveler* targets travelers and tourists, featuring in, for example, its October 2006 issue "59 Tours of a Lifetime," "Quebec City," "iPod Travel," "Wales Trip Planner," and "Local Outfitters." *National Geographic Adventure,* by contrast, reaches out to the armchair traveler as well with such articles (again in its October 2006 issue) as "Stalking Tanzania's Man-Eating Lions," "Alps Trail Dispatches," "Is Piracy on the Rise?" and "Where To Live and Play: 31 Amazing Adventure Towns."

The society promotes environmental awareness in other ways as well. For example, it publishes a variety of world, con-

Stanley D. Brunn

tinental, and regional maps for schools and offices. More specialized maps of various sorts—ranging from maps detailing the distribution of flora and fauna to those offering the latest satellite images of built or natural environments—appear as supplements in monthly issues of its magazines. Other materials—atlases, books about nature and adventure, and videos about the planet's natural environments, cultures and customs, and history—are also available. Most prominent among these are the PBS series *The Incredible Planet,* which the society launched in 1975, and the *National Geographic Explorer* television series, which debuted a decade later and continues today on the immensely popular National Geographic Channel. These varied means spread the message about such issues as the disappearance of plant and animal species, the human struggle for cultural survival, and the short- and long-term consequences of climate change.

Other programs and initiatives signal the continued commitment of the NGS to environmental awareness. In 1988, in order to emphasize the importance of geography's role in the modern world and to improve its instruction, especially in the public schools, the society initiated the National Geographic Education Foundation,[3] funds from which are allocated to train elementary, middle, and high school teachers nationwide in such subjects as the physical environment, resource use, human use and misuse of the earth, conservation, and sustainability. A related, more-high-profile program is the National Geographic Bee—sponsored by the society every year since 1989—in which fourth through eighth graders are tested on their knowledge of geography. Also in the late 1980s, the society organized the Committee for Research and Exploration, whose objective was to assist the efforts of scholars and scientists around the world in increasing knowledge of our planet, its environment, and its inhabitants. Also relevant to our concerns here are

the Conservation Trust, "dedicated to the conservation of the world's biological and cultural heritage" and "support[ing] innovative solutions to issues of global concern";[4] the Expeditions Council, funding "exploration and adventure around the world" in order to "foster a deeper understanding of the world and its inhabitants";[5] and the Explorers in Residence program, whose goal is "to highlight and enhance [the society's] long-standing relationships with some of the world's preeminent explorers and scientists," many of whose initiatives are conservation related.[6]

From its inception more than a century ago, the NGS has advanced environmental awareness, promoting both research and educational efforts. It has been particularly innovative in its use of visual technologies, its ventures ranging from the early print version of *National Geographic,* to the later televisual Discovery Channel, to the most recent hyperspatial NationalGeographic.com. Students and teachers of all ages and at all levels benefit regularly from the broad range of programs and services the society offers. The NGS has been, and remains, an important agent of change, a tireless advocate of the earth in the face of ever-increasing environmental assaults.

Notes

1. See http://dsc.discovery.com.
2. See http://www.nationalgeographic.com.
3. See http://www.nationalgeographic.com/foundation /index.html.
4. See http://www.nationalgeographic.com/conservation /index.html.
5. See http://www.nationalgeographic.com/council.
6. See http://www.nationalgeographic.com/explorers-program/eir.

References

Bendavid-Val, Leah. 2001. *Stories on Paper and Glass: Pioneering Photography at National*

Geographic. Washington, DC: National Geographic Society.

Carter, Alice A. 1999. *The Art of National Geographic: A Century of Illustration*. Washington, DC: National Geographic Society.

Lutz, Catherine, and Jane L. Collins. 1993. *Reading National Geographic*. Chicago: University of Chicago Press.

McCarry, Charles, ed. 1997. *From the Field: A Collection of Writings from National Geographic*. Washington, DC: National Geographic Society.

National Geographic Society (NGS). 1989. *National Geographic Index, 1888–1988*. Washington, DC: National Geographic Society.

———. 1999. *National Geographic Photographs: The Milestones: A Visual Legacy of the World*. Washington, DC: National Geographic Society.

Newman, Cathy. 2001. *National Geographic Fashion*. Washington, DC: National Geographic Society.

———. 2002. *Women Photographers at National Geographic*. Washington, DC: National Geographic Society.

Poole, Robert M. 2004. *Explorers House: National Geographic and the World It Made*. New York: Penguin.

Raven, Peter H. 2000. *National Geographic Expeditions Atlas*. Washington, DC: National Geographic Society.

Rothenberg, Tamar Y. 1994. "Voyeurs of Imperialism: The *National Geographic* Magazine before World War II." In *Geography and Empire*, ed. Anne Godlewska and Neil Smith, 155–72. Oxford: Blackwell.

Steet, Linda. 2000. *Veils and Daggers: A Century of National Geographic's Representation of the Arab World*. Philadelphia: Temple University Press.

Chapter 4

Going Global

The Use of International Politics and Norms in Local Environmental Protest Movements in Japan

Kim Reimann

Over the past several decades, the environmental protest movement in Japan has undergone various ebbs and flows. After very high levels of activism at the local level in the 1960s and early 1970s, it entered a period of relative quiet in the late 1970s through the mid-1980s. From the late 1980s, however, a new wave of environment-related activism appeared that continues to the present and is richly detailed in this volume. In this most recent wave, new forms of activism have blossomed, and the environmental movement in Japan appears to have entered a new phase. When reviewing the protest campaigns that have been the most successful in the past decade in terms of organizing and capturing the attention of the public and policymakers, two interesting characteristics about the new movements stand out: (1) increasing numbers of *national-level networks* connecting local movements throughout the country and (2) the increasing use of *postmaterialist protest frames* that go beyond the victim-compensation and not-in-my-backyard (NIMBY) arguments of the 1960s and 1970s and call attention instead to the priceless value of nature and the need to protect fragile ecosystems.

This chapter explores the emergence and growth of these new movements in the last twenty years and focuses on one important source of the recent changes in movement organization and framing: international factors and the connection between local and global social movements. Building on the new literature in political science and sociology on transnational advocacy networks and social movements, I examine the ways in which grassroots activists and nongovernment organizations (NGOs) in Japan were influenced by global trends and seized new international opportunities presented to them to advance their cause. Since the early 1990s, international allies, norms, and institutions provided local activists in Japan new resources, frames, and incentives with which to organize at the national level. Although international factors represent only one influence shaping the new wave of environmentalism in Japan, they have often been overlooked in previous studies and play a more crucial role than one might expect.

This chapter is divided into three main sections. First, I present general background information on the new wave of environmental protest in Japan in the late 1980s and 1990s and highlight the characteristics that distinguish this wave from previous ones. Next, drawing on the new literature on transnational social movements, I outline in a theoretical section the ways in which international factors potentially influence movements at the national and local levels. Finally, I examine two well-known protest campaigns of the 1990s—the movement to stop the construction of the Nagara River Estuary Dam and the campaign to halt the Isahaya Bay Land Reclamation Project—and trace how international factors aided local activists by encouraging them to form more extensive national networks and providing them with new resources, new frames of reference, greater legitimacy, better media coverage, and more "voice" at the national level.

Background: The New Environmental Activism of the 1990s

In the 1960s and early 1970s, Japan had one of the most active environmental protest movements in the world with as many as three thousand different protest campaigns and ten thousand local disputes recorded in 1973 (Krauss and Simock 1980, 187). Protests during this period were mainly local in their focus, and, unlike that in the United States and several other industrialized countries, the movement never coalesced into a strong national-level movement. Until relatively recently, most environmental groups in Japan were small, local organizations that tended to have a NIMBY focus emphasizing health-related issues, pollution-victim compensation and stopping development projects that threatened the economic livelihood of local residents (Krauss and Simock 1980; McKean 1981; Schreurs 2002, chaps. 2–3; Broadbent 1998).

By the end of the 1970s, the environmental protest movement began to lose steam as the government rose to the challenge of dealing with the various serious environmental problems facing Japan. Tough antipollution legislation and measures enacted by the government in the early 1970s started to have their intended effect, and, by the late 1970s, Japan's exceedingly high levels of pollution receded to tolerable levels. State-mediated arbitration and compensation schemes set up at this time also provided new institutional channels through which to address local environmental problems, simultaneously discouraging protests, lawsuits, and other contentious tactics used by frustrated citizens in the 1960s and early 1970s (Upham 1987, chap. 2). In the late 1970s and early 1980s, thus, the number of environmental protests declined, and the movement in general became less visible (Schreurs 2002, chap. 3; Mason 1999, 189). Although there were movements for recycling and promoting organic produce, for example,

these sorts of consumer movements were of a very different nature than the more explicitly political and confrontational environmental movements of the 1960s and early 1970s that emphasized pollution victims.

Since the late 1980s, however, new or revitalized protest campaigns have appeared in numerous parts of Japan, and a second wave of environmental protest that continues to this day has spread across the country. Environment-related citizen activism that has captured national media attention and contributed to the general public debate on the environment from the late 1980s through the first decade of the twenty-first century includes antidam protests, campaigns to save wildlife habitats such as wetlands and coral reefs, anti–nuclear power protests, the anti–golf course movement, and campaigns to stop global warming (Niikura 1999; Fukatsu 1997; Cameron 1996, chap. 4; Sugai 1992; Reimann 2001).

Compared with the first wave of environment-related protests in the 1960s and early 1970s, these more recent movements have two distinguishing characteristics. First, in contrast to the very local focus of the earlier protest movements, environmental protest campaigns since the late 1980s—while still local in their immediate goals—are more likely to include national networks and involve cooperation between local and national groups. The antidam movement in Japan, for example, is composed of a set of very active local antidam campaigns that are now linked at the national level through the National Coalition on Water Resource Development Issues [Suigen kaihatsu mondai zenkoku renrakukai].[1] Similar national-level networks that bring together local campaigns include the Japan Wetlands Action Network [Nihon shitchi nettowaku], the Global Network for Anti–Golf Course Action, the Kiko Forum [Kiko foramu], the Save the Ozone Network [Sutoppu furon zenkoku renrakukai], and the Biodiversity Network Japan [Seibutsu tayosei

Japan]. In addition to these networks, national NGOs based in Tokyo such as the World Wildlife Fund–Japan, Greenpeace-Japan, the Friends of the Earth–Japan, the Wild Bird Society, and the Nature Conservation Society of Japan have also supported and participated in campaigns waged at the local level, helping them obtain higher levels of national attention.

In addition to this new national dimension, protest movements of the last twenty-five years also differ from earlier ones in terms of the sort of issues championed and the type of framing used. As previously noted, most of the protest movements of the 1960s and 1970s tended to focus on victims of severe pollution and/or development projects. The goals of such movements were to stop the most blatant and dangerous forms of environmental degradation and often stopped short of questioning the larger Japanese national goal of industrialization and development. In contrast to this, the movements of the second wave have tended to define their goals in broader and more postmaterial terms such as protecting the environment for quality-of-life and cultural reasons. Although many of the recent campaigns have involved materialist issues facing local communities, movement leaders from the 1990s on have often downplayed local politics, attempting instead to frame their causes and goals in national or universal terms. Campaigns to prevent local development projects, for example, are now more likely to adopt the postmaterialist frame of preserving local natural habitats for their own sake as precious natural settings whose destruction would be a loss for all Japanese.

This reshaping of environmental protest movements is in some ways quite remarkable considering the various obstacles facing activists who seek to organize nationally and question the prodevelopment status quo. As numerous political scientists and sociologists have noted, outsiders who challenge the state and the entrenched power elite in Japan face great institutional and cultural obstacles to

organization. Institutionally, legal and tax regulations for nonprofit organizations have until recently been very strict and have made it very hard for advocacy groups—especially those critical of the government—to incorporate, raise money, and establish a large membership base (Pekkanen 2000, 2006; Holliman 1990). Other institutional barriers include limited access to public information and a cumbersome legal system that discourages the use of the courts (Upham 1987, chap. 1). Cultural norms emphasizing harmony and cooperation have also discouraged challenges to authority and changes from below, and the state has often exploited such norms by isolating protestors in a way that makes their demands appear selfish (Pharr 1990; Broadbent 1998).

International Movements and Organizations as Political Opportunities

To understand these new patterns of protest and environmental activism, one needs to place Japan in the larger context of global environmental movements and evolving international political institutions and norms. Although domestic factors—for example, overdevelopment and environmental degradation, rising concerns among citizens about environmental issues, and negative socioeconomic impacts on local communities—have been important causes for the reemergence of the movement, they alone cannot fully account for the new patterns. Institutional and cultural barriers to organizing national movements and challenging the status quo in Japan remained fairly constant from the 1960s through the mid- to late 1990s, and, given this fact, one might have expected the recent movements to closely resemble those of the earlier period.

The connection of local protest movements to global movements and international factors is important to investigate since it is precisely in the late 1980s and the 1990s that global environmental issues emerged as an important focus of international relations. During this period, new ideas and opportunities at the international level not available in the 1960s and 1970s appeared and provided local activists with new tools for mobilizing and organizing protest. The second wave of environmental protest in Japan, thus, is part of a larger international trend that started in the 1980s and saw local and national movements in many parts of the world "going global" and tapping into new international resources and ideas.

A general survey of the new literature in political science and sociology on norms, transnational advocacy networks, and transnational social movements reveals that there are three main categories of international factors potentially providing new opportunities and resources for activists working at the local and national level: international norms, international allies, and international institutions. The rest of this section outlines these three factors and how they potentially transform domestic politics.

International Norms

In the past decade, scholars have become interested in the role of norms in international politics (Finnemore 1996; Florini 1996; Klotz 1995; Axelrod 1986; Risse, Ropp, and Sikkink 1999). Commonly defined as "shared expectations about appropriate behavior held by a community of actors" (Finnemore 1996, 22), norms are agreed-on standards or rules that can both regulate state behavior and serve as the basis for state identity. As a norm becomes accepted by a critical number of states and is institutionalized in international agreements and laws, it becomes increasingly difficult for states to ignore or violate it and remain a member of good standing in the international community promoting it (Finnemore 1996, chap. 1; Finnemore and Sikkink 1998).

International norms provide two possible types of resources for domestic-level activists and social movements. First, local and national activists can adopt an international norm as a new frame and mobilizing theme for their movement in order to universalize their cause and gain a new sort of legitimacy. Since international norms often represent universal ideals and aspirations, the incorporation of an international norm into the frame of a local or national social movement can allow that movement to overcome particularistic concerns and reach out to a wider audience. As movement leaders adopt new international norms, they often start to look at local problems from a different perspective and see their struggles in a different light. International norms also potentially provide social movements legitimacy vis-à-vis the state since it is much harder for states to alienate and criticize movements that successfully frame themselves as universal ideals.

In addition to reshaping movement frames, international norms also provide activists a tool with which to criticize the state and bolster their position. That is, such norms provide standards for state policies, and, if activists are able to show that the state is not living up to international agreements or accepted norms, they can expose it to international criticism and gain a source of legitimacy for their own critical positions.

In the area of the environment, international environmental treaties and conferences in the late 1980s and the 1990s collectively brought with them the new normative ideal of "sustainable development" that comprised both specific policy measures, such as environmental impact assessments, and a more ideational goal of respecting the natural environment (Held, McGrew, Goldblatt, and Perraton 1999, chap. 8). In addition to being a form of development that "meets the needs of the present without compromising the ability of future generations" (World Commission on Environment and Development

1987, 43), sustainable development was also a norm that emphasized the importance of democratic decisionmaking processes, the inclusion of local communities and citizens in policymaking, and state accountability in the area of the environment. Although the ideal of sustainable development first appeared in the international arena in the 1970s, it was only in the late 1980s and the 1990s, with advancements in scientific knowledge about global environmental degradation and the increasing international institutionalization of environmental regulations, that proenvironment norms gained an international ideational power that was exploitable by domestic actors. By 1992 and the commencement of the UN Conference on the Environment and Development (UNCED), the concept of sustainable development was one that few industrialized states would publicly challenge. With the creation of new international environmental treaties, it was now possible for NGOs and other societal actors to hold states accountable for environmental policies by referring to international standards and appealing to international norms.

International Allies

The second important international resource and opportunity for domestic social movements is the existence and support of international allies: international NGOs (INGOs), private foundations, foreign government officials, officials from international organizations, celebrities, etc. As the work of Sikkink, Keck, Brysk, and others has shown, transnational networks of activists have allowed "blocked" activists to circumvent obstacles at the local and national level and turn to the international arena for support (Keck and Sikkink 1998; Brysk 1994). International allies provide activists and NGOs with resources that may be lacking in the domestic context. These resources are both material and nonmaterial. Materially, foreign actors

can provide needed financing for NGO projects and activities that are unlikely to secure funding domestically. In terms of nonmaterial resources, international allies can help groups acquire international attention that, in turn, provides various benefits at home, such as greater domestic legitimacy and increased media coverage. When they involve powerful foreign governments that are sympathetic to the group or movement, these alliances can provide groups with the external political pressure needed to influence their own government (Martin and Sikkink 1993; Sikkink 1993; Keck and Sikkink 1998; McAdam 1998). Finally, international allies also act as agents for the transnational diffusion of ideas, and it is often through their contacts with international actors that activists are exposed to international norms and new forms of activism.

International Institutions

Finally, international institutions—both regimes and organizations—offer another set of important political opportunities for national groups excluded from the policy process domestically. As an international layer of political institutions, separate from but interconnected to domestic political institutions, international organizations offer a separate potential channel of access to decisionmakers. International conferences and meetings of international institutions (UN agencies, the World Bank, conferences of the parties to various treaties, etc.) are alternative political spaces where groups can voice their concerns and attain international exposure that can be used as a political resource to gain access to domestic policymaking processes from which they are otherwise excluded (Risse-Kappen 1995; McAdam 1998). Lobbying opportunities at international conferences, formal mechanisms set up by some governments to consult with NGOs at international meetings, and even more elaborate arrangements such as NGO representation on official delegations or the establishment of preconference policy "dialogues" at home all provide incentives for groups to organize or go global. Meetings of international institutions tend to be covered extensively by the press, so participation in them—especially when they take place in an organization's home county—offers public relations opportunities for an NGO, especially when its cause can be portrayed as conforming to international standards or approaches.

UN and other official international conferences also often serve as focal points around which national and local groups mobilize, coordinate activities, and otherwise work together. By providing activists with a unified target, venue, or basis for common action, international organizations and treaties and their conferences have stimulated new connections between local and national groups that previously had worked separately. The increasing number of international conferences from the late 1980s and the emergence of international environmental treaties in the 1980s and 1990s were important developments that helped domestic actors interested in similar environmental issues find each other and join forces. As groups prepare for an international conference, for example, they often create new networks that stimulate new forms of activism and collaboration at the national level as well.

The Cases: The Nagara River Estuary Dam and Isahaya Bay Land Reclamation Project

The new wave of environmental activism in Japan in the last twenty-five years or so is a useful case to examine for the influence and presence of international factors since the country contains several conditions that would seem to promote the use of international resources and opportunities. As previously mentioned, the domestic conditions

there in the 1980s were not favorable, and the fact that environmental activists in Japan were blocked domestically gave them strong incentives to go global and pursue international strategies. In the late 1980s, activists in Japan became more aware of international opportunities and factors since it was around this time that Western INGOs and foundations came to Tokyo seeking partners. Some of these INGOs and foundations started programs supporting Japanese NGOs (Greene 1990). Activists in Japan were, thus, aware of international resources and more likely to use them than activists in countries that were internationally isolated.

Japan is also a favorable location for the use of international tactics since statesmen there have tended to be sensitive to the country's international image and its place in international society. Since Japan's entrance into modern international politics in the nineteenth century, its leaders have been concerned about its role in the world and about gaining acceptance for it as a world power (Tamamoto 1993; Dore 1979–1980). This concern has made Japan sensitive to *gaiatsu* [outside pressure], and that sensitivity has been strategically used by a variety of actors both within and outside Japan (Gurowitz 1999; Schoppa 1997).

For environmentalists in Japan, international norms and *gaiatsu* became more potentially useful tools in the late 1980s and early 1990s since it was during this time period that the country emerged as an economic superpower looking for ways to increase its role and prestige in the international community. Limited by article 9 of the Japanese constitution to nonmilitary international contributions, leaders in Japan chose the environment as an area in which it would show international leadership through new global policy initiatives (Ohta 2000, 100–107). Such international aspirations made Japan more sensitive to normative pressure and criticism by environmentalists. If activists could show that Japan's domestic environmental policies did not live up to international standards, for example, the country's ability to be a global leader in the environment could be called into question.

The two cases chosen for this chapter are landmarks in the history of contemporary Japanese environmental movements because of the national networks of groups created and the high levels of media attention, and, consequently, the widespread public support, attracted. Although unsuccessful in that they did not attain their immediate objectives—stopping the construction of the Nagara River Estuary Dam and the Isahaya Bay Land Reclamation Project—these two campaigns did succeed in publicly calling into question state policies that privileged public works construction projects and economic development over environmental preservation and biodiversity. They also had an enormous influence in raising public awareness in Japan about environmental issues, and their highlighting of the failures of the state to adequately consider environmental and citizen concerns, in turn, paved the way for the success of other, later campaigns. The unsuccessful fight against the Nagara dam, for instance, was followed by the successful campaign to

halt the construction of a dam on the Yoshino River. The failure to save the wetlands of Isahaya Bay led to the more recent success of similar campaigns to preserve the Fujimae wetlands in Nagoya Bay and the Sanbanze tidal flats in Tokyo Bay. The two campaigns also revealed the need for the Japanese government to increase public participation in and the transparency of the decisionmaking process and to make itself more accountable to the public—norms that were being promoted through international conferences as integral aspects of the new paradigm of sustainable development.

The Nagara River Estuary Dam

The national campaign against the construction of an estuary dam on the Nagara River had its roots in the collapse of a long local struggle against the dam that started in 1968 and ended in 1988 when the final lawsuits by the fishing industry were settled (see figure 4.1). As the Ministry of Construction (MOC) and the Water Resources Development Corporation [Mizu shigen kaihatsu kodan] began making plans with local authorities to implement the dam project in 1988, a new move-

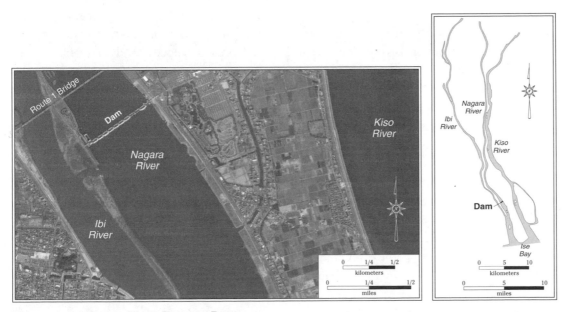

Figure 4.1. Nagara River Estuary Dam.

ment quickly emerged and spread the protest to the national level with the formation of SANREDC, the Society against the Nagara River Estuary Dam Construction [Naga-ragawakuchi seki kensetsu wo yamesasera shimin kaigi]. A network that eventually included sixty-three local and national groups and sixteen thousand members, SANREDC and its supporters represented a wide cross section of people from all parts of Japan, including fishermen, canoeists, biologists, writers, photographers, academics, local residents, national environmental NGOs, politicians, and celebrities (Cameron 1996, 147). In addition to its lobbying activities in Tokyo, SANREDC successfully mobilized large-scale events such as the annual festival-like Nagara River Day, which drew a crowd of up to fifteen thousand people and thousands of canoes as well as sparking protest rallies in Tokyo and a hunger strike that involved 114 people (Cameron 1996, 149; Niikura 1999, 100; Kyodo News Service 1992c; *Japan Environmental Monitor* [*JEM*], no. 32 [vol. 3, no. 10] [March 1991]: 18). The dam was completed in 1996, and this chapter examines the opposition movement from its conception in 1988 to its high point in 1996.

Norms and Frame Shift　Although the battle against the construction of the dam began at the local level in 1968, SANREDC's formation in 1988 marked a clear shift in the type of frames used by activists protesting the dam. Before 1988, the main concerns voiced by local protestors related to flood threats and how the loss of fish catch would hurt local fishermen. From 1988 and the spread of the campaign to the national level, the frame of the movement shifted to a broader, more postmaterial one emphasizing the natural and cultural heritage value of the river and the right of citizens to conserve natural resources for all Japanese. As is reflected in the campaign's slogan, "Preserve the Nagara River,

the Last Free-Flowing River in Japan!" activists focused on the loss of biodiversity and recreational pleasure that would result from dam construction and framed the issue as the loss not merely of one river but of clean rivers all over Japan (Niikura 1999, 100; Cameron 1996, 172–74). The movement also framed itself in national terms as an attempt to stop political corruption in public works projects, bring greater accountability and transparency to the policymaking process, and allow for greater participation of citizens in that process.

This new frame fit into the larger norm of sustainable and participatory development that was being promoted at the international level at the time, and, from the early years of the campaign, its leaders tapped into international norms and ideas emerging in antidam movements in other parts of the world. As the following sections show, the participation of the Nagara campaign in a larger global movement allowed it to articulate this new, post-material frame as a universal one that gave it greater legitimacy.

International Organizations: UNCED　The 1992 UN Conference on Environment and Development in Rio de Janiero was an important early influence on the Nagara antidam movement in several ways. First of all, UNCED helped SANREDC expand the movement from being a local one to being one with greater national-level participation in its early stages. The official preparatory process for this major international conference started in 1989 and became a focal point for NGOs and environmentalists all over Japan, who coordinated their efforts for the conference. SANREDC leaders were part of these preparations in 1989 and, through them, networked and met with activists from all parts of Japan. This process helped SANREDC greatly in its efforts during its early campaign years to mobilize national-level membership and support.

Participation in preparations for UNCED also exposed SANREDC organizers to anti-dam movements in other parts of the world and helped symbolically link the movement in Japan to the larger global antidam movement. In particular, SANREDC linked its own movement with the movement in India against the Sardar Sarovar Dam Project on the Narmada River, leading to a joint appearance of the leaders of both campaigns at an UNCED-related NGO conference held Japan in 1992 to talk about the similarities in the two movements. This linkage to the Narmada dam campaign was strategically useful for the Nagara activists since the Japanese government had withdrawn foreign aid financing to the Narmada dam project in 1990. Given the commonalities in the two cases, Nagara activists publicly argued that Japan should apply the same standards at home as it did abroad.

UNCED itself was also a public relations opportunity for the movement. The Japanese NGO country report complied for UNCED included a section on the Nagara River and the dam project, and Japanese NGOs chose SANREDC's leader, Amano Reiko, to speak about dams at the conference's Japan Day ('92 NGO Forum Japan 1992). Although Japanese government officials decided to cut Amano out of the program, her exclusion itself became a news item that drew attention to the Nagara case when most Japanese NGOs boycotted the event in protest (Cameron 1996, 178; Yokota and Iiyama 1992). Extensive Japanese press coverage of UNCED, thus, gave SANREDC additional public exposure.

Finally, it was at UNCED that Amano made important connections with major international environmental NGOs and famous Western environmentalists such as David Brower, the former executive director of the Sierra Club and founder of Friends of the Earth–US. As a follow-up to UNCED, she went to San Francisco to develop links between Japan's domestic and international policies, network and strategize with American NGOs, and generate media interest in the Nagara case (Cameron 1996, 150).

International Allies and Alliances International allies and alliances provided the Nagara campaign with numerous important resources, including normative support and legitimization, better press coverage, and a source of external political pressure. Immediately after UNCED, the participation of international actors in the Nagara movement noticeably increased. Starting in 1992, SANREDC invited prominent foreign speakers and experts to its annual Nagara River Day event and organized speaking tours in Japan for these international visitors as part of their trip. Speakers who appeared at SANREDC events in the early to mid-1990s include David Brower, Robert Herbst of the Tennessee Valley Authority (TVA), the ecological expert Robert Goodland from the World Bank, the freshwater biodiversity expert Jan Abramovitz from the WorldWatch Institute, the activist leader Dai Qing from the campaign to stop the Three Gorges Dam Project in China, the British author Fred Pearce, and the commissioner of the U.S. Bureau of Reclamation, Daniel P. Beard.

During their public appearances in Japan, all these international actors presented impressive cases for the environmentally destructive effects of dams and gave the Nagara movement normative support that legitimized its goals by relating them to the changing international view of dams. Reflecting the basic principles of sustainable and participatory development, the new emerging consensus on dams in the early 1990s in industrialized countries was that large-scale dam construction was environmentally destructive and that dam development had to be carefully planned to account for local concerns. Two figures who were particularly meaningful in presenting these new views in Japan were

Robert Herbst and Daniel Beard, both well known in Japan for their historical role in dam construction in the United States. The TVA, for example, was a model for dam and infrastructure development in Japan in the 1960s and 1970s and was highly respected by bureaucrats there. By the time Herbst came to Japan as a guest of SANREDC in 1992, however, the TVA's policies on dams had evolved considerably to a new position questioning the environmental sustainability of large-scale dams. Herbst warned Japan that TVA dam promotion was not a model to follow and that all new dam proposals should take environmental considerations and costs into account. Two years later, Daniel Beard arrived in Japan with a similar message, stating in his various speeches, television appearances, and presentations: "The era of dam construction is over" (Kyodo News Service 1992b; Beard 1996). Because Beard was highly respected, his support of the Nagara movement provided it legitimacy and made it much more difficult for state officials in Japan to portray the movement as purely selfish and antidevelopment. Beard's active promotion of sustainable development norms in Japan also had the effect of depicting the Japanese government as an international laggard in the industrialized world in terms of the new emerging norm on dam construction.

In addition to aiding the Nagara movement's normative framing and legitimacy, international allies also generated media coverage for the movement. As soon as foreign experts and figures started to appear in Japan in 1992 and support the antidam cause, media attention in Japan on the case also started to increase (Cameron 1996, 150, 178). SANREDC and its Japanese supporters also found the Western media sympathetic and useful in generating interest in their campaign and placed ads in and wrote letters to the *New York Times* as yet another strategy for attracting international attention and increasing media coverage in Japan (Cameron 1996, 150).

Finally, international allies also made efforts to lobby Japanese officials and elites on behalf of SANREDC, and their outside pressure made it much harder for the Japanese government to ignore the case. During their visit to Japan in 1992, for example, Herbst, Brower, and other foreign allies held a press conference in Tokyo when MOC officials refused to meet with them. In addition to pointing out problems with the proposed dam and suggesting that an independent team be set up to review the project, Herbst publicly criticized the MOC for its lack of consultation with NGOs and citizens about the project, pointing out how the TVA made conscious efforts to involve the public in decisionmaking processes (Cameron 1996, 164; Kyodo News Service 1992a). When SANREDC staged hunger strikes in 1993, forty foreigners from fourteen countries sent faxes and letters to the MOC, urging it to meet with protesters and cancel the dam project. Another example of how foreign allies attempted to engage the Japanese political elite came in 1996 when the U.S. Bureau of Reclamation hosted and organized a dam study tour in the United States for Japanese Diet members interested in current U.S. public works and dam policy (*JEM,* no. 83 [June 1996]: 10–11).

Despite these and many other efforts, however, Amano and SANREDC were, ultimately, unable to achieve the protest movement's goal, and the dam was completed in 1996. Since then, after other failed attempts to get rid of the dam, the project has become a well-known symbol in Japan of both a new burst of citizen activism in the 1990s and wasteful public works projects.

The Isahaya Bay Land Reclamation Project

Sponsored by the Ministry of Agriculture, Forestry, and Fisheries (MAFF), the Isahaya Bay Land Reclamation Project involved the reclamation of a portion of Isahaya Bay on

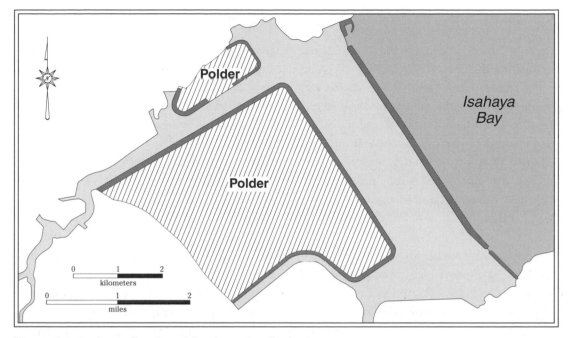

Figure 4.2. Isahaya Bay Land Reclamation Project.

the Ariake Sea in Nagasaki Prefecture (see figure 4.2). Initially proposed to increase agricultural production, then later justified as necessary for flood control, the project threatened the existence of one of the biologically richest wetlands areas in Japan.[2] As was the case with the Nagara dam project, the Isahaya project also involved decades of resistance by local fishermen whose economic livelihood it threatened. In 1989, after the last local fishing unions finally gave in to pressure by local government authorities and MAFF and planning on the project began, a new campaign to stop the project was launched. Led by the local marine biologist Yamashita Hirofumi, a new national network called JAWAN, the Japan Wetlands Action Network, formed in 1991, bringing together wetlands activists from other parts of Japan. By the mid-1990s, the campaign to save Isahaya Bay became one of the most well-known environmental battles in Japan and generated a great deal of public support through dramatic media images such as the writhing death throes of mudskippers [*mutsugoro*] in the drained

bay. Although the battle was lost when land reclamation commenced in 1997, the movement remained active through the late 1990s and into the first decade of the twenty-first century. This section examines transnational aspects of the campaign from its inception in the late 1980s to it high point in 1997–1998.

Norms and Frame Shift As was the case with the Nagara dam protest, the end of the local phase of protest against the Isahaya land reclamation project in 1988–1989 and the emergence in 1990–1991 of a new protest campaign with a more national-level organization and approach brought with it a clear shift in protest frames. Instead of being a local problem threatening the livelihoods of local fishermen, the frame adopted by protestors from 1990 presented the project in sustainable development terms as an environmental disaster with few real development benefits. Appealing to postmaterial nature lovers, the movement focused on such environmental costs as loss of biodiversity, threats to endangered species, and rapidly shrinking wetlands areas in Japan for bird-watching. Since 51 percent of Japan's birds species are waterfowl, activists argued, the loss of the wetlands had serious implications for wild birds in Japan. Furthermore, since Japan was a major East Asian flyway for migratory birds, wetlands loss there had larger regional and global environmental impacts. This new frame consciously built on normative arguments and frames promoting sustainable and participatory development found in the international arena at the time.

International Organizations International organizations, conferences, and treaties were important to Isahaya activists in numerous ways: as focal points for organizing nationally, as lobbying venues, as sources of media coverage, and as locations where contacts could be made with international allies. The international conference that proved to be particularly significant for JAWAN was the Fifth Conference of the Parties to the Ramsar Convention in Kushiro, Japan, in 1993. Formally called the Convention on Wetlands of International Importance Especially as Waterfowl Habitat, the Ramsar Convention is an international system for the designation and protection of important wetlands sites. The parties to the convention meet every three years, and its convening in Japan provided unique international opportunities for the movement.

First of all, the conference proved to be a focal point around which wetlands activists in Japan started to organize at the national level in a more effective way. The creation of JAWAN in 1991 was, in part, a reaction by Yamashita and the movement to the announcement in 1989 by the Japanese government that it would compete to host the international conference (*JEM*, no. 35 [vol. 4, no. 3] [June 1991]: 4–7; *JEM*, no. 18 [vol. 2, no. 6] [October 31, 1989]: 3). Prior to 1991, national networking on wetlands had, according to one activist, "not been particularly en-

couraging" (*JEM*, no. 26 [vol. 3, no. 4] [July 1990]: 9), but, with the goal of organizing for the Ramsar conference in Kushiro, JAWAN began organizing events around Japan that led to a much more organized national network of local wetlands groups. In October 1992, national organizing progressed a step further when JAWAN joined forces with three national environmental NGOs—the Wild Bird Society of Japan, the Nature Conservation Society of Japan, and Friends of the Earth–Japan—to form the Wetlands Coalition '93 in order to present a united front of Japanese environmental groups at Kushiro. By the time of the conference in 1993, a fairly organized national movement for wetlands protection existed within Japan, providing a new organizational resource for individual protest campaigns throughout Japan. The conference, thus, was an important focal point that aided the consolidation of many local protests into a single, more coordinated national network and movement (*JEM*, no. 54 [vol. 5, no. 2] [May 1993]: 16–17).

The Ramsar conference also provided a lobbying opportunity for Yamashita and other wetlands activists that was particularly well suited for applying international pressure on Japan using new arguments drawing on sustainable development norms. As the host of the conference, the Japanese government soon came under pressure to increase wetlands protection in Japan to show that it supported the ideals of the convention. In 1992, Japan had the lowest number of Ramsar sites among industrialized nations—it ranked fifty-fourth in the world in terms of actual covered acreage—and activists could point to the country's laggard international status to pressure the government to stop public works projects in biologically valuable wetlands areas such as Isahaya. In the various official meetings leading up to the 1993 conference as well as during the conference itself, these activists were able to get observer status and used this access to lobby the government to designate Isahaya and other areas in Japan as Ramsar sites and to start a more meaningful dialogue with citizens (Finkle 1993a; *JEM* 5, no. 3 [June 1993]: 6–7, 20).

Participation in the Ramsar convention also provided Japanese activists with a new perspective on their movement as a global one and helped them redefine its framing from being a local battle to being a more universal one of preserving wildlife nationally and globally. At symposia and seminars connected with the conference, international experts worried about what they called a "wetlands crisis" in Japan and warned that the country's regional importance in East Asia for migratory birds meant that this crisis concerned the global environmental community (*JEM,* no. 41 [vol. 4, no. 9] [February 1992]: 14–15; *JEM,* no. 42 [vol. 4, no. 10] [March 1992]: 18–19). One of the final outcomes of the conference was a general recommendation that tidal flats on the East Asian flyway be targeted for protection as part of regional efforts at wetlands protection. Such international comparisons and the placing of Japan in the larger picture of global environmental concerns gave Yamashita and his campaign to save Isahaya Bay an external source of legitimacy and credibility that it had previously lacked.

Since the conference took place in Japan, it also provided the movement extensive coverage in the Japanese press, coverage that tended to be supportive of the movement's goals. One newspaper provided Yamashita space to write a "special report," in which he described in detail how Japan lagged behind most countries in terms of wetlands protection as well as the biodiversity that would be lost if the Isahaya Bay project proceeded as planned (Yamashita 1993).

Finally, the Ramsar conference provided Japanese activists with the opportunity to make international allies and, through them, exert pressure on the Japanese government during the conference process. Western INGOs attending official Ramsar-related meetings in Japan in 1992 and 1993 met with JAWAN and Wetlands Coalition '93 and offered Japanese activists their influence and support. On learning how difficult it was for NGOs to gain access to government officials and the policymaking process in Japan, for example, INGOs used the international conference to pressure the Japanese government to adopt more participatory practices. At an official Ramsar symposium in Japan in 1992, INGOs raised the topic of NGO-government relations in Japan and asked the Japanese government what mechanisms were in place to include citizen participation in the process of making environment policy (*JEM,* no. 48 [vol. 5, no. 6] [October 1992]: 7). International connections between Japanese and American activists made during the conference would lead to other transnational alliances for wetlands campaigns in Japan in later years.

International Actors In addition to the general support given to Japanese wetlands activists by Western INGOs at the Ramsar conference, international actors were important allies of the campaign to save Isahaya Bay, and the use of foreign pressure was a very conscious and intentional strategy on Yamashita's part.[3] From the start, one of the goals of JAWAN was to create an international and regional network, and, since its founding meeting in 1991, JAWAN had organized an annual international wetlands symposium in Japan that included experts and activists from other countries (*JEM,* no. 35 [vol. 4, no. 3] [June 1991]: 4–7). These symposia created solidarity among wetlands campaigns and activists in Asia, the United States, and Latin America and, within Japan, helped create a new regional and global frame and context for local wetlands campaigns. In addition, they also provided the wetlands movement in Japan media attention since they were usually covered by the press.

More direct participation by Western INGOs in the Isahaya Bay campaign was activated in 1997 when it became clear that the process of land reclamation for the project would, in fact, begin. In April 1997, the gates cutting off water to Isahaya Bay were closed, and the Isahaya tidal flat was drained. In response to these events, a transnational campaign was quickly mobilized as an emergency last-ditch effort to save the bay.

In early 1997, before the closing of the gates, a coalition of American NGOs centered around the NGO members of the U.S. Ramsar Committee formed the American NGO Alliance to Save Isahaya Bay (Segawa 1997).[4] Throughout the spring and summer of 1997, this alliance joined forces with JAWAN, the World Wildlife Fund–Japan, and the Wild Bird Society of Japan in an all-out campaign to stop the project. First, the transnational coalition organized a letter-writing campaign directed at government officials and international organizations and meant to stop the project. Appealing to international norms, letters were sent to Prime Minister Hashimoto, President Clinton, the U.S. secretary of state, other foreign governments, and several UN convention secretariats urging the halt of the project and stating that it violated international agreements such as the Ramsar Convention, the Convention on Biodiversity, and bilateral migratory bird agreements that Japan had signed with the United States, Russia, and China (Segawa 1997; *JEM,* no. 91 [June–July 1997]: 5; Kyodo News Service 1997). When the gates were closed off in April 1997, over 250 organizations submitted formal protests to MAFF and the Environment Agency, and, by May, MAFF's Internet bulletin board was inundated by e-mails protesting the project (Fukatsu 1997, 30; *JEM,* no. 90 [April–May 1997]: 3; "Foes of Isahaya Project" 1997). In June 1997, Japanese protestors went to the United States and demonstrated in front of the UN during Hashimoto's attendance at the UN General Assembly special

session follow-up to UNCED. Western media coverage of the Isahaya campaign overlapped with these protests.

Transnational efforts to stop the project and support Yamashita continued in 1998. In April 1998, Yamashita received the Goldman Environment Prize, one of the top international prizes for individual work done in the area of environmental protection, and became an internationally recognized environmental hero (Katayama 1998; *JEM,* no. 96 [April–May 1998]: 1, 4–5, 7). Although further efforts by JAWAN to lobby U.S. government agencies to pressure the Japanese government did not bear fruit, transnational efforts targeting the Ramsar Convention Bureau did. In 1998, while in Japan for a wetlands conference, the secretary general of the Ramsar Convention Bureau, Delmar Blasco, put international pressure on the government by publicly criticizing Japan for not doing enough to protect its wetlands and set an example for the world as an environmental leader. Blasco also complained about the condition of the Isahaya Bay tidal flats and noted that the Ramsar Convention Bureau had been flooded with reports and criticisms of the environmental harm caused by the land reclamation project there.

These various transnational efforts in 1997 and 1998 provided important resources to Yamashita's campaign to save Isahaya Bay. First of all, foreign participation in the campaign generated greater media interest and helped make Isahaya one of the most controversial public works projects of the decade. Overall, media coverage of the conflict tended to be favorable to the protestors and had the general effect of raising public support for the movement. By the end of the 1990s, Isahaya Bay became a national symbol for useless environmental destruction and lack of accountability in Japan's "iron triangle"–controlled public works machine.

In addition to media coverage, Yamashita's conscious turn to international allies also

Kim Reimann

helped provide the movement a normative frame that was internationally recognized but also resonated with the Japanese public and could provide the movement legitimacy vis-à-vis its critics. Sustainable development and the value of protecting regionally important wetlands had potential appeal to nature lovers and bird-watchers in Japan. Evidence provided by INGOs of endangered bird species and migratory birds that fed at Isahaya Bay and their framing of the wetlands situation in Japan as an international crisis allowed Yamashita to frame the movement as concerned not merely with a local problem but with one that affected Japan and the East Asia region as a whole.

Postscript In 2000, JAWAN and the movement to stop the Isahaya Bay Land Reclamation Project suffered a leadership and personal loss with the sudden death of Yamashita. Since then, JAWAN and local activists have continued the fight and, through lawsuits filed in 2002 focusing on environmental damage to the Ariake Sea caused by the project, were able to temporarily halt the project in 2004. In 2005, however, these court rulings were overturned, and in October 2007 the reclamation project was completed. Despite this, the movement to save Isahaya Bay has by no means been in vain. Together with the campaign to stop the Nagara River Estuary Dam, this movement played an important role in reshaping the public debate in Japan in the 1990s on the environmental cost of development and public works and in activating greater interest in local environmental issues among the general public.

Conclusion

Since the late 1980s, a new wave of protest and grassroots movements focusing on the environment has swept Japan. As this chapter has shown, environmental movements in the last twenty years or so in Japan are different phenomena than previous protest movements. In contrast to the primarily local and NIMBY focus of the environmental protest movements of the 1960s and 1970s, the new movements are multilevel and multilayered ones that focus more on postmaterial values and often consciously link local, national, and international forces and ideas.

An international framework for analyzing movement dynamics offers interesting insights and clues as to how and why local environmental movements have changed over time. Changes in the international context in the 1980s and 1990s have led to new opportunities that local environmental activists in Japan and other countries have taken advantage of. The thickening of international regimes in the area of the environment, the rise of new transnational networks of support, and the establishment of new international norms in the 1980s and 1990s all contributed to the creation of a new international context that could be strategically used by activists blocked at the national and local level. The cases of protest movements against the Nagara River Estuary Dam and the Isahaya Bay Land Reclamation Project are very good examples of how these new international opportunities can affect and influence activists working at the grassroots level. With the help of international organizations, alliances, and norms, organizers of both campaigns were able to create a new type of local activism that built up national networks of support, consciously used foreign pressure, and shifted movement frames of reference from local, materialist ones to more universal, postmaterialist ones.

Although both campaigns failed to achieve their goals, they were landmarks for the most recent wave of environmental protest in Japan and have served as important models for many other local campaigns. The role and importance of international factors in aiding these two campaigns has, thus, a special significance. International factors—by generating extensive media coverage, by legitimizing movement goals, by shaming Japan in the international arena, and by providing universal ideals that most Japanese could understand—helped activists make their causes famous in Japan and contributed to an important shift in public awareness of environment issues. As these campaigns and many others struggle on, placing their actions in the wider global context provides a good historical perspective as well as a new analytic tool to understand new movements.

Notes

I am indebted to Richard Forrest for providing invaluable assistance in the case-study portions of this chapter. As a participant in many of the events described, he supplied "practitioner" insights and information that only an insider could know. Forrest was the East Asian representative of the National Wildlife Federation for most of the 1990s.

1. For more information, see www.geocities .co.jp/NatureLand-Sky/4094/suigen.htm (accessed May 9, 2003).

2. The bay provided feeding ground for as many as 232 species of birds and 300 species of marine life (see Fukatsu 1997).

3. Yamashita noted to one journalist: "As amply demonstrated in economic issues, the Japanese government is really weak against foreign pressure. We thought the same would hold true for an environmental issue. We sent out information to those foreign groups, asking them to lodge protests" (Fukatsu 1997, 30).

4. The coalition included the American Bird Conservancy, the Citizens Committee to Complete the Refuge, the Friends of the Earth, the International Rivers Network, and the National Wildlife Federation. The U.S. Ramsar Committee is a national advisory body that is led by NGOs and recognized by the U.S. government.

References

Axelrod, Robert. 1986. "An Evolutionary Approach to Norms." *American Political Science Review* 80: 1095–1111.

Beard, Daniel. 1996. "Creating a Vision of Rivers for the 21st Century." Speech delivered at the Nagara International Dam Summit, September 14. Available online at http://www.irn.org/index.asp?id=/basics/beard.html (accessed May 9, 2003).

Broadbent, Jeffrey. 1998. *Environmental Politics in Japan: Networks of Power and Protest.* Cambridge: Cambridge University Press.

Brysk, Alison. 1994. "Acting Globally: Indian Rights and International Politics in Latin America." In *Indigenous Peoples and Democracy in Latin America,* ed. Donna Lee Van Cott. New York: St. Martin's.

Cameron, Owen Kyle. 1996. "The Political Ecology of Environmentalism in Japan: Protest and Participation: 1983–1995." Ph.D. diss. Cambridge University, Trinity College.

Dore, Ronald. 1979–1980. "The Internationalization of Japan." *Pacific Affairs* 52, no. 4 (Winter): 595–611.

Finkle, James R. 1993a. "Conservationists Ready for Ramsar; Activists Poised to Put World Spotlight on Plight of Japan's Wetlands." *Daily Yomiuri,* June 8.

———. 1993b. "Ramsar Meeting May Be Turning Point; First Coalition of Japanese Green Forces Asserts Self." *Daily Yomiuri,* June 17.

Finnemore, Martha. 1996. *National Interests in International Society.* Ithaca, DC: Cornell University Press.

Finnemore, Martha, and Kathryn Sikkink. 1998. "International Norm Dynamics and Political Change." *International Organization* 52, no. 4 (Autumn): 887–917.

Florini, Ann. 1996. "The Evolution of International Norms." *International Studies Quarterly* 40, no. 3 (September): 363–89.

"Foes of Isahaya Project Protest with E-Mail Blitz." 1997. *Mainichi Daily News,* May 19.

Friends of the Earth and Japan Wetlands Action Network (JAWAN). 1993. "Japan's Wetlands and the Ramsar Convention." Tokyo: JAWAN, February.

Fukatsu, Hiroshi. 1997. "Tideland Project Brings Waves of Controversy." *Japan Quarterly* 44, no. 4 (October–December): 26–37.

Greene, Stephen G. 1990. "U.S. Non-Profits Help Groups in Japan—Yes, Japan." *Chronicle of Philanthropy* 27 (November): 6–16.

Gurowitz, Amy. 1999. "Mobilizing International Norms: Domestic Actors, Immigrants and the Japanese State." *World Politics* 51, no. 3 (April): 413–45.

Held, David, Anthony McGrew, David Goldblatt,

and Jonathan Perraton. 1999. *Global Trans-formations: Politics, Economics and Culture.* Stanford, CA: Stanford University Press.

Holliman, Jonathan. 1990. "Environmentalism with a Global Scope." *Japan Quarterly* 37, no. 3 (July–September): 284–90.

"Japan Urged to Set Example." 1998. *Yomiuri shimbun,* March 7.

Katayama, Mikiko. 1998. "Isahaya Activist Hopes International Aware[ness] Will Help Cause." *Yomiuri shimbun,* May 4.

Keck, Margaret E., and Kathryn Sikkink. 1998. *Activists beyond Borders.* Ithaca, NY: Cornell University Press.

Klotz, Audie. 1995. "Norms Reconstituting Interests: Global Racial Equality and U.S. Sanctions against South Africa." *International Organization* 49, no. 3 (Summer): 451–78.

Krauss, Ellis S., and Bradford L. Simock. 1980. "Citizens' Movements: The Growth and Impact of Environmental Protest in Japan." In *Political Opposition and Local Politics in Japan,* ed. Kurt Steiner, Ellis S. Krauss, and Scott C. Flanagan. Princeton, NJ: Princeton University Press.

Kyodo News Service. 1992a. "International Conservationists Meet to Urge Halt to Nagara Dam." October 3.

———. 1992b. "Lawyers Group Says Big Dam Era in Japan Is Over." October 17.

———. 1992c. "12,000 People Demonstrate against Nagara Dam Project." October 4.

———. 1996. "International Group to Gather to Voice Opposition to Nagara Dam." September 9.

———. 1997. "Clinton Urged to Press Japan to Drop Isahaya Project." May 22.

Martin, Lisa L., and Kathryn Sikkink. 1993. "U.S. Policy and Human Rights in Argentina and Guatemala, 1973–80." In *Double-Edged Diplomacy, International Bargaining and Domestic Politics,* ed. Peter Evans, Harold K. Jacobson, and Robert D. Putnam. Berkeley and Los Angeles: University of California Press.

Mason, Robert J. 1999. "Whither Japan's Environmental Movement? An Assessment of Problems and Prospects at the National Level." *Pacific Affairs* 72, no. 2 (Summer): 187–207.

McAdam, Doug. 1998. "On the International Origins of Domestic Political Opportunities." In *Social Movements and American Political Institutions,* ed. Anne N. Costain and Andrew S. McFarland. Lanham, MD: Rowman & Littlefield.

McKean, Margaret A. 1981. *Environmental Protest and Citizen Politics in Japan.* Berkeley and Los Angeles: University of California Press.

Niikura, Toshiko. 1999. "Campaigns against Dams in Japan and the Nagara River Estuary Dam." *Organization and Environment* 12, no. 1 (March): 99–104.

'92 NGO Forum Japan. 1992. *People's Voice of Japan—I Have the Earth in Mind, the Earth Has Me in Hand.* Tokyo: '92 NGO Forum Japan.

Ohta, Hiroshi. 2000. "Japanese Environmental Foreign Policy." In *Japanese Foreign Policy Today,* ed. Takashi Inoguchi and Purnendra Jain. New York: Palgrave.

Pekkanen, Robert. 2000. "Japan's New Politics: The Case of the NPO Law." *Journal of Japanese Studies* 26, no. 1: 111–43.

———. 2006. *Japan's Dual Civil Society: Members without Advocates.* Stanford, CA: Stanford University Press.

Pharr, Susan J. 1990. *Losing Face: Status Politics in Japan.* Berkeley and Los Angeles: University of California Press.

Reimann, Kim. 2001. "Building Networks from the Outside In: International Movements, Japanese NGOs and the Kyoto Climate Change Conference." *Mobilization* 6, no. 1: 69–82.

Risse-Kappen, Thomas. 1995. "Bringing Transnational Relations Back In: Introduction." In *Bringing Transnational Relations Back In: Nonstate Actors, Domestic Structures and International Relations,* ed. Thomas Risse-Kappen. Cambridge: Cambridge University Press.

Risse, Thomas, Stephen C. Ropp, and Kathryn Sikkink, eds. 1999. *The Power of Human Rights, International Norms and Domestic Change.* Cambridge: Cambridge University Press.

Scheerer, Jay. 1999. "Transnational Advocacy Networks and the Struggle to Save Isahaya Bay: Transnationalism with a Japanese Characteristics?" A.M. thesis, Harvard University.

Schoppa, Leonard. 1997. *Bargaining with Japan: What American Pressure Can and Cannot Do.* New York: Columbia University Press.

Schreurs, Miranda. 2002. *Environmental Politics in Japan, Germany and the United States.* Cambridge: Cambridge University Press.

Segawa, Shiro. 1997. "American NGOs Take Up Isahaya Bay Issue." *Mainichi Daily News,* May 23.

Sikkink, Kathryn. 1993. "Human Rights, Principled Issue-Networks, and Sovereignty in Latin America." *International Organization* 47, no. 3 (Summer): 411–41.

Sugai, Masuro. 1992. "The Anti-Nuclear Power Movement in Japan." In *Energy Politics and Schumpeter Dynamics,* ed. Helmar Krupp. Berlin: Springer.

Tamamoto, Masaru. 1993. "The Japan That Wants to Be Liked." In *Japan's Emerging Global Role,* ed. Danny Unger and Paul Blackburn. Boulder, CO: Lynne Rienner.

United Press International (UPI). 1996. "International Dam Confab Opens in Japan." September 14.

Upham, Frank. 1987. *Law and Social Change in Postwar Japan.* Cambridge, MA: Harvard University Press.

Watanabe, Teresa. 1997. "Tide of Anger at Japan's Parched Wetlands; Asian Nation Is Active Environmentally Abroad but Critics Say It Needs to Mend Its Ways at Home." *Los Angeles Times,* June 23.

World Commission on Environment and Development. 1987. *Our Common Future.* Oxford: Oxford University Press.

Yamashita, Hirofumi. 1993. "Japan Lags behind in the Preservation of Disappearing Wetlands." *Daily Yomiuri,* April 21.

Yokota, Hiroyuki, and Masahi Iiyama. 1992. "'Japan Day' Held in Rio de Janeiro." *Daily Yomiuri,* June 6.

Part II

Protesting the Effects
of Nuclear Radiation
and Chemical Weapons

Chapter 5

Citizen Activism and the Nuclear Industry in Japan
After the Tokai Village Disaster

Nathalie Cavasin

By the 1990s, several accidents at Japan's nuclear power plant facilities had occurred, all posing serious threats. International criticism of the management of Japan's nuclear program became pronounced. In Japan, the public concern about the safety of nuclear power has risen particularly after the two major 1990s accidents: the December 1995 sodium leak at the Monju (Fukui Prefecture) fast-breeder reactor and the September 1999 accident in Tokai Village (Ibaraki Prefecture) involving the uranium reconversion plant (see Cavasin 2002; and figure 5.1). Several other accidents occurred in Japan in nuclear plants after the Tokai disaster (Masuzoe 2000; Mukaidani 2000). For example, the August 2004 accident at the Mihama nuclear power plant (Fukui Prefecture) remains Japan's worst nuclear accident. It has led nuclear activists and even supporters of the nuclear industry to the conclusion that the deregulation of the market for electric power in Japan has resulted in an increase in nuclear accidents (Johnston 2005; Yoshii 2003).

After the Tokai accident, cover-ups of problems at nuclear plants have been revealed, further eroding the already diminishing popular support for nuclear energy (French 2000; Yoshida 2000). In 2002, the newspaper *Nikkei* broke the story of cover-ups of damage at nuclear plants (e.g., cracks in coolant pipes) by the Tokyo Electric Power Company [Tokyo denryoku kabushiki kaisha] ("More Cover-Ups" 2002). Other media revelations include the falsification of data ("Nuclear Safety at Stake" 2007; "Tepco Must Probe"

2007) and the poor work ethic characterizing power plant workers (Hisane 2007; "Weak Works Ethic" 2006).

Hasegawa (2004, 11) has written: "The stereotypical image of the obedient Japanese who silently follows the traditional order is still strong internationally, but the voices of people interested in environmental movements and environmental NGOs [nongovernment organizations] in the new public sphere are loud and diverse." Still, the literature on antinuclear activism in Japan is lacking in many respects. I follow Hasegawa (2004, 132) in defining antinuclear activism as being "opposed to the civil use of nuclear energy." And I deal in particular with the Tokai accident because it has created new forms of local activism. Data were collected through interviews with several citizens involved in antinuclear activism (Klandermans and Staggenborg 2002). The report that follows gives an overview of how in the 1990s such activism was locally focused.

Antinuclear Movements in Japan

As noted by Mario Diani (2003), it is difficult to gauge the nature of social movements. They are more than just protests. In Japan, it is possible to recognize four types of antinuclear social movements (Hasegawa 2003).

The first type of movement is one whose supporters live in the area where a power plant is planned. Often those supporters are farmers and fishermen since power companies tend to

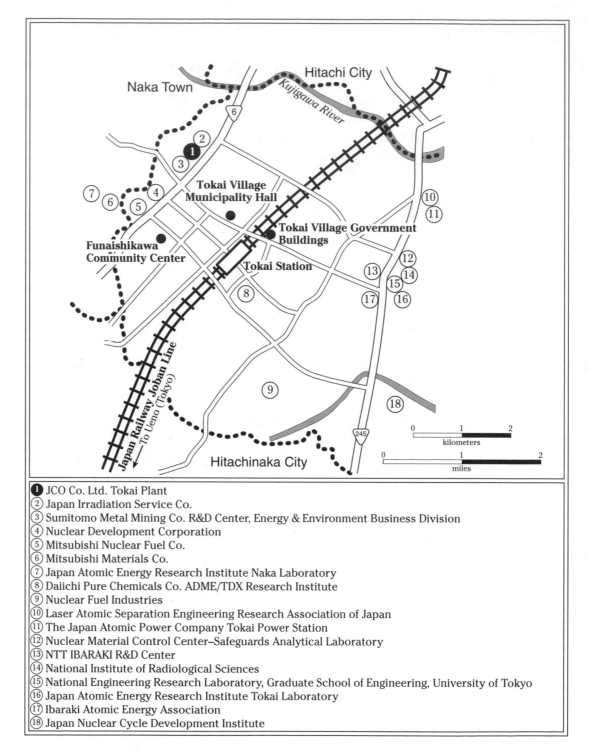

❶ JCO Co. Ltd. Tokai Plant
② Japan Irradiation Service Co.
③ Sumitomo Metal Mining Co. R&D Center, Energy & Environment Business Division
④ Nuclear Development Corporation
⑤ Mitsubishi Nuclear Fuel Co.
⑥ Mitsubishi Materials Co.
⑦ Japan Atomic Energy Research Institute Naka Laboratory
⑧ Daiichi Pure Chemicals Co. ADME/TDX Research Institute
⑨ Nuclear Fuel Industries
⑩ Laser Atomic Separation Engineering Research Association of Japan
⑪ The Japan Atomic Power Company Tokai Power Station
⑫ Nuclear Material Control Center–Safeguards Analytical Laboratory
⑬ NTT IBARAKI R&D Center
⑭ National Institute of Radiological Sciences
⑮ National Engineering Research Laboratory, Graduate School of Engineering, University of Tokyo
⑯ Japan Atomic Energy Research Institute Tokai Laboratory
⑰ Ibaraki Atomic Energy Association
⑱ Japan Nuclear Cycle Development Institute

Figure 5.1. Tokai Village.

target depopulated farming and fishing villages when scouting locations for plants. Feeling that they have a right to live and work where they please, supporters strongly resist the sale of land to the power company. The key strategy employed by power companies in circumventing such opposition and carrying through a nuclear location plan [*genpatsu ricchi keikaku*] is securing the support of local elected officials. This tended to be successful because, until about 1980, the most influential organizations in Japan were political parties or associations. In recent years, however, residents have been increasingly consulted (Hasegawa 2003). For example, in response to opposition by a local citizens' group to the planned construction by the Tohoku Electric Corporation [Tohoku denryoku kabushiki kaisha] of a nuclear power plant in Maki, a small farming town on the Sea of Japan (Niigata Prefecture), a referendum on the issue was held, 60 percent of residents voting against and, thus, blocking construction (Kotler and Hillman 2000).

The second type of movement is one whose supporters are affiliated with political parties—such as the former Socialist Party [Kyu shakai to] or the Communist Party [Kyo san to]—or labor unions. Most of the people involved in this type of movement were born in the *dankai no sedai,* or Japanese post–World War II baby boom period, 1947–1949. Most are also university educated and opposed to the industrial sector on the grounds of its heavy involvement in the war effort. Their movement involvement focuses mostly on the dissemination of information not widely available through conventional media sources about the dangers associated with nuclear power, although, with the advent of the Internet, such information has become much more widely available. Some, however, employ the alternative strategy of becoming power company stockholders, partly in an attempt to influence the company from within, and partly in an attempt to obtain useful inside information.

The third type of movement is one whose supporters are women, mainly housewives, who live in metropolitan areas. Such a movement is a recent phenomenon, few women in Japan being involved in antinuclear activism until spurred by the 1986 Chernobyl accident. The first concerted effort of this movement was the circulation of a petition in 1990 and 1991 in support of a "nuclear-free law" [*kogenpatsu ho*]. Even though more than 3.3 million signatures were collected, the law was never debated in the Diet. After that experience, the movement has tended to focus its attention on other issues, such as the environment, education, and social problems, sometimes partnering with nonprofits and NGOs. While it has staged individual antinuclear protests that have gained media attention, it has had little long-term effect on government policy.

The fourth type of movement is one whose supporters are professionals—lawyers, university professors, etc.—opposed to government policy on nuclear energy. They disseminate their ideas in the newspaper *Hangenpatsu shimbun* [Against nuclear newspaper], founded in 1978. Some are active members of antinuclear organizations such as CNIC, or the Citizens' Nuclear Information Center [Denshiryoku shiryo joho shitsu], the research committee on nuclear power problems of the Japan Scientists' Association [Nihon kagakusha kaigi], the People's Research Institute on Energy and Environment [Shimin enerugi kenkyujo], Greenpeace–Japan, and the committee for the protection of the environment of the Japan Federation of Bar Associations [Nihon bengoshi rengokai]. Affiliated with the movement are NGOs promoting natural energy sources such as the Citizens' Alliance for Saving the Atmosphere and the Earth [Chikyu kankyo to daikioson wo kangaeru zenkoku shimin kaigi], the Renewable Energy Promoting People's Forum [Shizen enerugi suishin shimin foramu], and the Green Energy "Law" Network [Shizen enerugi shokushin ho]. These NGOs are mainly involved in collecting and disseminating information on power companies employing nuclear energy in Japan and on nuclear accidents in Japan and around the world.

The tendency of all antinuclear activity in Japan is to start strong and campaign actively against the construction of a new nuclear power plant but, should the battle be lost, to gradually accept and even rely on nuclear power. The CNIC describes the situation as follows: "So far three nuclear power plant installation plans in three different regions have been cancelled as a result of strong campaigns led by local activists, including medical doctors, monks, teachers, etc. However, local anti–nuclear energy movements tend to lose strength once nuclear facilities are installed. That's why there are many reactors at the same plant. Public resistance to nuclear energy shows itself most clearly in the reluctance of local citizens to accept nuclear facilities of any type. It requires very substantial subsidies to get local agreement for nuclear facilities. However, once communities become dependent on the nuclear industry, they are more inclined to accept additional facilities, because they want to keep the subsidies coming" (Philip White, CNIC, e-mail to author, January 12, 2007).

Antinuclear activism in Japan is now focused on two major issues: the disposal of high-level radioactive waste and the threat posed to nuclear power plants by earthquakes. Concern about the disposal of radioactive waste has been intensified by the search being conducted by the Nuclear Waste Management Organization of Japan [Denshiryoku hatsuden kankyo seibi kiko], founded in October 2000, for communities willing to host waste-storage facilities. Local opposition is strong and widespread, especially because of concern that poorer communities will have little economic choice but to accept the state-funded projects ("Hosting Radioactive Waste Sites" 2006). Concern about earthquake-related safety issues is not nearly so great at the grassroots level, with only occasional local movements appearing, such as that centered in Hamaoka in Shizuoka Prefecture. New government guidelines meant to reduce the risk to power plants of earthquake damage were promulgated in 2006, guidelines that did not take account of public opinion (Fujino and White 2007).

The Tokai Accident and Local Antinuclear Activists

The worst nuclear accident to date in Japan occurred on September 30, 1999, at a uranium processing plant of the JCO Corporation [JCO kabushiki kaisha] in Tokai (see figure 5.1 above and photograph 5.1). The plant began operations in 1973, and, as the demand for fuel increased, operations were expanded. The criticality accident (or accidental nuclear chain reaction) in question—in which radiation leaked into the atmosphere—was rated as level 4 on the seven-level international scale (see photograph 5.2).

Tokai Village is located in the northern part of Ibaraki Prefecture along the coast of the Pacific Ocean near the prefectural city of Mito, 140 kilometers northeast of Tokyo. The village of thirty-four thousand has been a center of nuclear energy development in Japan. It encompasses thirty-seven square kilometers, an area dotted with nuclear facilities. One-third of all jobs there are at the thirteen industry-related plants and institutes. Tokai has, over the years, benefited from tax revenues coming from the nuclear energy industry, revenues that have financed such public works projects as the resurfacing of roads and the rebuilding of the train station (see photograph 5.3). Support for the nuclear industry ran high in the village—until the accident, at which point many began questioning the government's nuclear energy policy.

The main criticism is the slow response

Photograph 5.1. JCO Corp. (Courtesy Nathalie Cavasin.)

Photograph 5.2. Location of the accident site at the JCO plant. (Courtesy CNIC, Tokyo.)

time. Approximately seventy residents of Tokai were exposed to dangerous levels of radiation before any warnings were issued. The government's inadequate approach to crisis management reminded people of its equally inadequate responses to the Great Hanshin Earthquake of 1995 and the massive oil spill just off the coast of the Sea of Japan in 1997. The feelings of residents of Tokai were mixed, however, and some considered the subject of the criticality accident taboo. Still, according to Fujino Satoshi, a public relations officer with the CNIC in Tokyo: "The attitude of local people in Tokai Village is changing. . . . However, there is elsewhere a clear, vivid example of change found in Fukushima Prefecture, where the governor is taking a clear stance against action nuclear policy" (Fujino 2003).

Antinuclear activism in Tokai took various forms after the accident. Oizumi Soichi, the owner of an auto parts factory located one hundred meters from the JCO plant, formed the first group of radiation victims and filed suit in September 2002 demanding compensation from JCO and its parent company, Sumitomo Metal Mining [Sumitomo kinzoku kozan kabushiki kaisha], for the effects on his, his wife's, and other affected residents' health. As of December 2006, nineteen trials had been held.

Another approach is that taken by Yatabe Yuko (see photograph 5.4), a homemaker [shufu] living in the area of Tokai Village

Photograph 5.3. Bullet train station, Tokai Village. (Courtesy Nathalie Cavasin.)

Photograph 5.4. The Tokai antinuclear activist Yatabe Yuko. (Courtesy Nathalie Cavasin.)

who became actively involved in antinuclear activism after the accident (Ban 2002). As she reports: "The people in Tokai had lived for more than forty years with the nuclear energy and didn't talk about it. . . . However, after the accident, the thinking of the people changed a little bit." She added that, immediately after the accident, no one explained the danger posed by the radioactive rain and that "nobody wanted to take responsibility for the accident" (Yatabe 2003). This is confirmed by a CNIC study indicating that the final report of the Nuclear Safety Commission committee investigating the accident included no information on the effects of radiation exposure (see Hasegawa and Takubo 2001). People were left to educate themselves about nuclear energy. And Yatabe did so, collecting materials from different sources on the topic and becoming very knowledgeable in the process.[1]

Immediately after the accident, Yatabe and other local residents organized the screening in Tokai of *Village of Naja* [*Nadja no mura*], a 1997 film on the subject of Chernobyl directed by the photographer Motohashi Seiichi. The success of the screening, which was attended by more than six hundred people, encouraged the organizers to pursue the goal of further informing ordinary citizens about the risks posed by nuclear energy. To that end, they formed a group named the Circle of Naja [Nadja no wa]. Yatabe herself has written of her own concerns, as well as on those of children, after the Tokai accident in *Ground Zero* [*Gurando zero*], a journal published by the Japan Chernobyl Foundation [Nihon cherunobiri rentai kikin], a Japanese NGO (see Yatabe 2008). She also formed the Association on the ITER Project [ITER so shiru sai], a study group related to the ITER (International Thermonuclear Experimental Reactor) project in Naka, located close to Tokai. A 2001 survey canvassing people living within three kilometers of the proposed ITER project site that she conducted with the aid of the study group found that none of the

Photograph 5.5. The Tokai antinuclear activist Fujii Gakusho. (Courtesy Nathalie Cavasin.)

respondents knew that the project involved a nuclear reactor. The promoters and developers of the project had discussed only its contribution to local economic development and scientific progress generally.[2]

While she has since taken a job in another prefecture and is, therefore, unable to engage in antinuclear activism as fully as before, Yatabe does still follow the issue in Tokai and attends group meetings when she is able. She also continues to encourage local residents to have regular checkups (Yatabe 2007).

Another antinuclear activist in Tokai is Fujii Gakusho, a priest (see photograph 5.5). Unlike Yatabe Yuko, Fujii was born in Tokai. In his opinion, people have already begun to forget about the accident, a situation not helped by the fact that the local government has aligned itself behind the nuclear power industry, even going so far as to construct various information centers—Information Plaza Tokai [Infomeshon puraza tokai], Tokai Atom World [Atomu warudo (Tokai tenjikan)], Tokai Terra Park [Tokai tera paku], and the Science Museum of Atomic Energy [Denshiryoku kagaku kan]—that simply tout the merits of nuclear power without outlining the dangers involved. It is for these reasons that he promotes further demonstrations, thus maintaining a continued "awareness of the future risks of the nuclear power" (Fujii 2003).

Until the Tokai accident, village politics

had been dominated by the proponents of nuclear energy. But that changed when the antinuclear activist Aizawa Kazumasa was elected to the village assembly to serve a term that ran from February 2000 to January 2004.[3] Aizawa has been involved in antinuclear movements since the 1960s. He had been a researcher in the modern history of Japan at the Ibaraki Prefectural Museum of History [Ibaraki ken ritsu rekishi kan] before becoming a member of the assembly.[4] He was also representing the plaintiffs in a lawsuit initially filed in 1974 against the Tokai II plant (the number two atomic power station in Tokai and the first large-scale power plant in Japan to generate eleven hundred megawatts of electricity), which began commercial operation in 1978 (Aizawa 2003). (The Tokai II case and a similar suit brought against the Ikata nuclear power plant in Shikoku were in the 1970s the only such cases in Japan.) The suit dragged on for twenty years. But, despite the fact that Aizawa's election to the assembly gave him easy access to information that before he had been able to obtain only with great difficulty, if at all, it failed, the court dismissing it in 2004 (the Ikata case had been dismissed in 1992).

From the perspective of his long experience as an activist, Aizawa reports: "In the 1970s, after the student movements, environmental problems reached a peak, with an increase in citizen movements, which was finally followed by the passage of a new environmental law at the national level." The same period also saw the rise of citizen antinuclear movements, movements supported by a number of prominent nuclear scientists. According to Aizawa: "Citizens started to think about nuclear energy after the Chernobyl accident in 1986, but it was not until the Tokai accident that they were really conscious of the safety issues raised by nuclear energy" (Aizawa 2003). Still, in the immediate aftermath of the JCO accident, he was the only candidate to discuss the danger of

Photograph 5.6. The Mito antinuclear activist Nemoto Gan. (Courtesy Nathalie Cavasin.)

nuclear power and what its future role in Tokai should be.

Nemoto Gan (see photograph 5.6), another antinuclear activist, lives north of Tokai in Mito, the capital city of Ibaraki Prefecture. His activities concern us here because he is a member of the network Ibaraki Action Coalition against Nuclear Power [Han genshiryoku ibaraki kyodo kodo], which opposes the further development of nuclear power and related facilities in Tokai (Nemoto 2004). A longtime activist, he served as a plaintiff in the Tokai II case and is currently involved in the campaign to prevent the dismantling of the magnox nuclear reactor owned by the Japan Atomic Power Company [Nihon denshiryoku hatsuden kabushiki kaisha] and located in Tokai Village. (Opposition to the dismantling is based on the potential exposure of workers to radiation and the problems involved in the management of the radioactive waste [Nemoto 2004].)[5] In these and other ways, he has long sought to bring local concerns to national attention.

The movements developed by all these antinuclear activists appear to have expanded because they are part of a larger activist network facilitating the exchange of information. It is particularly interesting that the exchange of information proceeds so efficiently because, at least in 2003, only one of the four activists I interviewed had Internet access. (Until recently, Japan lagged far behind the United

States in terms of ownership of personal computers and the availability of Internet access.) The others resort to more traditional means of communication, such as phoning, faxing, or exchanging letters. The situation is, however, changing. More and more activists are creating Web pages, running blogs, and diffusing information through e-mail mailing lists. The impact of the Internet on such activism is clearly an important research area to pursue.

The Role of Nonprofit Organizations

Nonprofit organizations play an important role in antinuclear movements in Japan. A good example is the CNIC—cofounded in 1975 by Takagi Jinzaburo (1938–2000), a former nuclear scientist turned antinuclear activist—believed to be Japan's largest antinuclear group, with more than two thousand registered members (Takagi and CNIC 2000). Independent from the government and industry, the CNIC supports itself financially by tapping into established networks of activists, whether through membership fees, donations, or the sale of its publications.[6] It sees itself mainly as a provider of information on nuclear energy, a function it performs in several ways. One way is through its publications. Another is through the organization of symposia. Most recently, in 2006, it was involved in the organization of a symposium in Yoyogi (a Tokyo district) commemorating the twentieth anniversary of the Chernobyl accident and meant to raise public awareness about the issues surrounding nuclear energy. Other issues confronted by the CNIC include the danger posed to nuclear power plants by earthquakes and the impact on local residents of the Tokai accident. In order to address the latter, CNIC has, along with the Japan Congress against A- and H-Bombs [Gensuikin], created the JCO Criticality Accident Comprehensive Assessment Committee [JCO rinkai jiko sogo kyoka kaigi].[7]

Conclusion

Clearly, local antinuclear movements have attracted a dedicated following and have built the beginnings of a national network. But they still remain fragmented, their long-term stability has not yet been demonstrated, and no nationwide antinuclear group has as yet emerged. Certainly, the election of Aizawa, a declared antinuclear activist, to a term in the Tokai assembly suggests a promising change of attitude at the grassroots level. But it is far too early for even a preliminary conclusion as to the long-term effect of the antinuclear movement in Japan.

Notes

1. It should be noted that those working in the nuclear power and related industries are contractually prohibited from participating in any public discussion of the industry (see Suk 2000).

2. For reports of Yatabe's other activities, see *Riron Sensen* [The ideology front] 66 (Autumn 2001): 84–91.

3. Although in 2004 he sought a second term and lost, he has decided to run again in 2008 (Aizawa 2007).

4. He had been living in Mito and moved to Tokai to be closer to his job.

5. There are no others cases anywhere in the world of the dismantling of such a reactor (of the thirty-seven worldwide in 2004, twenty-four were no longer in operation).

6. According to Broadbent (2003, 206), such established networks offer activists the opportunity "to mobilize larger groups of actors without convincing each member of the virtues of the case."

7. Hasegawa and Takubo (2001) reports on several effects.

References

Aizawa, Kazumasa. 2003. Interview with the author. June 27. Waseda University, Tokyo.
———. 2007. Fax interview with the author. January 18.
Ban, Hideyuki. 2002. "Yuko Yatabe: An Activist

with Mother's Eyes." *Nuke Info Tokyo,* no. 87 (January/February): 10.

Broadbent, Jeffrey. 2003. "Movements in Context: Thick Networks and Japanese Environmental Protest." In *Social Movements and Networks: Relational Approaches to Collective Action,* ed. Mario Diani and Doug McAdam. New York: Oxford University Press.

Cavasin, Nathalie. 2002. "Tokai-mura Disaster." In *Encyclopedia of Modern Asia,* ed. David Levinson and Karen Christensen, 5:505. New York: Scribner's Reference.

Diani, Mario. 2003. "Introduction: Social Movements, Contentious Actions, and Social Networks: From Metaphor to Substance." In *Social Movements and Networks: Relational Approaches to Collective Action,* ed. Mario Diani and Doug McAdam. New York: Oxford University Press.

French, Howard W. 2000. "Accident Makes Japan Re-Examine A-Plants." *New York Times,* January 13.

Fujii, Gakusho. 2003. Interview with the author. July 29. Tokai Village.

Fujino, Satoshi. 2003. Interview with the author. May 21. CNIC office, Tokyo.

Fujino, Satoshi, and Philip White. 2007. Interview with the author. January 16. CNIC office, Tokyo.

Hasegawa, Koichi. 2003. *Kankyo undo to atarashii kokyoen—kankyo shakaigaku no pasupekutibu* [Environmental movements and the new public sphere—the perspective of environmental sociology]. Tokyo: Yuhikaku.

———. 2004. *Constructing Civil Society in Japan: Voices of Environmental Movements.* Melbourne: Trans Pacific Press.

Hasegawa, Koichi, and Yuko Takubo. 2001. *JCO Criticality Accident and Local Residents: Damages, Symptoms and Changing Attitudes.* Translated by Gaia Hoerner. Tokyo: Citizens' Nuclear Information Center.

Hisane, Masaki. 2007. "Japanese Nuclear Power Steams Ahead." *Asia Times Online,* February 10. http://www.atimes.com/atimes/Japan/IB10Dh01.html (accessed March 3, 2008).

"Hosting Radioactive Waste Sites a Chance to Balance the Books." 2006. *Asahi shimbun,* October 7.

Johnston, Eric. 2005. "Japan's Nuclear Nightmare." Japan Focus, article no. 278. http://japanfocus.org/products/details/1694 (accessed January 10, 2007).

Klandermans, Bert, and Suzanne Staggenborg, eds. 2002. *Methods of Social Movement Research.* Minneapolis: University of Minnesota Press.

Kotler, Mindy L., and Ian T. Hillman. 2000. *Japanese Nuclear Energy Policy and Public Opinion.* Report prepared in conjunction with an energy study sponsored by the Center for International Political Economy and the James A. Baker III Institute for Public Policy, Rice University. http://www.rice.edu/energy/publications/docs/JES_NuclearEnergyPolicyPublicOpinion.pdf (accessed January 19, 2007).

Masuzoe, Yoichi. 2000. "Japan's Nuclear Disaster." *Japan Echo* 27, no. 1 (February): 16–18.

Miller, Brian A. 2000. *Geography and Social Movements: Comparing Antinuclear Activism in the Boston Area.* Minneapolis: University of Minnesota Press.

"More Cover-Ups of Nuclear Plant's Faults Revealed." 2002. *Nikkei Interactive Net,* September 20.

Mukaidani, Susumu. 2000. "The Terror of Tokaimura." *Japan Echo* 27, no. 1 (February): 19–24.

Nemoto, Gan. 2004. "We Oppose Removal and Dismantling of Tokai Reactor." *Nuke Info Tokyo,* no. 103 (November/December).

"Nuclear Safety at Stake." 2007. *Japan Times,* August 17.

Suk, Sarah. 2000. "Tokai Caught between a Rock and Nuclear Plant." *Mainichi shimbun,* September 23.

Takagi, Jinzaburo, and the Citizens' Nuclear Information Center (CNIC). 2000. *Criticality Accident at Tokai-mura: 1 Mg of Uranium That Shattered Japan's Nuclear Myth.* Translated by Gaia Hoerner, Akiko Fukami, and Taeko Miwa. Tokyo: Citizens' Nuclear Information Center.

"Tepco Must Probe 199 Plant Check Coverups." 2007. *Japan Times,* February 2.

"Weak Works Ethic Blamed for Errors." 2006. *Asahi shimbun,* October 19.

Yatabe, Yuko. 2003. Interview with author. July 29. Naka.

———. 2007. Telephone interview with author. January 18.

———. 2008. E-mail interview with author. January 4.

Yoshida, Yasuhiko. 2000. "Nuclear Safety Issue Jars Japan to Reconsider Its Policy Program." *Japan Quarterly* 47, no. 1 (January–March): 56–64.

Yoshii, Hidekatsu. 2003. "Denshiryoku enerugi isonsei kara no tenkan" [Nuclear energy: From dependence to change]. *Keizai,* no. 96:43–53.

Chapter 6

Citizen Advisory Boards and the Cleanup of the U.S. Nuclear Weapons Complex

Public Participation or Public Relations Ploy?

John J. Metz

[Nuclear weapons production has been] a secret operation, not subject to laws. . . . No one was to know what was going on. . . . This is our business, it's national security. Everybody else, butt out.

> W. Henson Moore,
> Deputy Secretary of Energy,
> June 15, 1989, at Rocky Flats

By 1990, the environmental and safety failures of its past operations had come to haunt the U.S. Department of Energy (DOE) and its complex of nuclear weapons–producing factories (see figure 6.1).[1] The 1980s had been a busy time at the complex. The military buildup of the Reagan administration had doubled production in an aging and technologically complicated system (Cochran, Arkin, Norris, and Hoenig 1987a, 3–4; OTA 1991; Makhijani, Hu, and Yih 1995; DOE 1995). The total absence of outside oversight and a stunning lack of concern about the disposal of nuclear and toxic wastes during four decades of operation had allowed nuclear and toxic materials to contaminate all the major sites and to spill out into neighboring civilian areas (Makhjani, Hu, and Yih 1995; DOE 1995, 1996, 1997b). A cascade of horror stories about the operations of the complex had been discovered by the grassroots groups that had formed at almost every site, stories that were reported in local and national news me-

dia (e.g., Schneider 1988a, 1988b, 1989a, 1989b, 1989c, 1989d, 1989e, 1989f, 1989g). The FBI had raided one major facility, Rocky Flats, and seized all available records, accusing the contractor of violating a court order by secretly burning radioactive waste in an incinerator at night. At another facility in Fernald, Ohio, the DOE had for four years allowed people living in the vicinity to drink well water contaminated with uranium and had settled a class action suit for $78 million to compensate the victims and monitor their health for the rest of their lives. The DOE and predecessor organizations had performed radiation experiments on prisoners, children with disabilities, and pregnant women.

As grassroots activists publicized these conditions and news organizations published exposés about them, enough citizens became aware of and concerned about the problems to force congressional representatives to act. Congressionally mandated blue-ribbon committees reviewed the problems and recommended that citizens be involved in the "cleanup" of the DOE facilities (OTA 1991; DOE/EPA 1996).[2]

Expanding citizen participation in resolving environmental and other disputes has gained increasing favor within the DOE during the last forty years. The process invariably seeks to involve stakeholders in decisionmaking by bringing contending groups together to promote dialogue and compromise, as the

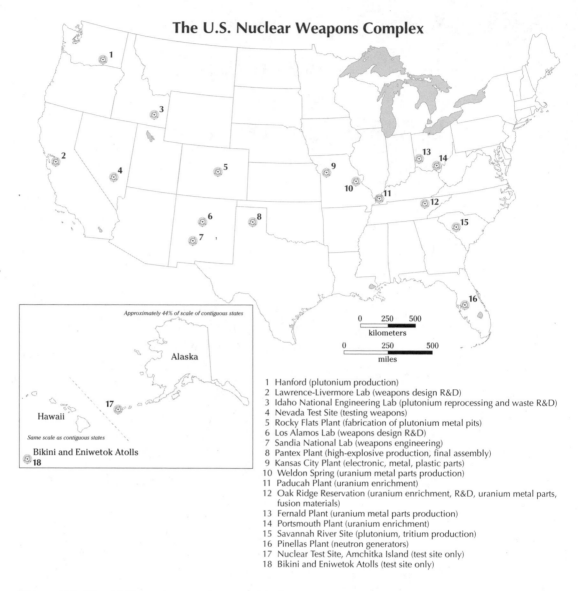

The U.S. Nuclear Weapons Complex

Approximately 44% of scale of contiguous states

Alaska

Hawaii

Same scale as contiguous states

Bikini and Eniwetok Atolls
18

1 Hanford (plutonium production)
2 Lawrence-Livermore Lab (weapons design R&D)
3 Idaho National Engineering Lab (plutonium reprocessing and waste R&D)
4 Nevada Test Site (testing weapons)
5 Rocky Flats Plant (fabrication of plutonium metal pits)
6 Los Alamos Lab (weapons design R&D)
7 Sandia National Lab (weapons engineering)
8 Pantex Plant (high-explosive production, final assembly)
9 Kansas City Plant (electronic, metal, plastic parts)
10 Weldon Spring (uranium metal parts production)
11 Paducah Plant (uranium enrichment)
12 Oak Ridge Reservation (uranium enrichment, R&D, uranium metal parts, fusion materials)
13 Fernald Plant (uranium metal parts production)
14 Portsmouth Plant (uranium enrichment)
15 Savannah River Site (plutonium, tritium production)
16 Pinellas Plant (neutron generators)
17 Nuclear Test Site, Amchitka Island (test site only)
18 Bikini and Eniwetok Atolls (test site only)

Figure 6.1. The U.S. nuclear weapons complex.

groups create mutually acceptable solutions to the problems over which they have previously contended (Renn, Webler, and Wiedemann 1995; Applegate 1998; Susskind, Levy, and Thomas-Larmer 2000; Webler and Renn 1995). The DOE also wanted to modernize its aging nuclear complex. To get the funds to update its facilities, especially because the end of the cold war made it harder to justify nuclear warhead production,[3] the DOE realized that it had to improve its public image and demonstrate that it had become a "good citi-

zen" by cleaning up the radioactive and toxic contamination at the cold war sites and by including the public in the process (D'Antonio 1993, 260–62).

Within this operational and public relations nightmare, the DOE radically altered its relationship to citizens by establishing a variety of public participation programs, including citizen advisory boards at twelve of its nuclear weapons–production sites (see tables 6.1–6.2). This move to incorporate citizens in the decisionmaking process was a complete

John J. Metz

reversal of the previous five decades of operation, transforming a bureaucracy that had been totally closed to the public into one apparently committed to democratic participation.[4] In the decade since these changes were begun, nine so-called site-specific advisory boards (SSABs) have met regularly and participated in the environmental remediation programs at their respective sites.[5]

In this chapter, I review and evaluate this process. First, I describe the nuclear weapons complex of the cold war era and the processes by which the nuclear weapons were made. Second, I sketch a few of the environmental and health problems besetting the weapons complex and draw some conclusions about how the nuclear establishment related to citizens during the cold war era. Third, I describe the attempts of the DOE to involve the public in planning and implementing the cleanup of weapons sites via the citizen advisory board program. Fourth, I evaluate the success of the SSABs. During the 1990s, the performance of the SSABs was mixed, with some sites having effective boards and others stymied by conflicts. During the second Bush administration, the DOE seems to be withdrawing support from the SSABs and exhibiting many of the characteristics of the cold war nuclear establishment. Finally, I conclude by trying to answer the question of whether the citizen advisory boards represent true public participation or are simply a public relations ploy.

The Nuclear Weapons Complex of the Cold War Era

After the Soviet Union detonated its fission bomb in August 1949, President Truman decided in January 1950 to develop thermonuclear fusion weapons (Clarfield and Wiecek 1984, 116–48). This led the civilian-controlled Atomic Energy Commission (AEC), which had been created by the 1946 Atomic Energy Act, to plan and build a vast expansion of the fa-cilities it had inherited from the Manhattan Project.[6] The high-pressure Manhattan Engineering District (MED), as the Manhattan Project was formally designated, had built three major facilities—Oak Ridge, Tennessee (to develop nuclear materials and prototypes for plutonium production), Los Alamos, New Mexico (to design and build the bombs), and Hanford, Washington (to produce and extract plutonium)—but had relied on scores of private factories and machine shops to build and supply nuclear and nonnuclear materials and parts for the first weapons (Cochran et al. 1987a, 1987b; Makhijani, Hu, and Yih 1995; DOE 1995, 1996, 1997b; "The Sites" 2001). Many of these private factories became contaminated and are now part of the over three hundred sites in thirty-four states that are part of the cleanup (DOE 1997b; "The Sites" 2001). In the early 1950s, the AEC set about building thirteen major new facilities to bring the entire weapons-building process into government-owned facilities (for a map of the complex, see figure 6.1 above; for a summary of the major sites, see table 6.1 below).

These plants consumed huge amounts of land: the Idaho National Engineering Lab took 2,314 square kilometers, the Savannah River Site (SRS) 806 square kilometers, and Hanford 1,450 square kilometers (DOE 1996). At its largest, the weapons complex occupied more than 9,300 square kilometers of territory, had 11 million square meters of building space (DOE 1995, 2), and employed more than 100,000 workers.[7] In an assessment of the environmental damage caused by the system, the DOE claims that the total cost of nuclear weapons production up to 1995 had been $300 billion, in 1995 dollars (DOE 1995, 2). However, a comprehensive independent assessment of weapons and delivery-vehicle planning, production, and cleanup costs puts the total bill at thirteen times the DOE estimate: $3,900 billion (Schwartz 1995, 1998).

Table 6.1 Summary of Major Facilities of the U.S. Nuclear Weapons Production Complex

Site	Location 1	Size	Functions	Workers Mid-1980s (N)	Budget Mid-1980s ($ million)	Future Use	Contractor	Cleanup Cost, 1996 Estimate ($ million)
Los Alamos Laboratory	North central New Mexico	112 km²	Design, develop, test nuclear weapons; other nuclear weapons research, e.g., laser weapons for space	7,368	999	Ongoing weapons development	1943–2004, University of California	4,081
Lawrence-Livermore Laboratory	80 km E of San Francisco; site 300–100 km E of SF	2.6 km²; 28 km²	Design, develop, test nuclear weapons, space-based lasers; explosives testing	8,541	937	Ongoing weapons development	1951–2004, University of California	2,506
Idaho National Engineering Laboratory	Southeastern Idaho, 67 km W of Idaho Falls	2,314 km²	New reactor R&D; recycle highly enriched uranium from navy spent fuel to SRS reactors	5,064	366	DOE activities to end; possible future industrial uses	University of Chicago, EG&G, now Bechtel-BWXT	18,622
Oak Ridge National Laboratory	16 km W of Oak Ridge TN, 48 km W of Knoxville	1,160 ha of 14.3 km² site	Develop plutonium reactor and reprocessing; experimental reactors, research	5,045	391	Ongoing research site	1948–1984, Union Carbide; 2004, BWXT	9,351
Argonne National Laboratory	20 km SW of Chicago	689 ha	R&D reactors; energy research	2,965	251	Ongoing research site	1951–2004, University of California	843
Sandia National Laboratory	10.4 km E of Albuquerque	1,128 ha of 307-km² site	R&D nonnuclear parts of weapons; energy research	8,480	1,102	Ongoing research site	1953–2004, University of California	1,591
Oak Ridge K-25 Enrichment Plant	21 km W of Oak Ridge city	600 ha within 14.3-km² site	Produce highly enriched uranium	3,900	1,388	Converted to private industrial production	1943–1984, Union Carbide; 2004, Bechtel-Jacobs	7,286

(continued)

Table 6.1 Summary of Major Facilities of the U.S. Nuclear Weapons Production Complex (continued)

Site	Location 1	Size	Functions	Workers Mid-1980s (N)	Budget Mid-1980s ($ million)	Future Use	Contractor	Cleanup Cost, 1996 Estimate ($ million)
Paducah Enrichment Plant	8 km W of Paducah, KY	300-ha plant on 1,369 ha	Produce low-enriched uranium	1,300	1,388	Leased to semi-private United States Enrichment Corporation	1951–1984, Union Carbide; 2004, Bechtel-Jacobs	4,830
Portsmouth Enrichment Plant	32 km N of Portsmouth, OH	1,483-ha site	Produce low- and highly enriched uranium	1,483	1,388	To be closed and decommissioned from 2007 on	1951–1984, Union Carbide; 2004, Bechtel-Jacobs	3,960
Fernald Feed Materials Plant	27 km NW of Cincinnati	420-ha site	Produce uranium metal for plutonium targets and for other weapons parts	1,083	119	Being closed by 2006; recreation	1951–1986, National Lead; 2004, Fluor Fernald	3,017
Hanford Plutonium Production Complex	Along the Columbia River in southeastern Washington	1,450-km² site	Plutonium production	13,650	986	Being closed by and cleaned by 2070	1946–1964, GE; 1964–1994, multiple; 2004, Fluor Hanford	50,208
Savannah River Site	West central South Carolina, 30 km W of Augusta, GA	806-km² site	Produce plutonium, deuterium, tritium for weapons	15,480	1,200	Ongoing nuclear weapons production	1953–1989, Duport; 2004, Westinghouse	48,769
Rocky Flats Plutonium Processing Plant	26 km NW of downtown Denver	155-ha industrial site within 2,486-ha site	Made plutonium metal core of fission triggers of thermonuclear weapons	5,991	485	Being closed by 2006	1951–1975, Dow Chemical; 1975–1989, Lockheed Martin; 2004, Kaiser-Hill	17,319

(continued)

Table 6.1 Summary of Major Facilities of the U.S. Nuclear Weapons Production Complex (continued)

Site	Location 1	Size	Functions	Workers Mid-1980s (N)	Budget Mid-1980s ($ million)	Future Use	Contractor	Cleanup Cost, 1996 Estimate ($ million)
Oak Ridge Y-12 Plant	3.2 km SW of downtown Oak Ridge	324 ha within 14.3-km^2 site	Make fusion package, uraniumn metal parts	7,213	602	Ongoing weapons production	1947–1984, Union Carbide; 2004, BWXT-Y12	6,168
Mound Plant	16 km SSW of Dayton, OH	306 ha	Detonators, nonnuclear part; tritium recovery	2,364	216	Being closed by 2015	1947–1988, Monsanto; 2004, CH2M-Hill	1,356
Kansas City Plant	19 km S of Kansas City, MO	120 ha	Electrical, electromechanical, plastic parts for weapons	7,853	532	Ongoing weapons work	1949–2004, Honeywell	447
Pinellas Plant	9.6 km NW of St. Petersburg, FL	40 ha	Neutron generators, other electronic and mechanical parts	1,926	139	DOE closed; future nonweapons work	1957–1996, GE	437
Pantex Plant	27 km NE of Amarillo, TX	6,500 ha	Assesmble and dissemble weapons	2,749	198	Ongoing weapons work	1956–2000, Mason & Hanger; 2004, BWXT	683
Nevada Test Site	104 km NW of Las Vegas	3,510-km^2 site	Test weapons	8,414	484	Ongoing weapons work	EC&G until 1999; 2004, Bechtel	3,644

Sources: "Workers, Mid-1980s" and "Budget, Mid-1980s," taken from Cochran et al. (1987b). All other data taken from DOE (1996).

Table 6.2. Operational Arrangements of SSABs

	Fernald	Rocky Flats	Hanford	Paducah	Savannah River	Oak Ridge	Idaho	Nevada	Los Alamos
Members (N)	15	15–25	31	10–15	25	15–20	15	18	12–17
How select original members	Consultant choose individual from stakeholder groups: local residents preferred	EPA and state of Colorado chose six candidates to be on the board and to select the remaining members	Structured into 3 interest groups by a paid independent team; each group chooses its representatives	Members chosen by working group and facilitator to represent 15 categories	7-member panel	Blind selection by an independent individual along with a 7-member panel	7-member panel	Chosen by the state of Nevada Division of Environmental Protection, approved by the DOE	Led by a formation committee, approved by the DOE
How select new members	Entire board elects	New members are chosen by the advisory board itself	Same as above	No official process; through personal contacts with current board members	The board is responsible for recruiting and selecting new members	Blind selection by a 4-member selection panel	Same as above	Chosen by the advisory board	Chosen by the Los Alamos area office
Decisions by consensus?	Not officially, but in practice	Yes	Yes	Necessitated, but little member commitment	No, majority rule	No, majority rule	Yes	No, majority rule	No, majority rule
Use facilitator?	Yes	Yes	Yes	Yes	Chairperson, no outsiders	No	Yes	Yes	No
Categories of stakeholders	Educators, environmentalists, local government	Educators, peace activists, former DOE employees, environmentalists, local government	Educators, peace activists, former DOE employees, Native Americans, Hispanics, local government	Educators, environmentalists, local government	Educators, African Americans, former DOE employees, local government	Educators, former DOE employees, peace activists, environmentalists (until 1998), local government	Environmentalists, peace activists (until 1996), former DOE employees, Native Americans, local government	Educators, peace activists, Native Americans, former DOE employees, local government	Educators, former DOE employees, Native Americans, Hispanics, local government
DOE support in 1990s	Strong	Moderate/weak	Strong	Mixed	Strong	Strong	Strong	Strong	Strong

Source: Unless otherwise indicated, all information is taken from Bradbury and Branch (1999).

The Production of Plutonium and Enriched Uranium

Modern nuclear weapons result from a combination of fission and fusion reactions. Nuclear reactions convert matter to energy according to Einstein's famous formula $E = mc^2$, where E is the energy released, m is the matter converted to energy, and c is the speed of light. Fission reactions occur when either uranium-235 (U-235) or plutonium-239 (Pu-239) is hit with a neutron, destabilizing it and causing it to split into pieces. In the process, some matter is converted to energy, and two or three neutrons, plus radiation and heat, are ejected. Fusion reactions join two forms of hydrogen into helium but require temperatures of several million degrees for ignition. The first bombs, those dropped on Hiroshima and Nagasaki, were fission bombs. Presented with designs that used a fission bomb to create the conditions to trigger the fusion reaction, President Truman made the decision to develop fusion, or hydrogen, bombs. Protests by scientists and others that such a weapon was too horrible to be built were overshadowed by the period's fear of communism, the victory of the Communists in China, the Korean War, and the irresistible power that such weapons gave to the U.S. president and military. In May 1950, Truman accepted the option presented by a National Security Council report calling for a massive rearming of the United States, including the development of the "super," a fusion bomb.[8] The report became the blueprint for the administration's decision to meet "every communist threat throughout the world" (Clarfield and Wiecek 1984, 115–30), even when the perceived threats were anticolonial civil wars, like Vietnam; nationalistic attempts to control resources by elected governments, like Iran in 1953 and Chile in 1973; or reform-minded democratic governments, like Guatemala in 1954.

The key part of nuclear weapons, therefore, is fuel for the fission reaction, U-235 or Pu-239. U-235 is extremely rare, about 0.7 percent of all uranium, but to make a practical bomb the percentage must be concentrated, or "enriched," to 93 percent. The MED had developed two processes to enrich uranium and chosen one, combining uranium and fluorine into uranium hexafluoride, which vaporizes at temperatures above 64°C (Makhijani and Saleska 1995, 39), and sending the gas through thousands of filters in a gigantic plant at Oak Ridge. In the early 1950s, the AEC built two similar facilities, one at Paducah, Kentucky, and the other at Piketon, Ohio, north of Portsmouth (DOE, 1995, 1996, 1997b; see also figure 6.1 above).

Plutonium is a human-created element with ninety-four protons in its nucleus. It is created when U-238, the isotope that is 99.2 percent of all uranium, absorbs a neutron into its nucleus in a reactor and goes through two disintegration stages. The Pu-239 is embedded in the uranium and so must be chemically separated, or "reprocessed," to be made into bomb parts. The Manhattan Project had built three plutonium-production reactors at Hanford, but the AEC added five more between 1948 and 1955 and then a ninth, the N reactor, in late 1963 (Makhijani, Ruttenber, Kennedy, and Clapp 1995, 216; DOE 1995). The expanded complex of the cold war called for another plutonium-production facility to be built along the South Carolina–Georgia border, on the Savannah River. The SRS built five production reactors in the 1950s expansion. These were to produce both plutonium and tritium, a gaseous isotope of hydrogen that is placed in the center of the fission trigger of the bomb to "boost" it with a small fusion reaction. Tritium gas is created by irradiating lithium-6 targets, which splits the lithium-6 into tritium (DOE 1995; Makhijani and Saleska 1995, 56–58; Makhijani, Ruttenber, Kennedy, and Clapp 1995, 246–48).

Extracting plutonium from irradiated fuel requires remote-control equipment because the fuel remains highly radioactive.[9]

The process involves dissolving the fuel in nitric acid and chemically separating out the Pu-239. This process leaves large volumes of highly radioactive acidic waste. Rather than convert this waste to a safer form, the MED, AEC, and DOE put it into storage tanks. At Hanford, they added, but sometimes failed to keep records on, substances to neutralize the acids in order to avoid the expense of stainless steel tanks. By 1990, all the waste produced since 1943 remained in 177 tanks, except the approximately 1 million gallons that have leaked into the soil from 77 of the tanks (DOE 1997b, 35). Hanford also poured 200 million gallons of "slightly" radioactive high-level wastewater into the ground (DOE 1997b, 75). Another 440 billion gallons of radioactive and hazardous liquids were dumped (Alvarez 2003, 31), so plumes of radioactive water are moving toward and into the Columbia River (for maps, see DOE 1995, 73). The SRS and the Idaho National Engineering Lab also have huge volumes of reprocessing waste dating from their first production in the 1950s. Overall, 382,000 cubic meters (about 100 million gallons) of high-level waste containing 960 million curies of radioactivity await treatment (DOE 1997b). The plan has been to dry the waste and mix it with molten glass, creating large glass logs to be buried in an underground repository, but the process will take until 2050, and Hanford's total remediation cost was estimated at $50 billion (DOE 1996, 3, Washington sec.: 4). The SRS will continue to produce weapons material, but its cleanup of previously generated waste was estimated to cost $48.8 billion and to last until 2050 (DOE 1996, 3, South Carolina sec.: 4).

Uranium Mining and Processing

Before the uranium can be enriched or the plutonium created and reprocessed, the ore must be mined and the uranium processed into reactor fuel and special U-238 targets for plutonium reactors. The AEC was intent on establishing a U.S.-based source of uranium, so it set up an incentive program in 1948,[10] which led to the industry growing rapidly and supplying about half the uranium used by 1960 (Yih, Donnay, Yassi, Ruttenber, and Saleska 1995, 110). The remainder was imported from Canada, Australia, France, Niger, Gabon, Namibia, and South Africa.

The uranium industry was located in the intermountain Western states, especially on the Colorado Plateau (see figure 6.1 above). Radon gas is a decay product of the progeny of U-238, so in underground uranium mines radon accumulates to lethal levels if the mines are not ventilated. The AEC knew but ignored this problem as it began promoting uranium mining in the 1940s. Not until 1971 were national standards of air quality in uranium mines enacted, after the AEC had secured all the uranium it needed (Yih et al. 1995, 117).[11] Many, perhaps half, of the miners subsequently contracted lung cancer or other respiratory diseases.

Beginning in the late 1970s, the miners, legally represented by Stewart Udall, who had been Kennedy's secretary of the interior, began a class-action suit seeking compensation. Udall also filed a separate suit on behalf of people who lived downwind from the Nevada Test Site and so had been exposed to fallout from atmospheric nuclear tests.[12] Despite compelling evidence of AEC culpability, the courts rejected any government responsibility on the grounds of the government's sovereign immunity. However, the stories of the suffering of the miners and "downwinders" were so compellingly presented in the media that, in 1990, Congress passed and President Bush signed a law guaranteeing future health care to the miners and payments of $100,000 to miners and up to $50,000 to downwinders who contract one of twelve types of cancer. The government has been slow to make these payouts, especially to Navajo widows who have no government record of their marriage

(Schneider 1993a, 1993b; Janofsky 2001; Makhijani 2001; Alvarez 2001).

The uranium mills produce "yellow cake" (uranium oxide), and this was converted to uranium hexafluoride at civilian facilities in Gore, Oklahoma, and Metropolis, Illinois. The Metropolis plant is just across the Ohio River from the Paducah Gaseous Diffusion Plant, to which it sends its uranium hexafluoride for enrichment. The Paducah plant did the initial enrichment to 1–2 percent U-235. The slightly enriched uranium was then sent either to Oak Ridge's K-25 complex, which could produce 4 percent fuel, or to the Portsmouth plant, which could concentrate U-235 to 93.5 percent (Cochran et al. 1987b, 126–32). In 1962, the K-25 complex closed its two largest plants, and, in 1985, the entire K-25 enrichment area was closed. The DOE leased the Paducah and Portsmouth plants in 1992 to the semiprivate United States Enrichment Corporation, which continues to produce 3–5 percent U-235 fuel for civilian power plants at Paducah (DOE 1997b, 20).

Both the enriched and the "depleted" uranium (material from which most of the U-235 has been filtered) were important "feed materials" for weapons components. Highly enriched uranium was made into fuel rods for navy submarines or for SRS plutonium-producing reactors. Depleted uranium was formed into targets for plutonium-producing reactors or into uranium metal for the parts of bombs that direct neutrons into the warhead's fusion package (DOE 1997b, 22). Fernald Feed Materials Plant, eighteen miles northwest of downtown Cincinnati, was the primary site where these conversions were made (Cochran et al. 1987b, 7–12).[13] Both the depleted and the enriched uranium were in the form of uranium hexafluoride, so one of the three main tasks Fernald performed was to convert the uranium hexafluoride to uranium metal. Depleted uranium metal was shaped into either targets for Hanford's plutonium-production reactors or "derbys," cake-shaped masses, to be shipped to Oak Ridge's Y-12 plant or to Rocky Flats to be made into bomb parts. Fernald also recycled uranium scraps from Y-12, Rocky Flats, Hanford, and other sites back into uranium hexafluoride for further enriching. Some of this had been irradiated and so contained bits of plutonium, but neither Fernald nor the enrichment plants had special procedures to protect workers from this most deadly of materials or from fission products contaminating the uranium.

In fact, managers at the Fernald plant were extremely careless about controlling the radioactive and toxic materials that the plant processed. Uranium dust was produced in several areas and was supposedly filtered by dust collectors and scrubbers, but these safeguards were often worn out or not functioning, allowing dust within the breathing zone of workers to exceed the maximum allowable concentration (MAC) by factors of one hundred or even one thousand. In the 1950s and 1960s, the collectors were emptied by pouring the dust down a shoot from the second floor into a barrel on the first floor, a cloud of radioactive dust often wafting into both floors in the process. In the 1950s, workers were not issued respirators until the dust concentration reached ten times the MAC (Makhijani 1996a). The scrubbers that were to prevent the dust from escaping were often broken, so how much uranium escaped remains uncertain. The most recent estimates are that between 340 and 450 metric tons were released in the air and 80 tons into surface waters (Makhijani 1996b, 7). Most of Fernald's escaped radioactivity has come from radon gas emitted by highly concentrated ore wastes left over from the Manhattan Project and stored in several concrete silos at the Fernald site. Radon escaped from 1951 until 1991, and this produced most of the 170,000 curies estimated to have been released (Makhijani 1996b). Disposal of this waste has proved to be the most difficult and contentious aspect of Fernald's cleanup (Makhijani and Makhijani, 2006).

John J. Metz

Uranium releases combined with carelessly dumped wastes contaminated the groundwater under the plant and the drinking water in the plant's vicinity. The DOE knew in 1980 that neighboring wells were contaminated, but it did nothing to fix the problem. When a local resident, Lisa Crawford, discovered in 1984 that she had been feeding her baby uranium-contaminated water and that the DOE knew about the contamination, she began a class-action suit against the contractor and the DOE. Crawford soon became the president of Fernald Residents for Environmental Safety and Health (FRESH) and became with other members of FRESH key drivers of the Fernald Citizen Advisory Board. In 1989, the DOE settled the suit, paying $78 million for health monitoring and compensation.

Assembling the Bombs

The actual components of the weapons were manufactured at Oak Ridge's Y-12 complex, which made the uranium and fusion package, and at Rocky Flats, which produced plutonium metal and then molded and machined the plutonium into the hollow shell "pits" that are the cores of the bombs. Three other large plants made the nonnuclear components. The Pinellas plant, halfway between Clearwater and St. Petersburg, Florida, made neutron generators to initiate the fission reaction. The Mound plant, near Dayton, Ohio, made actuators, ignitors, and detonators. The Kansas City plant produced the plastic, mechanical, and electrical parts. All these parts were sent to the Pantex plant, near Amarillo, Texas, for final assembly.

Sacrificing Citizens for National Security

The discussion to this point has already suggested that the government was intent on manufacturing nuclear weapons regardless of the health or financial costs to nuclear industry workers and the general population. The more deeply one delves into the functioning of the AEC/DOE nuclear complex, the more disturbing the information one finds (see tables 6.3–6.5). Perhaps most shocking are the human experiments (table 6.4), which clearly violate international human rights standards. Yet, as disturbing as the experiments were, they are consistent with the other practices of the nuclear establishment. For example, exposing the entire population to radioactive fallout (table 6.3), placing military personnel within the radioactive blast zone during weapons tests, allowing contaminated working conditions within the factories (table 6.5), and disposing of wastes irresponsibly all display a consistent pattern of actions that threaten the well-being of citizens. The apologists for the AEC/DOE regularly claim that standards were different in earlier times, that people did not see the dangers that seem so obvious today, that the rush to defend the nation from an aggressive Soviet threat left the decisionmakers no choice but to act as they did. However, when we read the detailed studies that have emerged as independent citizens have investigated DOE operations, we see that the nuclear establishment did know that its actions would cause harm but acted anyway. A report to the AEC in April 1948 stated: "The disposal of contaminated waste in present quantities and by present methods, if continued for decades, presents the gravest of problems" (DOE 1995, 8). In 1948, the committee selecting the site for nuclear tests was warned by an air force meteorologist that it should select an East Coast site to avoid fallout (Ortmeyer and Makhijani 1997). Documents from Fernald's files show that management knew of dangerous working conditions: a 1960 memo noted that dust within the breathing zone of some workers was ninety-seven thousand times the MAC; a 1955 document noted that measured uranium releases were "far below true stack releases"; and another memo sixteen years later called the releases "inherently deceptive" (Makhijani 1996b, 2). The failure to notify local residents that their drinking water was contaminated with uranium was so outrageous a public relations disaster for the DOE that it contributed to the agency's decision to pay the Fernald victims $78 million. The Fernald suit reminds us that AEC/DOE behavior has cost the agency a great deal of money. The costs include not only the compensation paid (or promised) to miners, fallout victims, soldiers intentionally exposed to radiation, and nuclear workers (Flynn 2001), but also the $250–$500 billion that the DOE expects to spend cleaning up the messes it created.

A 1988 *New York Times* story suggested part of the institutional cause of the waste and safety problems (Schneider 1988a). Although the AEC and the DOE own the weapons-complex real estate and buildings, they do not run the facilities themselves. Private corporations were hired to do that. The corporate contractors had all their operational costs paid and were guaranteed a base payment as

Table 6.3. Nuclear Testing Impacts	
Tests	Atmospheric tests within continental United States: 106[a] Atmospheric or underwater tests in Pacific: 106[a] Atmospheric tests in South Atlantic: 3[a] Underground tests in Nevada: 725 (156 vent to atmosphere)[a] Space tests: 9[a]
Nevada test site	Site selected because close to Los Alamos; despite fallout it will drop throughout United States; AEC warns Kodak before tests because fallout destroying newly made film [b] Consistently heavy fallout closest to site (downwind in Nevada, Utah) [b] Occasional heavy fallout when radiation cloud encounters storms, located randomly in time and space; vast majority never identified (Albany, NY, a 1953 exception) [b]
Pacific tests	Destroy home islands of indigenous people [c] 1954 H-bomb test covers Ronglap Island with radioactive "snow"; not evacuated for 50 hours, high doses; 1957 returned to island; AEC wants to monitor radiation impact on people: they are "more like us than the mice" [c] 1946 test contaminates virtually entire Pacific Fleet and thousands of sailors [c]
Protests	During the 1950s growing opposition in United States and world [d] AEC responds by "reeducating" public with false propaganda: radiation called "sunshine units" [d] AEC attacks critics, labels them "Communist sympathizers," secretly undermines careers [d]
Impacts	National Cancer Institute study finds all United States exposed to radiation, producing 25,000–50,000 thryroid cancers, 10 percent of which will be fatal [e] AEC knew of iodine-131 in milk, failed to order disposal, though did warn film manufacturers of tests [f] Nevada underground tests create high-level nuclear waste totally lacking any geologic or manufactured containment [f]
Compensation	1990, G. H. Bush signs legislation to provide health care and compensation to uranium miners and "downwinders" who have one or more of 12 types of cancer; government processing of claims slow, rejecting many, like Navaho miners' widows who have no legal marriage documents [g] In 2000, Congress passes Energy Employees Occupational Illness Compensation Act for 600,000 workers who may have been affected by working at complex

[a] IPPNW/IEER (1991, 15–17, 61–68). [b] IPPNW/IEER (1991); Ortmeyer and Makhijani (1997). [c] IPPNW/IEER (1991, 69–88); Weisgall and Moore (1994). [d] Clarfield and Wiecek (1984, chap. 8). [e] Ortmeyer and Makhijani (1997). [f] Makhijani (2001); Wald (1997); Waldman (1997). [g] Alvarez (2001); Schneider (1993a, 1993b); Stieber (1995).

John J. Metz

Table 6.4. Human Experiments by the AEC, the ERDA,[a] and the DOE

"Atomic soldiers"	250,000–500,000 military purposely placed 2–7 km from nuclear tests, taken onto detonation site within hours of test; want to see if can fight on radioactive battlefield [b]
Deliberate release of radioactivity	1949 "Green Run" at Hanford—to calibrate sensors to measure Soviet Union plutonium production [c] 1940s and 1950s, radioactive lanthanum-140 and tantalum-182 over Oak Ridge and Los Alamos; seemed to be testing for radiation dispersal bombs [c]
Plutonium injections	18 people at Los Alamos, Oak Ridge, University of California, San Francisco, University of Chicago, University of Rochester given plutonium injections that are 2.4 to 14.7 times larger than limit for plutonium workers; 22 others injected with other radioisotopes; e.g., MIT injects with radium and thorium [d]
Inject women	200 women, many pregnant, given "cocktails" of radioactive iron at Vanderbilt University hospital; at least 3 fetuses die of radiation [e]
Whole body irradiation	For 30 years patients at Los Alamos, Oak Ridge, University of Cincinnati Hospital given "full-body radiation" to observe impacts; early claimed to be therapeutic, but later known to be useless treatment; contributes to death of 8–20 patients [f]
Irradiate mentally handicapped	Treat children at Fernald State School in Massachusetts with radioactive substances late 1948–1961; treated children belong to "Science Club," given special treats for joining [g]
Irradiate prisoners' testicles	131 prisoners in Oregon and Washington have testicles irradiated with gamma radiation equal to 20,000 X-rays in exchange for possible reduced sentences; must also have vasectomy; to measure radiation impact for weapons complex workers [h]
Insert radium into nasal passages	0.5–2.3 million children in experiments inserting radium into nasal passages to treat adenoid and sinus problems; 2,300–10,000 extra deaths In 1940s, 7,000 submarine and airplane personnel treated with radium for ear pressure problems
Compensation	Sovereign immunity absolves United States of all claims by these victims In 1990, Bush sets up fund to compensate verified victims, but few qualify In 1993, Secretary O'Leary establishes Advisory Committee on Human Experiments; its 1996 report recommends that only a few hundred people should get compensation, medical notification, or even an apology [i]

[a] ERDA, the Energy Research and Development Administration, replaced the AEC briefly before being converted to the DOE. [b] IPPNW/IEER (1991); Minutaglio (1994); Weisgall and Moore (1994). [c] Makhijani (1994, 25). [d] Makhijani (1994). [e] Gordon (1996). [f] Makhijani (1994). [g] Makhijani (1994). [h] Makhijani (1994). [i] Gordon (1996).

well as a bonus. The bonuses were based on cost containment, quality assurance, meeting production goals, and keeping workers and the environment in general safe. However, only 10 percent of the bonus was based on health and safety, while 45 percent was based on cost containment and 45 percent on meeting production goals. Hence, the contractors received millions of dollars in bonuses for operating in ways that contaminated workers and the environment, leaving the taxpayer to pay for the cleanup and the plant's neighbors to pay with compromised health (Schneider 1988a). Reinforcing this perverse system, the DOE managers who measured contractor performance also received bonuses based on the same criteria. In his detailed study of the Rocky Flats plant, Ackland (1999b, 143–45) describes how the push for more production led to high levels of employee contamination—in 1968, Rocky Flats had more than 80 percent of all DOE employees who exceeded their maximum radiation-exposure limits— and to the two major fires that came very

Table 6.5. Health Impacts on Workers at Nuclear Facilities

AEC/DOE position before 1999	Factories operate safely; people have threshold below which radiation does not affect them; AEC/DOE monitor worker exposures to prevent injuries
Hanford workers study, 1960s	Preliminary results of study find excess cancer rates; AEC cancels study, confiscates data, hires new researchers to complete study: they find no excess of cancer [a]
Fernald conditions	Uranium dust filters wear out, not replaced for years; 370–600 tons vented 1951–1989; in 1966, canister leaks 1.82 tons of uranium hexafluoride onto dairy farm neighbor; some dust contaminated with plutonium fission products No internal monitoring of workers; simple film badges used, data often lost Uranium contaminated Miami River aquifer, drinking water of neighbors
Rocky Flats conditions	Make plutonium metal parts; plutonium burns in air; one-millionth of an ounce in lungs can kill; small fires every day In 1957, fire burns uncontrolled for 4 hours; workers turn on exhaust fans, which carry fire into ducts and filters; filters burn for 13 hours, depositing plutonium into neighboring communities, how much highly disputed; no warning to local officials; explosion cuts electricity, shutting fans off In 1969, fire burns out of control for hours, almost breaches roof; 1 ton of plutonium burns; DOE claims none escapes; firetruck accidentally knocks out electric power, shuts fans off, allowing fire to be controlled Precipitation runoff from plant carries plutonium into drinking water reservoirs of local communities and contaminates with plutonium; one a superfund site [b]
Oak Ridge conditions	In 1990s, cluster of unexplained disease besets workers and neighbors of high-efficiency incinerator at K-25 enrichment area; workers claim the TSCA incinerator, which burns mixed radioactive and toxic wastes, is releasing toxins; DOE claims it captures 99.999 percent of emissions; *Tennessean* newspaper does long series on complaints in 1996–1997; in June 1997, governor convenes "expert" panel, but it finds incinerator safe, though incinerator is only monitored for special tests [c]
Paducah conditions	July 1999 *Washington Post* story revealed that the plant had enriched recycled uranium contaminated with plutonium for 23 years, exposing workers to high levels of plutonium [d]
Government payouts	In 2000, Congress includes in 2001 defense bill a fund to compensate workers who suffer from beryllium and silica poisoning or radiation-induced cancer and "special cohort workers" (those who worked in three enrichment plants); in 1990, Radiation Exposure Compensation Act gives downwinders and uranium miners compensation for special cancers 1984 Veterans Dioxin and Radiation Exposure Act sets standards for Veterans Administration to accept claims of veterans affected by radiation Government slow to pay compensation, denies many claims [e]

[a] D'Antonio (1993, 46–47). [b] Ackland (1999a, 1999b). [c] Flynn (2001). [d] Flynn (2001). [e] Alvarez (2001); Flynn (2001); Makhijani (2001).

close to contaminating much of Denver with plutonium (Ackland, 1999a, 1999b). As the contractor Dow Chemicals noted in a 1967 memo: "Over the past several years there has been a strong emphasis on seeking more efficient production methods in an effort to reduce the cost of producing plutonium for weapons. This emphasis has resulted in some remarkable savings in cost in the face of more stringent processing requirements, but some of the changes have contributed to exposure increases. It is not possible to revert to old processing techniques without causing large additional capital expenditures and gross increases in the operating costs associated with plutonium processing" (Ackland 1999b, 144–45). Another aspect of the contracts that encouraged irresponsible production methods was that contractors were absolved from any liability no matter how they operated. As information about the practices became known, the government spent large amounts of money ($40 million in 1994 alone [Flynn 2001]) defending contractors from lawsuits. With these kinds of incentive structures operating, practices that endangered workers and local residents were inevitable. Similar incentives in the environmental cleanup emerged after 2000 as the DOE established payment bonuses and penalties for contractors based solely on the date they completed their work (Alvarez 2003; see also below).

The "New" DOE: Cleaning Up with Citizen Guidance

The contract system described above partially explains the problems of the nuclear complex, but it is also a symptom of a broader issue: the lack of outside control over the AEC/DOE and the atmosphere of national security–justified secrecy that permeated the entire enterprise. When Congress set up the AEC, it created it as a five-member civilian-controlled organization, but the AEC gradually ceded decisions about the types and specifications of weapons to the Department of Defense (DOD), with the result that, by 1967, the AEC had become a mere supplier of "engineering support to meet DOD's demands" (Cochran et al. 1987a, 2–3). The AEC established a decentralized structure that gave local field offices considerable power, making them semi-independent of headquarters (Fehner and Gosling 1996, 8). As part of the decentralization process, the AEC/DOE never established systemwide internal standards for safety or waste management, effectively turning the issues over to the contractors, who, as has been noted,

had no liability. Congress also created a special committee to oversee the AEC, the Joint Committee on Atomic Energy (JCAE) (Fehner and Gosling 1996, 8). Although conceived as a watchdog, the committee only occasionally questioned AEC actions, becoming essentially an uncritical promoter and defender of the AEC (Ford 1982, 62–84, 166–69).

It was not that the AEC had no warnings of problems but that it consistently ignored outside reviews recommending changes. In 1947, the AEC organized the Safety and Industrial Health Advisory Board, the panel of outside experts that issued the report quoted above about waste being the "gravest of problems." As civilian nuclear power was being developed in the late 1950s, questions of waste disposal arose and led the JCAE to hold hearings during which the uncontrolled dumping of wastes into the ground at Hanford and into rivers at Oak Ridge was revealed, but ignored. In 1960, a National Academy of Sciences panel of experts assembled to explore high-level waste-disposal sites notified the AEC that its waste-disposal practices were unsatisfactory, leading the AEC chair, Glenn Seaborg, to label the panel's report "uninformed" (Fehner and Gosling 1996, 14). After the Three Mile Island partial meltdown in 1979, the DOE established an in-house panel known as the Crawford Committee to review reactor safety within the weapons complex: the committee found it seriously deficient, but its recommendations were ignored. The doubling of production during the Reagan build-up of the 1980s so worsened conditions that whistleblowers at Hanford and Rocky Flats went public with their concerns. At Hanford, Casey Rudd found that, while equipment throughout the site had been modified to repair breaks or correct bad designs, neither the changes, the calculations justifying the changes, nor evidence of the approval of the changes by peer engineers had ever been included in the blueprints (D'Antonio 1993, 168–97). At Rocky Flats, Jim Stone found the exhaust ducts full of plutonium, plant equipment antiquated, and no efforts at reform, so he continuously wrote reports critical of conditions. After he was fired, he took his reports to the FBI (Ackland 1999b, 204).

Although the AEC was not changing, society was. The social reforms of the 1960s led to new environmental regulations and agencies. The Atomic Energy Acts gave the AEC and the DOE authority over radioactive materials in their possession, but whether the DOE's handling of toxic and hazardous wastes was subject to the laws and rules of the new federal regulatory agencies and the states was unclear. An affirmative judgment in 1984 in a suit brought by two environmental groups over 2 million pounds of leaked mercury at Oak Ridge's Y-12 complex resolved the question and opened the door to challenges from grassroots and national groups all around the country (table 6.5 above). The National Environmental Policy Act of 1969 had already mandated federal agencies to submit environmental impact statements before beginning new construction, and this process revealed information that had previously been unavailable. It also required hearings at which citizens could voice objections to the plans. Grassroots groups and reporters used these new tools to pry information out of the DOE, often confirming their suspicions about DOE activities.

From 1984 to 1990, scores of revelations about lax standards, dangerous operations, and irresponsible waste-management practices were revealed, leading to the shutdown of all the major facilities, effectively ending the ability of the United States to manufacture nuclear weapons (DOE 1995, 1997b). Of course, the United States had more than thirty thousand warheads stockpiled and cannibalized old warheads to maintain an active arsenal. Congress and the DOE established numerous review boards and committees to evaluate the condition of the facilities and suggest how to clean them up (see table 6.6). Virtually all these review committees

suggested that the DOE include citizen participation in the cleanup process, and two of them suggested citizen advisory boards (OTA 1991; DOE/EPA 1996). Since most of the key facilities were placed on the Superfund "National Priorities List," they were required to include public participation in any cleanup. By the early 1990s, the DOE reversed its previous five decades of secrecy by publishing and making available in reading rooms the torrent of "remedial assessments/feasibility studies," environmental reports, and complex-wide studies that began emerging on the current conditions at the various sites and possible remediation methods. By 1995, most sites had reading rooms filled with tens of thousands of pages of reports.

As the courts ruled that the DOE must observe environmental laws, including public participation, the DOE decided to make citizen advisory boards a major component of its public participation program. Advisory boards were first suggested by the Office of Technology Assessment in 1991; and in 1993 the Keystone Committee,[14] established by the Environmental Protection Agency (EPA) to set cleanup priorities, strongly recommended them (DOE/EPA 1996, app. C). The first boards were established at Fernald and Hanford in 1993, with guidelines for them published the next year (see DOE 1994), and, over the next several years, ten other weapons-complex sites set up SSABs.

Officially, there was only one board for the entire complex, with members at each site. In practice, the various sites were given considerable freedom to establish the ground rules to fit their own situations. By the time the SSABs were being established, the Clinton administration's Hazel O'Leary had become the secretary of energy. O'Leary appointed former critics of the DOE to some senior positions (Alvarez 2000) and appears to have been genuinely shocked by the breadth and depth of environmental, health, and safety problems that were exposed before and

Table 6.6. Processes Leading to the Establishment of Citizen Advisory Boards

1984: *LEAF v. Hodel* decision rules Oak Ridge Y-12 mercury spills subject to Resources Conservation and Recovery Act

1984: Fernald air and groundwater contamination revealed; Ohio sues DOE; FRESH forms; Fernald citizens sue DOE

1984: Hanford Environmental Action League (HEAL) forms to focus on environmental and health issues related to Hanford [a]

1985: Reagan Secretary of Energy Herrington establishes new assistant secretary for environment, safety, and health; begins to negotiate with EPA and states on cleanup of sites

1986, February: Hanford DOE office releases 19,000 pages of information; HEAL finds evidence of many releases of radioactivity, including "Green Run"

1986: After Chernobyl, Herrington has National Academy of Sciences panel evaluate DOE reactors; it recommends an external oversight committee to monitor

1988, January: Herrington sets up Advisory Committee on Nuclear Safety with J. Ahearn as head (Ahearn Committee) to report to DOE secretary

1988, February: Hanford's N Reactor shut down, ending plutonium production at Hanford [b]

1988, August: Savannah River reactor surges; Ahearn Committee finds ingrained complacency; Herrington investigator finds 19-page memo outlining three decades of accidents, near meltdowns of which DOE had no record; Savannah River shutdown [c]

1988, September: Congressional hearings publicize problems; Congress establishes 5-member Defense Nuclear Facilities Safety Board to review safety and environmental condition of weapons sites

1988, October: DOE reveals thousands of tons of uranium released at Fernald; Fernald shut by strike

1988, October: 6 *New York Times* stories reveal SRS reactors had emergency shutdowns at twice the rate of civilian reactors; supervisor happens to be present to stop meltdown

1988, October: Rocky Flats temporary shutdown when 3 workers contaminated

1988, December: Herrington approves dose reconstruction study at Hanford; Congress approves study of thyroid illness in downwind populations

1989: James Watkins secretary of energy

1989, June: FBI raids Rocky Flats for violating a court ban on burning radioactive waste; Rocky Flats shut down

1989: Fernald shuts down; DOE settles suit by Fernald residents for $78 million

1989: Congressional hearings call for DOE to set priorities for cleanup; 10 governors, 49 state attorneys general, EPA ask DOE to set cleanup priorities

1989, December: EPA asks Keystone Center of Colorado to set up a national dialogue to set cleanup priorities for DOE

1990–1995: Keystone Committee (Federal Facilities Environmental Restoration Dialogue Committee) meets, sets priorities, recommends Citizen Advisory Committees at major sites [d]

1991: Office of Technology Assessment reviews weapons complex environmental conditions; recommends citizen advisory boards [e]

1992, March: Rocky Flats grand jury investigation of conditions leading to FBI raids ends with deal between prosecutors and DOE that seals all records; grand jury members outraged, want investigation results revealed, but prohibited [f]

1992, April: U.S. Supreme Court rules DOE subject to state laws

1992, October: Congress passes and Bush signs Federal Facility Compliance Act, based on which DOE, EPA, and states can regulate cleanup of weapons complex

1992–1995: DOE, EPA, and states establish cleanup agreements

1993: Keystone Committee presents interim report outlining plans for SSABs [g]

1993: Fernald initiates SSAB; other sites follow

1994, May: Office of Management and Budget approves charter establishing SSAB for DOE

[a] D'Antonio (1993, 43). [b] D'Antonio (1993, 232). [c] D'Antonio (1993, 250–55). [d] DOE/EPA (1996, app. C).
[e] OTA (1991). [f] Ackland (1999b, 232). [g] DOE/EPA (1996, app. C).

during her tenure. O'Leary's DOE accepted many of the recommendations of the Keystone Committee. Some of the crucial ones were that all groups of stakeholders were to be represented, especially grassroots critics of the DOE; that decisions were to be made by consensus, if possible, rather than by simple majority votes; that senior DOE personnel were to attend the boards' meetings and must explain how the DOE has responded to boards' recommendations; that the boards were to concern themselves with "big-picture" issues and should not second-guess the DOE and the contractors on the details of their actions; that the boards could occasionally hire outside specialists to review DOE actions but that these consultants could not duplicate work that DOE contractors were doing; and that consultants could be hired to help establish boards and their rules of operation and to facilitate meetings (for a more complete list, see table 6.7).

The SSAB guidelines (DOE 1994; DOE/EPA 1996) state that all stakeholder groups are to be included and list the following as examples: local residents, including minorities and low-income communities; local reuse committees seeking employers to use old DOE facilities; local government officials; the business community; school districts and universities; DOE and contractor employees; local environmental groups; civic/public interest organizations; the religious community; local homeowners; the medical community; Native American tribes; and labor organizations.

Several of these, like the business community, labor organizations, and local government, were already organized with easily identified goals. Three kinds of grassroots groups had organized to influence the DOE

Table 6.7. SSAB Guidelines

To be credible, the selection process of SSAB members must be fair and open; EM (the Environmental Management office of the DOE) recommends that the board be balanced and representative of all stakeholders who see themselves affected by environmental restoration, waste management, and other activities

The board should have between 15 and 30 members, depending on the size of the site, and not include alternates

DOE managers and stakeholders should set up meetings to educate the managers, contractors, and stakeholders about the SSAB concept and the Federal Advisory Committee Act and to agree on a process for membership selection

After members are selected, they should select a chair and a vice chair, develop a mission statement, set ground rules for operating the board

The SSAB may use an independent convener to assist the board with possible strategies for drafting a mission statement, establishing operating procedures, drafting goals and objectives, and selecting major policy issues to address

The SSAB may use a facilitator to ensure that board members set and reach meeting objectives, maintain focus, work as a team, reach consensus, and operate with optimum efficiency

Senior DOE and contractor managers responsible for the cleanup should attend meetings to ensure that SSAB advice is heard by decisionmakers

SSABs should focus on the "big picture," the major policy issues, rather than specific cleanup details; the DOE and its regulatory partners need clear articulations of stakeholder principles, priorities, and values

Boards may request that the DOE fund independent technical review of key issues; funds could be used to support travel, per diem and compensation for technical experts and researchers from national public interest organizations, universities, or private consulting firms; technical funding should complement rather than duplicate technical programs of the DOE and its regulating agencies

The DOE must report to the boards what actions result from SSAB recommendations and communicate to the boards what parts of their recommendations were not accepted and why

Source: DOE/EPA (1996).

John J. Metz

facilities. Peace groups opposed the manufacture of nuclear weapons on moral and practical grounds and used environmental, health, and safety issues as evidence of the evil that nuclear weapons inevitably bring and as tools to block weapons production; examples include the Oak Ridge Environmental Peace Alliance and the Rocky Mountain Peace and Justice Center at Boulder. Environmental groups opposed the ways the DOE operated because of environmental and safety damage that nuclear production had caused; these groups avoided taking stands that opposed all nuclear weapons production in order to expand their appeal to those who believed nuclear weapons to be necessary. FRESH and the Hanford Education Action League are examples. The peace and environmental groups often joined forces, but they had sufficiently different goals that, at times, they disagreed on tactics and/or methods of protest.

The third kind of group, DOE support groups, arose at many sites (e.g., the Hanford Family [D'Antonio 1993] and the local oversight committee [LOC] at Oak Ridge) as information about the safety and environmental failures emerged and closing the facilities became a possibility. DOE support groups were filled with workers who worried that they would lose their good-paying jobs, with the local and regional political leaders who feared that their tax base would disappear, with retired workers who felt that their life work was being denigrated, and with local businessmen and real estate developers who feared that closing the plants and publicizing the environmental dangers would hurt their ventures. The communities surrounding DOE facilities were, in effect, company towns with a patriotic culture that tolerated no dissent. Much of this came from the AEC/DOE-imposed national security restrictions and from the government-sponsored and community-accepted belief that workers had been fighting and "won" the cold war. Among the strongest supporters of the DOE were retired nuclear-industry employees, especially the scientists and highly educated managers, and this group often became active on the SSABs.

National environmental groups also played crucial roles. The American Friends Service Committee supported organizers at Rocky Flats in the 1970s and early 1980s. The National Resources Defense Council (NRDC) funded a group of researchers who accumulated key information and published three volumes of detailed information on the complex in the late 1980s (Cochran et al. 1987a, 1987b); the NRDC also led coalitions in legal battles forcing accountability on the DOE. The Physicians for Social Responsibility and the Institute for Energy and Environmental Research (IEER) obtained grants in the late 1980s to prepare a massive study of the impacts of nuclear weapons production in all the nuclear weapons states (Makhijani, Hu, and Yih 1995). The IEER has also trained activists in the technical bases of DOE operations and provided DOE critics with a steady stream of scientifically based analyses of DOE practices. In 1987, the grassroots groups at the various sites linked themselves into the Military Production Network, which changed its name to the Alliance for Nuclear Accountability in 2000. This alliance of thirty environmental and peace groups meets three times per year, hosted by the various member groups at the different weapons sites, to develop collective positions on issues and cooperative strategies for organizing citizens, doing research, and taking legal actions.[15] These activists have been able to share their experiences and support one another while at the same time becoming an effective lobby.

The DOE's acceptance of advisory boards (see photograph 6.1) placed the opposition grassroots groups in a quandary: should they participate or not? All the DOE critics had been calling for more democratic control over the DOE, so here was an apparent opportunity to have some influence. Some activists and groups (e.g., Paula Elofson-Gardine of the Environmental Information Network at Rocky Flats [personal communication, 1997] and Greg Mello at Los Alamos [Mello 1997]) decided that the SSABs would be incapable of having any real impact and refused to participate. Other groups waited to see how the selection process unfolded before deciding. At all but SRS some activists decided to give it a try.

The Performance of the Advisory Boards

The SSABs have had a mixed performance record. How to measure their success/failure is a significant problem because it is difficult to identify objective and unbiased criteria by which to judge. There is a growing body of literature on citizen involvement in environmental and social decisionmaking (for reviews, see Bradbury, Branch, and Malone 2003, app. A; Webler and Renn 1995; and Webler 1995), but very few of these studies address how to evaluate success objectively. The resolution of dispute negotiations almost always conforms to the preferences of one or more of the contending stakeholder groups (Renn et al. 1995), and so to define the winners' position as "success" is to privilege their position. One way of evaluating is to ask the participants whether they consider the process to have been successful. The validity of this approach depends on whether all the stakeholder groups are included. Between 1995 and 2003, about half the DOE SSABs lost the members who were critical of the DOE (see table 6.2 above), so evaluations based on participant satisfaction from SSABs that have excluded all critics can be seen as biased, even when the participants do not see themselves as such.

One approach to evaluation that explicitly uses a subjective criterion is Arnstein's "ladder of participation." During the late 1960s, Arnstein studied federal poverty alle-

Photograph 6.1. Meeting of the advisory panel. (Courtesy U.S. Department of Energy.)

viation programs that included citizen participation components (see Arnstein 1969). She proposed "power" as the evaluative criterion and described eight "rungs" to the ladder of citizen power. The lowest rungs give citizens no power but seek to "cure" them through lectures. Middle rungs, which she called "informing, consulting, placating," give citizens information, and even ask their opinions, but keep decisionmaking power in the hands of authorities; this is the level of power we find in the DOE's SSABs. The highest rungs of the ladder actually give citizens the authority to decide issues, a situation that Arnstein sees as the goal of public participation.

Renn et al. (1995) suggest a way to escape subjectivity by proposing that public participation be evaluated, not by its outcomes, but by the *process* through which it functions. These scholars have adapted the concepts of Jürgen Habermas to define evaluative criteria-based processes that are "fair" and "competent" (Webler 1995). By *fair*, they mean that all participants will have an equal chance to put their concerns on the agenda; to approve and propose rules of discussion; to debate and critique proposals and rules; to put forth and criticize validity claims about language, facts, norms, and expressions; and to suggest a moderator, method of facilitation, and rules of discussion. The second criterion,

John J. Metz

competence, refers to how well the participatory process provides the procedural tools and knowledge needed to make good decisions. It demands that all participants be given clear definitions and the opportunity to challenge the validity of terms, concepts, and facts used by others in the discussion. All participants should have equal access to the available and relevant systematic knowledge about the objective world (Webler 1995). These few sentences only hint at the multiplicity of factors so far identified as ensuring that public participation discourses will actually be fair and competent. A key factor not included, however, is authority: will the groups be merely advisory, or will their decisions have legally sanctioned power? SSABs are only advisory.

Before we apply these concepts to the SSABs, we can make a preliminary judgment about the SSABs based on how well they have fulfilled the criteria established by the Keystone Committee and accepted by the DOE (table 6.7 above). Have the boards included DOE critics? Have they brought contending groups into constructive dialogue that has produced consensus positions on cleanup decisions? Has the DOE accepted the recommendations the boards have issued?

At sites that are being closed (Hanford, Idaho, Fernald, Rocky Flats), the boards functioned well, deciding issues by consensus, creating positive working relationships, and developing a pool of citizens with a sophisticated understanding both of the environmental problems of the sites and of the maze of laws regulating the cleanup (table 6.2 above). However, at sites with ongoing weapons missions, most community members, and especially business and elected leaders, have wanted to keep the good-paying jobs and to minimize the perceptions that their community is vulnerable to health hazards, so they have been strong supporters of the DOE. This has created an antagonistic attitude that has either kept DOE critics off the SSABs (SRS), pushed critics to resign (Oak

Ridge, Paducah), induced the DOE to remove dissident members (Los Alamos in 1997), or convinced the DOE to terminate the entire board (Pantex in 2000). Nevertheless, by 2000, boards at many sites, including those without any dissidents or without consensus decision structures, felt that they had made real contributions to the remediation process, and local DOE personnel agreed (Bradbury et al., 2003). That decisionmaking power was divided between the DOE, the EPA, and state regulatory agencies allowed the state agencies to demand that SSAB advice be heeded. The policies of the second Bush administration have, however, undermined the confidence of even the pro-DOE boards, so I briefly review those changes.

Management by the Second Bush Administration of the DOE Complex

The most important aspect of the policies of the second Bush administration is the decision to build new "tiny" (thousand-ton yield), "mini" (hundred-ton yield), and "micro" (ten-ton yield) nuclear weapons that can be used in conventional wars (Dowler and Howard 1991; Schwartz 2002; Deller 2002; Smith 2003; Speed and May 2005). This was part of a general move toward unilateral action, exemplified by withdrawal from or rejection of nine international treaties (Deller 2002) and the attack on Iraq. With regard to the weapons-complex cleanup, the Bush administration initiated a "top-to-bottom review," altered contracts according to a "performance-management plan," and instituted a "risk-based end-state" process to determine what levels of residual contamination can be left at the sites. The top-to-bottom review concluded that the sites could be cleaned more quickly and cheaply by altering the contracts to a performance-management plan, which specified that contractors would receive large bonuses if they completed their work within budget before the DOE established a closure date and would lose money for every month they exceeded the deadline. This policy provides financial incentives to do the remediation quickly and within budget, so it mirrors the incentive structure that created the contamination before 1990.

The risk-based end-state initiative argued that the final uses of the sites should determine the levels of residual waste that can be left. The future-user scenarios ranged from an occasional visitor, to industrial workers, to residents who work off-site, to subsistence farmers. In general, the DOE seeks to classify the sites as having low use levels, thereby allowing much higher levels of residual contamination and reducing cleanup costs greatly (Makhijani and Gopal 2002). For example, the DOE has converted Rocky Flats and the Hanford Reservation to wildlife refuges and claims that only occasional visitors will use these sites, allowing residual contamination levels to remain high. Since these sites will remain radioactive for thousands of years into the future, assuming that any use level less intense than that of a subsistence farmer can be institutionally maintained is likely to expose future generations to life-threatening risks (Makhijani and Gopal 2002). This dispute emerged over residual plutonium levels at Rocky Flats, as I discuss below.

The SSABs and the public were effectively excluded from evaluating all the Bush administration changes, leading many SSAB members to conclude that the DOE was withdrawing from public participation and intending to increase the amounts of contamination left at the sites (Bradbury et al., 2003; see also table 6.8). For example, at Paducah in 2003, the DOE lowered cleanup standards to cut costs (e.g., leave contaminated groundwater because the city water system had been extended to those affected) and finally got the EPA and the Kentucky Department of Environmental Protection to agree. The SSAB, however, had not even been aware that the final agreement was being decided, much less

had its views heard as part of the decision-making process, so seven members resigned (PGDPSSAB 2003; Mark Donham, personal communication, 2008). The DOE has also created the Office of Legacy Management to maintain the sites after remediation is complete, but what role the public will have with this new office remains so uncertain that the chairs of all the SSABs wrote a letter to DOE headquarters in January 2004 promoting the continued use of SSABs by the Office of Legacy Management and protesting the reductions in SSAB budgets at closure sites (Bierer et al. 2004). The changes are so egregious that the Los Alamos board, which the DOE reconstituted in 1997 after removing recalcitrant members, has complained that recent actions "bespeak a strong sense that EM [the Environmental Management division of the DOE] is backing away from its public participation commitments" (NNMCAB 2003a, 2003b). The Bush administration–initiated changes have severely undermined SSAB member confidence in the DOE's commitment to public participation, but, even in the "open" days of the 1990s, significant problems existed.

Inherent Problems with DOE SSABs

Advisory Boards Have No Authority The first issue is that the boards are only advisory and have no legally sanctioned power. Moreover, they can be terminated at any time by the DOE, as was the Pantex board in 2001. The Amarillo community was extremely antagonistic to DOE critics, but Secretary O'Leary required that its SSAB include DOE opponents, so members of several groups of DOE critics joined. The Pantex board split into "booster" and "critic" factions that never developed constructive ways of collaborating. When the board began discussing the environmental impacts of the storage of plutonium obtained from the disassembly of weapons, the DOE dissolved the board (McBride 2001).

Table 6.8. Conflicts between the DOE and SSABs during the Second Bush Administration	
Board	**Event**
Rocky Flats	In 1997, rejected first Rocky Flats cleanup agreement because will leave 15 times higher concentration of plutonium in soil than other sites; in 2003, protest revised cleanup agreement, which will reduce residual soil plutonium to levels of other sites only within 3 feet of soils and will alter surface water sampling protocols to mix water samples [a]
Hanford	DOE proposes to reclassify three quarters of "high-level nuclear waste" as "incidental" waste, allowing high-level waste to be dumped into shallow burial sites [b]
Paducah	SSAB rejects DOE accelerated cleanup plan [c] 7 of 18 members resign because public participation a "sham," excluded from access to decisions, information withheld by contractors [d] Board letter to DOE agrees with concerns of members who resigned [e]
Northern New Mexico	Citizen advisory board (CAP) cites recent events as sign DOE is abandoning public participation; recommends reversing this trend [f] CAB criticizes short DOE time lines for commenting on proposals as "absurd," leading to question the competency of those issuing it [g]
Fernald	Attempts to alter cleanup standards opposed emphatically by CAB [h]
Idaho	CAB recommends the DOE clean site to "residential" level of risk, not the "industrial" level the DOE proposes [i] CAB rejects DOE plan to renegotiate cleanup levels [j]
All sites	Chairs of all SSABs jointly recommend that DOE reverse dramatic cuts in SSAB budgets, collaborate with Legacy Management to transition SSABs of closed sites into Legacy Management, work out citizen involvement in long-term stewardship of closed sites [k]
Pacific National Lab study of SSABs	DOE seems to be abandoning its commitment to public participation in the view of many experienced SSAB members; distrust is reemerging; stakeholders have high expectations and will be angry if the DOE terminates public participation [l]

[a] RFCAB (2003); Moore (2002, 2005). [b] Alvarez (2003). [c] PGDPSSAB (2002). [d] PGDPSSAB (2003). [e] PGDPSSAB (2003). [f] NNMCAB (2003a). [g] NNMCAB (2003a). [h] Klepal (2003, 2004). [i] INEELCAB (2004a). [j] INEELCAB (2003, 2004b). [k] Bierer et al. (2004). [l] Bradbury et al. (2003); Makhijani and Ledwidge (2003).

In 1997, the DOE removed eleven dissident members of the Los Alamos SSAB (Hoffman 1997), and a year later a judge ruled the action legal (Hoffman 1998). The DOE can also set less stringent cleanup levels than a board wants if it can get the EPA and state regulators to agree, as has occurred at Rocky Flats. In 1996, the DOE, with the approval of the federal and state regulators, proposed a radionuclide soil action level (RSAL) of 651 picocuries/gram.[16] The RSAL is the level below which residual plutonium in soils would not be removed. This level, which was more than three times higher than the level approved at the Nevada Test Site and more than fifteen times higher than any other plutonium-contaminated site, was based on the assumption that the area would be a wildlife refuge forever. The SSAB and the community complained that the high levels of residual plutonium would be dangerous for more than 100,000 years, that no one knows how people will use this land even one thousand or ten thousand years from now, and that the RSALs must, therefore, be much lower. Their concerns led the DOE to fund an independent review of the issue, one of the few examples when SSAB pressure led to an independent scientific assessment. This review recommended 35 picocuries/gram as the RSAL. The DOE never formally responded to the review, but, in 2002, it proposed that the top three feet of soil have 50 picocuries/gram as a standard, that the three- to six-foot layer have a 1,000 picocuries/gram limit, and that soils below six feet have no limit, even though plutonium transport pipes and other plutonium hot spots remain buried (Makhijani and Gopal 2002; Moore 2002, 2005). Although 86 percent of public comments opposed this option, it was chosen and has been implemented. The DOE ignored the SSAB's recommendation that the subsurface have the same 50-picocuries/gram standard as the surface soils (RFCAB 2003). During the public participation process seeking to revise the 651-picocuries/gram RSAL, it was learned that, in 1995, the DOE, the EPA, and the Colorado Department of Public Health and Environment had made two secret deals: to complete the cleanup by the end of 2006 and to keep the total cost under $7 billion (despite the DOE's 1996 estimate that it would cost $17 billion). After this revelation, two DOE critics, who had been key members of the board, quit.

Weapons Sites Hostile to Critics   Because the weapons factories employ large numbers of people and pay high wages, the economic inputs to communities make town fathers and workers want to suppress all criticism, and this means that citizens critical of the DOE often endure social ostracism and hostility. The formation of the Oak Ridge SSAB illustrates the pattern. In 1991, Oak Ridge city and surrounding county governments got a DOE-funded grant from the Tennessee Department of Environment to create an LOC. The DOE was closing its K-25 uranium enrichment complex, so the DOE and the city fathers were trying to find new industries to occupy the empty factories. Not surprisingly, the LOC consistently took positions minimizing the health dangers from DOE operations and supporting the DOE. The LOC claimed to represent the public and so maintained that it should be the Oak Ridge SSAB. Activists critical of the DOE had to rely on the un-

dersecretary for environmental management to force the Oak Ridge DOE officials to create an independent SSAB that included DOE critics. This was done in 1995 (Hutchison 2003). However, the activists all resigned in early 1998 when board hostility blocked their efforts, even though the Oak Ridge Environment and Peace Alliance representative had been one of the most productive members of the board (Hutchison 2003; Sigal 2000).

Rely on DOE and Contractors for Information
Boards are explicitly instructed to rely on the DOE and contractors for all information and not to hire consultants to duplicate the work the DOE is doing. They are also enjoined to focus on the big picture and ignore the details. But the big-picture decisions often depend on details. For example, decisions about RSALs are, as we have seen, based on such details as the levels of risk involved and how site use may change in the future. The example of Fernald's attempt to mix radium-rich uranium wastes into glass, a process called vitrification, is another revealing example of how relying on the DOE and its contractors for information prevents SSAB members from obtaining any independent assessment of the choices that DOE contractors are making.

In the 1940s and 1950s, the United States obtained some of its uranium from mines in the Belgian Congo and Australia, and these ores had much higher concentrations of uranium and radium than other ores, up to 65 percent uranium versus the typically less than 1 percent found in most other ores. After these high-uranium ores were processed to remove the uranium, the waste was much more radioactive than ordinary uranium waste, and it contained high concentrations of radium-226, which decays into radon gas, and thorium-230. Because of the higher radioactivity and the radon threat, these materials were stored separately. The high-uranium ores were processed by the Mallinckrodt Corporation in St. Louis until 1956 and at Fernald until 1959 (Fioravanti and Makhijani 1997, 220). Mallinckrodt shipped about thirty-one thousand fifty-gallon barrels of its wastes to Fernald, where they and the Fernald-processed wastes were placed in two reinforced concrete "silos" between 1952 and 1959. The high radium content of the waste produces radon gas, so these silos have been the largest source of radioactive emissions at Fernald. Cracks appeared in the silos in 1964, and they were sealed and supported by an earthen berm that was placed against the outside walls to balance the pressure from within. Several efforts to reduce radon gas emissions in the 1970s and 1980s were ineffective. In 1991, 630 cubic meters of bentonite clay were placed on top of the wastes, and this has reduced radon emissions.

Treating these wastes has been the most difficult aspect of the Fernald remediation (see photographs 6.2–6.3). There were three silos of waste: silos 1 and 2 held the wastes from the Congo ores, and silo 3 held dried powdery wastes from other concentrated ores. The DOE and its contractor suggested either mixing the wastes with cement (cementation) or mixing them into glass, vitrification. Vitrification had the advantage that it resists disintegration much longer than cement, prevents radon gas from escaping, and produces smaller volumes of treated waste, which were to be shipped to the Nevada Test Site for disposal. Cementation did not prevent radon escape, but it was a technically easier procedure. Making glass is a complex process, so it was unclear whether the silos' materials could be vitrified. The Pacific Northwest National Lab had done tests on materials similar to those in the silos and suggested that they could be mixed into the glass but that it would be a tricky process because of the high sulfur and

Photograph 6.2. Fernald nuclear facility, 1988. (Courtesy U.S. Department of Energy.)

Photograph 6.3. Fernald nuclear facility after cleanup, 2006. (Courtesy U.S. Department of Energy.)

lead content of the material. The DOE and its contractor proposed, and the EPA and the Ohio-EPA approved, vitrification as the treatment in the legally binding record of decision in 1995 (Fioravanti and Makhijani 1997, 225–32). The contractor (Fluor Daniel Fernald) proposed building a pilot plant that would first vitrify nonradioactive material similar to the silo wastes and then vitrify small amounts of the actual waste. After consideration of the pilot plant results, it would then upscale to a production plant.

The contractor budgeted the pilot plant at $15.8 million, without including operational costs, maintenance costs, escalation costs, or construction and project-management costs (Fioravanti and Makhijani 1997, 233). The melter was an experimental model with no track record. The contractor violated elementary engineering principles by constructing the equipment that would interface with the melter according to the preliminary information provided by the subcontractor supplying the melter. When the subcontractor delivered a melter that differed significantly from the

preliminary specifications, Fluor Daniel Fernald had to make 225 design changes between May 1995 and May 1996 to get what it had built to match the melter. By the time the changes were made and the pilot plant was ready to be tested, the estimated cost had increased to $66 million. As the pilot plant ran, it encountered major problems with the system that feeds the waste material to be vitrified into the melter: the pipes were too small, design changes had led to sharp bends in the piping that restricted flow, and seals and pumps were abraded by the coarse material (Fioravanti and Makhijani 1997, 233–36). Running the melter led to sulfate froth accumulating on the surface of the glass-waste mix and a concentration of electricity and heat flow in this layer (Fioravanti and Makhijani 1997, 236–39). Finally, on December 26, 1996, during one of the test runs of the pilot plant, a small hole opened in the bottom of the melter and leaked the entire contents of the melter. Luckily, the test was employing a surrogate material, so no radioactive material escaped, but the melter was destroyed. The

John J. Metz

pilot plant had consumed about $50 million of the originally estimated $57 million cost of the entire project—and failed. In April 1997, Fluor Daniel Fernald estimated that vitrification would cost $476 million (Fioravanti and Makhijani 1997, 239–48).

The DOE and the contractor decided to have an independent review team examine the process and reconsider treatment. After considering the team's report, they went back to the regulators and the Fernald Citizen Advisory Board and asked to change the waste-treatment plan to the cementation of silos 1 and 2 and the placing of the silo 3 contents in plastic bags. The regulators agreed, so, in 2004, the contractor began the cementation process, completing it in 2006. In late 2006, the DOE designated the cleanup complete and the SSAB process over. The disposal of the silo wastes, however, remains problematic. The contents of silo 3 had been scheduled to be disposed of at the Nevada Test Site, but the state of Nevada successfully sued to block shipment, so the material was shipped to an Energy Solutions (formerly known as EnviroCare) facility in Utah, even though the concentration of thorium-230 is sufficient that, within fifty years, decay will produce radiation levels of radium-226 that exceed the facility's limits (Makhijani and Makhijani 2006, 7). The silo 1 and 2 wastes emit enough radiation that they require burial in a deep geologic repository. They were shipped for temporary storage to a facility in Texas run by Waste Control Specialists. Final disposal remains to be decided.

DOE Has Too Few Competent Managers to Oversee Contractor Operations
The Fernald vitrification process and many other examples (see Fioravanti and Makhijani 1997) illustrate the DOE's inability to manage its contractors. The absence of complex-wide safety rules allows contractors to set safety standards, even though in 1988 Congress ordered the DOE to establish a nuclear safety

enforcement plan that had teeth, like fines and criminal sanctions for violations. Eleven years later "the department's Office of Environment, Safety, and Health, which is responsible for hundreds of nuclear facilities, has a staff of six people and was almost totally dependent on contractors self-reporting" (Alvarez 2000, 29). In July 1998, a worker was killed at the Idaho National Lab when a high-pressure carbon dioxide fire-suppression system accidentally went off. Thirteen other workers were saved by a rescue team that entered the building without breathing equipment. Several similar accidents had previously occurred but were ignored. Secretary of Energy Richardson blocked the site manager's claim for bonus payments later that year, but, in September 1999, the manager, now in a senior position at headquarters, wrote a message opposing a safety rule because it would "place a significant burden on our contractors" (Alvarez 2000, 30). One week later, he went to work for the INEEL contractor who had failed to implement safety rules. The revolving door between the DOE and its contractors is common (Alvarez 2000). By January 2004, Congress was demanding that the DOE start fining contractors for violations, but the Bush administration is proposing to allow the contractors at federal nuclear facilities to write their own safety standards in order to prevent the implementation of congressional rules (Zuckerbrod 2004).

Failure to Remediate Large Volumes of Wastes
The DOE proposes and the SSABs have approved leaving large volumes of waste in the places where it was irresponsibly dumped in the past. The Environmental Management Programmatic Environmental Impact Statement states that 30 of 44.5 million cubic meters of low-level radioactive waste and 6.6 of 13 million cubic meters of mixed low-level and hazardous waste will be left in the ground (DOE 1997a; Fioravanti and Makhjani 1997, 33–39). Another 89,000 cubic meters of bur-

ied transuranic waste[17] and associated soils are never mentioned in the analysis (Fioravanti and Makhijani 1997, 38).

The DOE Institutional Culture of Secrecy and Control Continues The failure of the Fernald silos project outlined above is not unique. Similar violations of standard engineering principles are occurring at Hanford, as it plans to stabilize and store the reprocessing wastes of fifty years (Martin 2006), and in the inconsistent treatments of buried transuranic waste (see Fioravanti and Makhijani 1997). One example is the secret agreements made at Rocky Flats in 1995 to complete the cleanup by the end of 2006 and to keep the cost below $7 billion (DOE 1996; see also table 6.1 above). A second example is the failure to plan adequately for the treatment of Hanford's tank wastes. After courts blocked the DOE's 2003 attempt to redefine three-quarters of the high-level waste as incidental and, thus, buriable on-site, the vitrification plant costs have increased from $4.3 to $12 billion, and the start date has moved from 2011 to 2019 (Martin 2006).

The review of the DOE's environmental cleanup process by Fioravanti and Makhijani (1997) identified the agency's most basic problem as its institutional culture. This culture includes the lack of a sound process of internal scientific and technical peer review involved in decisionmaking, contractors who are not accountable for their failures, the tendency to rush into large projects without adequate scientific and engineering preparation, the lack of independent regulation of the DOE's nuclear activities, and a primary commitment to developing, producing, and testing nuclear weapons (Fioravanti and Makhijani 1997, 259). All these are the result of what Robert Alvarez calls "an isolated and privileged management culture" (Alvarez 2000, 31), which is itself justified as essential to national security. The national security–justified lack of accountability has given enormous power to DOE officials and huge profits to the corporate contractors. The opportunities to retire into high-paying positions with the contractors give DOE managers incentives to block any reforms (Alvarez 2000).

Conclusion

As the DOE weapons complex ground to a halt in the 1980s, the bad press, congressional pressure, and court rulings forced the agency to accept citizen participation in its cleanup. As far back as Reagan's second term, Secretary of Energy John Herrington had realized that, to obtain support for rebuilding the crumbling nuclear weapons complex, the DOE would have to create a new public image and that that new image would be enhanced by involving the public in cleanup decisions (D'Antonio 1993, 220).

The Clinton administration secretaries of energy came from outside the nuclear establishment and appear to have been genuinely sympathetic to the workers and citizens harmed by weapons production (Alvarez 2001; Flynn 2001). They pushed to make citizen participation a key component of environmental remediation and to ensure that citizen advisory boards conformed to the recommendations of the Keystone Committee, especially the inclusion of DOE critics. During the Clinton years, the accomplishments of the SSABs varied. At sites where weapons production was ending, the activists and other members constructively guided the remediation process (e.g., Fernald, Rocky Flats, Hanford, and Idaho). On other boards at sites with ongoing weapons production, either activists have concluded that they could not attain their goals and quit (e.g., Oak Ridge, SRS, and Paducah), or they have been removed when the SSAB made recommendations that the DOE found intolerable (Los Alamos, Pantex). At all sites, a significant number of citizens have learned a great deal about the problems of remediating the nuclear weapons–production complex and have offered large amounts of solid advice to the DOE. In many cases, individuals with opposing viewpoints have come to respect one another and to work constructively together. Nevertheless, if we apply the standards of fairness and competence defined by Renn et al. (1995), the SSAB experiment fails to provide the public with successful participation. A basic failure is that the boards had no real authority. Rather, power remains in the hands of the DOE and its contractors, and even the most successful board, Fernald, had too few resources to allow it to discover the violations of basic engineering practices that led to the silos fiasco.

DOE personnel and contractors are not accountable for their actions because, as they argue (and the rest of society accepts), national security demands that they be free to do whatever they think necessary to defend the country. During the cold war, the fear of communism effectively kept all but a few citizens from challenging this claim. The large flows of money to the workers and contractors (appropriately skewed toward the top) provided monetary incentives to reinforce the patriotic fervor of those involved.

The end of the cold war threatened the justifications of the nuclear and military establishments in the United States. In 1989, 150 nuclear weapons scientists and their corporate collaborators met at Los Alamos to find new missions for themselves. Several of the participants suggested targeting smaller developing nations that have biological, chemical, or nuclear weapons programs. By 1991, Secretary of Defense Cheney issued a top-secret policy requiring the military to begin planning nuclear strikes against China and against developing countries capable of producing chemical, biological, and nuclear weapons (Smith 2003, 4). A 1991 article by Los Alamos researchers argued that conventional weapons killed too many civilians to be usable and that new tiny, mini, and micro

nuclear weapons should be created to "deter" dictators running rogue countries, the "bunker buster" nukes being able to threaten them personally (Dowler and Howard 1991). The Clinton administration adopted the first Bush administration's nuclear policy, directing the military to plan nuclear strikes at countries that could conceivably produce chemical, biological, or nuclear weapons (Smith 2003, 5–6). The Clinton and first Bush administrations ignored the contradiction between their nuclear policy and the Nuclear Non-Proliferation Treaty (renewed in 1995), which prohibited the United States from using nuclear weapons against nonnuclear treaty members and mandated that the United States and other nuclear powers begin reducing their nuclear stockpiles. In the mid-1990s, the military modified the B-61 bomb into an earth-penetrating bunker buster, but it was unable to penetrate deeply enough to be effective. After taking power, the second Bush administration completed a "nuclear posture review" in late 2001. The parts of the review that were leaked to the press indicated that the United States is prepared to use nuclear weapons on China, Iraq, Iran, Syria, and Libya and would produce a "robust nuclear earth penetrator" (RNEP) warhead. The new warhead would require a new facility to make plutonium "pits" to replace Rocky Flats. President Bush's nuclear plans explicitly violate the nonproliferation treaty by targeting nonnuclear state signatories to the treaty. The subsequent evidence that the RNEP would also be unable to reach deep, hardened bunkers or to prevent massive nuclear contamination of the surrounding people led Congress to cancel the program in 2005.

Nevertheless, in October 2006, the DOE's National Nuclear Security Agency announced that it would publish a "Complex Transformation Supplemental Programmatic Environmental Impact Statement" outlining plans to build a "nuclear weapons complex for the 21st century" (NNSA 2006; D'Agostino

2007), and, in late 2007, it published it (NNSA 2007). In early January 2007, a story appeared that a design for a new "reliable replacement warhead" had been accepted and that it would combine designs from the two competing weapons labs at Los Alamos and Lawrence-Livermore, California. This hybrid design will keep scientists at both labs occupied and will require the rebuilding of the nuclear weapons complex (see Broad, Sanger, and Shaker 2007). The threat of the cold war "victory" has been met: the nuclear weapons builders are manifestly back in control.

Not surprisingly, the second Bush administration has decided to rush the environmental restoration of the old weapons factories to completion by increasing the levels of residual waste that can be left at the sites. It has initiated the risk-based end-state process, which defines future uses in ways that will minimize human occupancy in order to allow lower levels of remediation and higher levels of contamination. For example, defining Rocky Flats, INEEL, Hanford, and Fernald as wildlife refuges means that people will be exposed to dangerous radiation or toxic chemicals only on their infrequent visits to the sites. Guaranteeing these uses for the next 10,000–200,000 years may seem absurd, but such absurd assumptions are the basis of DOE policies. The DOE is also reducing public participation to the point that the chairs of the SSABs and even the DOE-installed board at Los Alamos are openly questioning the agency's commitment to public participation (NNMCAB, 2003a, 2003b; Bierer et al. 2004). In 2006, funding to the SSABs was cut dramatically and the cleanups of both Fernald and Rocky Flats completed.

So has public participation been a temporary public relations ploy to allow the DOE to rebuild its nuclear capability? I have no evidence that it was a planned strategy, but, whether consciously planned or an inadvertent product of DOE inertia, the result is the same: the weapons builders remain firmly in

power. It is hard to imagine that the United States will give up its ability to project nuclear weapons into whatever parts of the world it desires. The consistency of nuclear policies regardless of party or individual president in power seems clear. It also seems that the ongoing U.S. reliance on nuclear weapons is undermining the nonproliferation treaty and inducing more countries to develop nuclear weapons (Sanger 2005).

The relationship between the DOE and its contractors also seems safe from reform. There will be no accountability, as the transfer of government services to private contractors, along with ever-increasing billions of taxpayer dollars, will continue until the money runs out.

However, the SSAB process has created a much larger group of knowledgeable citizens. Even DOE supporters have become disillusioned by the turn of events under the second President Bush. And the citizens are more organized. At Fernald, the DOE proposed changing the cleanup level of the aquifer and increasing the radioactivity of materials being placed in the on-site disposal facility, but the people fought back, led by FRESH. As these changes came to light, FRESH mobilized both of Ohio's Republican senators, both local congressional representatives (who are conservative Bush loyalists), county supervisors, water district officials, local politicians, and private citizens to oppose the changes. Since the DOE's goal had been to close Fernald in 2006 and Fluor Fernald faced financial penalties if it did not reach the target date, FRESH let the DOE know that any changes would have to go through public hearings and that it would ensure that the hearings lasted long enough to make a 2006 completion impossible. The DOE backed down and completed Fernald in mid-2006, terminating the Fernald Citizen Advisory Board.

FRESH is unique in its ability to mobilize political power. At most other sites, the DOE will be able to do what it wants. Fears of terrorism have provided an effective substitute for fear of communism, so the public seems willing to stand back. Admittedly, the second Bush administration is extremist, but its philosophy corresponds well to the culture of the DOE. Even in the "open" 1990s, the citizen advisory boards had little ability to affect DOE plans. As the horror stories of AEC/DOE abuses fade from memory, and as the new weapons facilities are built, there will be few incentives to keep the SSABs functioning. DOE actions may, however, expand the group of DOE critics.

Notes

This chapter has benefited from the patient explanations provided by many people involved in public participation at the DOE complex. An inspiring amount of hard work and concern has come from a wide range of these citizens and from DOE and state regulatory personnel. I had considerable help from Eileen C. Slattery as a research assistant. My wife, Mary, and daughter, Rosa, have also helped by giving me the time to do this project.

1. The following acronyms are used throughout the text of this chapter: AEC = Atomic Energy Commission; DOD = Department of Defense; DOE = Department of Energy; EPA = Environmental Protection Agency; FRESH = Fernald Residents for Environmental Safety and Health; IEER = Institute for Energy and Environmental Research; JCAE = Joint Committee on Atomic Energy; LOC = local oversight committee; MED = Manhattan Engineering Project; NRDC = National Resources Defense Council; RNEP = robust nuclear earth penetrator; RSAL = radionuclide soil action level; SRS = Savannah River Site; SSAB = site-specific advisory board.

2. These facilities cannot ever be completely cleaned up because the wastes are distributed so widely and deeply in the environment. Indeed, a major dispute between citizens and the DOE is over what level of residual contamination is safe. The trade-off is, ostensibly, between cost and safety.

3. In 1989, 150 government and industry representatives met at Los Alamos to design a post–cold war role for themselves and decided that mininukes to deter rogue states was their answer (Smith 2003).

4. Williams (2002) and other commentators erroneously claim that the DOE and its predeces-

sor agencies followed the "decide, announce, defend" model of decisionmaking, but the cold war nuclear establishment just acted: it never bothered announcing or defending.

5. Three of the original twelve have been terminated (see below).

6. After a bitter debate, Congress created the AEC as a civilian-controlled organization (for the formation of the AEC, see Clarfield and Wiecek 1984, 108–16), but, as weapons were built and stockpiled, the AEC gradually turned control of nuclear weaponry over to the Department of Defense, with the result that, by 1967, the AEC had become a mere supplier of "engineering support to meet DOD's demands" (Cochran et al. 1987a, 2–3).

7. Employment in the complex peaked at 117,257 in 1964, but, in 1952, 84,608 construction workers were building the new facilities, while 58,101 production workers operated the factories (Cochran et al. 1987a, 14).

8. That report, NSC-68, first outlined an overdrawn account of the Soviet threat and then suggested four options for the United States: do nothing; withdraw into isolationism; preemptively attack the Soviet Union with nuclear weapons; rearm for a war to meet and defeat the Soviet Union throughout the world. After summarily rejecting the first two options, the administration seriously debated the attack on the Soviets, but most advisors believed that the United States had too few weapons to destroy the Soviet Union, so the fourth option was chosen (see Clarfield and Wiecek 1984, 120–48).

9. The AEC developed several ways to reprocess the plutonium from the U-238 targets of the production reactors. The MED built two reprocessing plants using a "bismuth-phosphate" process to extract plutonium for the first bomb, which was detonated in New Mexico, and for the Nagasaki bomb. In the late 1940s, a different aredox process was developed and scaled up to a production plant by 1951. The redox plant at Hanford reprocessed nineteen thousand metric tons of irradiated fuel before closing in 1967. A third purex process was developed in two "canyons" at the SRS, which opened in 1954 and 1955; a purex plant was also completed by 1956 at Hanford (Makhijani and Saleska 1995; DOE 1997b, 22–24).

10. The AEC offered a $10,000 reward for major finds and guaranteed high prices for ten years (Ringholz 1989).

11. Ringholz (1989) has an extended discussion of the AEC's recalcitrance and the failed attempts of Duncan Holloway and others to reduce the dangers.

12. Fallout from the ninety atmospheric nuclear

tests was more concentrated close to the test sites in the Western United States, but thunderstorms did result in radiation washout in relatively heavy doses throughout the United States, though the lack of radiation-detection equipment allowed many such washouts to remain undetected (IPPNW/IEER 1991, 60–63).

13. During the Manhattan Project, private factories in Michigan, New York, New Jersey, Ohio, and Missouri processed and fabricated uranium metal. These factories are now among the three hundred sites needing remediation (see DOE 1996; "The Sites" 2001). In 1952, the AEC asked the Mallinckrodt Corp. of St. Louis, which had done much of the MED uranium processing, to build a plant similar to Fernald in 1958 at Weldon Spring, Missouri, thirty-two kilometers west of St. Louis. In 1967, the AEC decided that Fernald was adequate and closed Weldon Spring. The cleanup of the Weldon Spring Site was completed in 2002. See www.weldonspring.org.

14. The EPA asked the Keystone Center of Colorado in late 1990 to establish a committee to set priorities for the weapons-complex cleanup. This committee was officially called the Federal Facilities Environmental Restoration Dialogue Committee. It published an interim report in 1993 and a final report in 1995 (DOA/EPA 2006).

15. For more information on the Alliance for Nuclear Accountability, see http://www.ananuclear.org.

16. A curie is a measure of radiation in which 37 billion disintegrations occur each second; a picocurie is one one-trillionth of a curie. Hence, a picocurie means 0.037 disintegrations/second or 2.2 disintegrations/minute.

17. Transuranic wastes are materials contaminated with plutonium and other human-made elements with atomic numbers above uranium's 92. Until 1970, these materials were buried in shallow dumps, often stored in cardboard boxes. In 1970, the AEC created the TRU category, defining it as waste having more than ten nanocuries/gram; this material was to be placed in retrievable storage. In 1984, the DOE raised the definition cutoff to one hundred nanocuries/gram.

References

Ackland, Len. 1999a. "The Day They Almost Lost Denver." *Bulletin of the Atomic Scientists* 55, no. 4:58–65.
———. 1999b. *Making a Real Killing: Rocky Flats and the Nuclear West.* Albuquerque: University of New Mexico Press.

Alvarez, Robert. 2000. "Energy in Decay." *Bulletin of the Atomic Scientists* 56, no. 3:24–35.
———. 2001. "Making It Work." *Bulletin of the Atomic Scientists* 57, no. 4:55–57.
———. 2003. "The Legacy of Hanford." *Nation* 277, no. 5:31–35.

Applegate, John S. 1998. "Beyond the Usual Suspects: The Use of Citizen Advisory Boards in Environmental Decision Making." *Indiana Law Journal* 73:903–57.

Arnstein, Sherry. 1969. "A Ladder of Citizen Participation." *American Institute of Planners Journal* 35:216–24.

Bierer, James, et al. 2004. Letter to Robert Card, January 15, regarding funding of SSABs. Available at http://www.hanford.gov/hanford/files/HAB_Chairsletter.pdf (accessed January 2, 2008).

Bradbury, Judith A., and Kristi M. Branch. 1999. *An Evaluation of the Effectiveness of Local Site-Specific Advisory Boards for U.S. Department of Energy Environmental Restoration Programs.* Washington, DC: U.S. Department of Energy.

Bradbury, Judith A., Kristi M. Branch, and E. L. Malone. 2003. *An Evaluation of the DOE-EM Public Participation Programs.* Pacific Northwest National Laboratory Report no. 14200. Available at www.PNL.gov/main/publications/external/technical_reports/PNNL-14200.pdf (accessed November 8, 2007).

Broad, William J., David Sanger, and Thom Shaker. 2007. "U.S. Selecting Hybrid Design for Warheads." *New York Times,* January 7, 2007, A1.

Clarfield, Gerard H., and William M. Wiecek. 1984. *Nuclear America: Military and Civilian Nuclear Power in the United States, 1940–80.* New York: Harper & Row.

Cochran, T. B., W. M. Arkin, R. S. Norris, and M. M. Hoenig. 1987a. *Nuclear Weapons Databook.* Vol. 2, *U.S. Nuclear Warhead Production.* Cambridge, MA: Ballinger.
———. 1987b. *Nuclear Weapons Databook,* Vol. 3, *U.S. Nuclear Warhead Facilities Profiles.* Cambridge, MA: Ballinger.

D'Agostino, Thomas P. 2007. "Statement of Thomas P. D'Agostino Deputy Administrator for Defense Programs National Nuclear Security Administration before the House Armed Services Committee Subcommittee on Strategic Forces, April 5, 2006." Available at http://www.nnsa.doe.gov/docs/congressional/2006/2006-

04-05_HASC_Transformation_Hearing_
Statement_(DAgostino).pdf (accessed January
3, 2008).

D'Antonio, Michael. 1993. *Atomic Harvest: Hanford and the Lethal Toll of America's Nuclear
Arsenal.* New York: Crown.

Deller, Nicole, comp. 2002. "Treaty Overview: A
Summary of U.S. Compliance with Nine Secu-
rity-Related Treaties." *Science for Democratic
Action* 10, no. 4:8–9. Available at www.ieer
.org.

Department of Energy (DOE). 1994. *Office of
Environmental Management Site Specific Ad-
visory Board Guidance.* Washington, DC: U.S.
Department of Energy, Office of Environmental
Management.

———. 1995. *Closing the Circle on the Splitting
of the Atom.* Washington, DC: U.S. Department of Energy, Office of Environmental Management.

———. 1996. *The 1996 Baseline Environmental
Report.* 3 vols. Washington, DC: U.S. Department of Energy, Office of Environmental Management.

———. 1997a. *Final Waste Management Programmatic Environmental Impact Statement
for Managing Treatment, Storage, and Disposal of Radioactive and Hazardous Waste.* DOE/
EIS0200-F. Washington, DC: U.S. Department
of Energy, Office of Environmental Management.

———. 1997b. *Linking Legacies: Connecting the
Cold War Nuclear Weapons Production Processes to Their Environmental Consequences.*
DOE/EM-0319. Washington, DC: U.S. Department of Energy.

Department of Energy (DOE)/Environmental Protection Agency (EPA). 1996. "Site-Specific
Advisory Board Final Guidance." www.epa
.gov/swerffrr/documents/oem196.htm (accessed
November 8, 2007).

Dowler, Thomas W., and Joseph S. Howard II.
1991. "Countering the Threat of the Well-
Armed Tyrant: A Modest Proposal for Small
Nuclear Weapons." *Strategic Review* 19, no. 4
(Fall): 34–40.

Fehner, Terrence R., and F. G. Gosling. 1996.
"Coming in from the Cold: Regulating U.S. Department of Energy Nuclear Facilities, 1942–
96." *Environmental History* 1, no. 2:125–33.

Fioravanti, Marc, and Arjun Makhijani. 1997.
*Containing the Cold War Mess: Restructuring
the Environmental Management of the U.S.
Nuclear Weapons Complex.* Takoma Park,

MD: Institute for Energy and Environmental
Research.

Flynn, Michael. 2001. "A Debt Long Overdue."
Bulletin of the Atomic Scientists 57, no. 4:38–
48.

Ford, Daniel. 1982. *The Cult of the Atom: The Secret Papers of the Atomic Energy Commission.*
New York: Simon & Schuster.

Gordon, Danielle. 1996. "The Verdict: No Harm,
No Foul." *Bulletin of the Atomic Scientists* 52,
no. 1:31–40.

Hoffman, Ian. 1997. "Agency Takes Issue with
Citizen's Board." *Albuquerque Journal,* August
13.

———. 1998. "DOE Can Disband Citizen's Committee." *Albuquerque Journal,* July 31.

Hutchison, Ralph. 2003. Interview with author.
June 20. Oak Ridge, TN. Hutchison is the executive director of the Oak Ridge Environment
and Peace Alliance.

Idaho National Engineering and Environmental
Laboratory Citizens' Advisory Board (INEEL-
CAB). 2003. "Cleanup Driven by Risk-Based
End States." Recommendation 103. Avail-
able at http://www.inlemcab.org/Recommend/
pdf/103.pdf (accessed January 2, 2008).

———. 2004a. "Draft Idaho National Engineering
and Environmental Laboratory Risk-Based End
State Vision." Recommendation 109. Available
at http://www.inlemcab.org/Recommend/pdf/
109.pdf (accessed January 2, 2008).

———. 2004b. "Public Involvement in the Development of the Draft Idaho National Engineering and Environmental Laboratory Risk-Based
End State Vision Document." Recommendation 110. Available at http://www.inlemcab
.org/Recommend/pdf/110.pdf (accessed January 2, 2008).

International Physicians for the Prevention of
Nuclear War (IPPNW)/Institute for Energy and
Environmental Research (IEER). 1991. *Radio-
active Heaven and Earth.* New York: Apex.

Janofsky, Michael. 2001. "Ill Uranium Miners
Wait as Payments Lapse." *New York Times,*
March 27, A1.

Klepal, Dan. 2003. "Fernald Cleanup Change Proposed." *Cincinnati Enquirer,* October 10.

———. 2004. "Fernald Cleanup Changes Sought."
Cincinnati Enquirer, March 1, A1.

Makhijani, Annie, and Arjun Makhijani. 2006.
"Shifting Radioactivity Risks." *Science for Democratic Action* 14, no. 3:1–2, 4–8, 10–12. Available at www.ieer.org (accessed January 2007).

Makhijani, Arjun. 1994. "Energy Enters Guilty

Plea." *Bulletin of the Atomic Scientists* 50, no. 2:18–29.

———. 1996a. "Fernald Workers Radiation Exposure." *Science for Democratic Action* 5, no. 3:3–4. Available at www.ieer.org (accessed January 2007).

———. 1996b. "Health and Environmental Impacts of Nuclear Weapons Production: Radioactivity in the Fernald Neighborhood." *Science for Democratic Action* 5, no. 3:1–2, 5–7, 12. Available at www.ieer.org (accessed January 2007).

———. 2001. "The Burden of Proof." *Bulletin of the Atomic Scientists* 57, no. 4:49–54.

Makhijani, Arjun, and Sriram Gopal. 2002. "Setting Cleanup Standards to Protect Future Generations: The Scientific Basis of the Subsistence Farmer Scenario and Its Application to the Estimation of Radionuclide Soil Action Levels (RSALs) for Rocky Flats." *Science for Democratic Action* 10, no. 3:1–6. Available at www.ieer.org (accessed January 2007).

Makhijani, Arjun, Howard Hu, and Katherine Yih, eds. 1995. *Nuclear Wastelands: A Global Guide to Nuclear Weapons Production and Its Health and Environmental Effects.* Cambridge, MA: MIT Press.

Makhijani, Arjun, and Lisa Ledwidge. 2003. "Back to the Bad Old Days." *Science for Democratic Action* 11, no. 4:1, 9–16. Available at www.ieer.org (accessed January 2007).

Makhijani, Arjun, A. J. Ruttenber, E. Kennedy, and R. Clapp. 1995. "The United States." In *Nuclear Wastelands: A Global Guide to Nuclear Weapons Production and Its Health and Environmental Effects,* ed. Arjun Makhijani, Howard Hu, and Katherine Yih, 169–284. Cambridge, MA: MIT Press.

Makhijani, Arjun, and Scott Saleska. 1995. "Production of Nuclear Weapons and Environmental Hazards." In *Nuclear Wastelands: A Global Guide to Nuclear Weapons Production and Its Health and Environmental Effects,* ed. Arjun Makhijani, Howard Hu, and Katherine Yih, 23–64. Cambridge, MA: MIT Press.

Martin, Todd. 2006. "Re: Tank Waste Program Path Forward." HAB (Hanford Advisory Board) Consensus Advice no. 192. September 8. Available at http://www.hanford.gov/hanford/files/HAB_Adv-192.pdf (accessed November 8, 2007).

McBride, Jim. 2001. "Amarillo, Texas, Nuclear Plant's Board Hurt by Infighting, Members Say." *Amarillo Globe-News,* November 25.

Mello, Greg. 1997. "Re: Citizens Advisory Board (CAB) at Los Alamos National Laboratory (LANL)." July 31. Available at www.lasg.org/almletter_b.html http://www.lasg.org/archive/1997/almletter.htm (accessed November 8, 2007).

Minutaglio, Bill. 1994. "Boley Caldwell Wants an Apology." *Bulletin of the Atomic Scientists* 50, no. 3:35–38.

Moore, LeRoy. 2002. "Lowering the Bar." *Bulletin of the Atomic Scientists* 58, no. 2:28–39.

———. 2005. "The Bait-and-Switch Cleanup." *Bulletin of the Atomic Scientists* 61, no. 1:50–57.

National Nuclear Security Administration (NNSA). 2006. "Notice of Intent To Prepare a Supplement to the Stockpile Stewardship and Management Programmatic Environmental Impact Statement—Complex 2030." *Federal Register* 71, no. 202 (October 19): 61731. Available at http://www.nnsa.doe.gov/docs/newsreleases/2006/Complex_2030_NOI_10-19-06.pdf (accessed January 2, 2008).

———. 2007. "Complex Transformation Supplemental Programmatic Environmental Impact Statement." Available at http://www.nnsa.doe.gov/docs/ComplexTrans/SPEIS.pdf (accessed January 3, 2008).

Northern New Mexico Citizens' Advisory Board (NNMCAB). 2003a. "Recommendation to the Department of Energy no. 2003-3: DOE's Commitment to Public Participation." Available at www.nnmcab.org/recommendations/recommendation-2003-03.htm (accessed November 9, 2007).

———. 2003b. "Recommendation to the Department of Energy no. 2003-8: Unreasonable Time Constraints for Department of Energy Directives." Available at www.nnmcab.org/recommendations/recommendation-2003-08.htm (accessed November 9, 2007).

Office of Technology Assessment (OTA). 1991. *Complex Cleanup: The Environmental Legacy of Nuclear Weapons Production.* Washington, DC: U.S. Government Printing Office.

Ortmeyer, Pat, and Arjun Makhijani. 1997. "Worse Than We Knew." *Bulletin of the Atomic Scientists* 53, no. 6:46–50.

Paducah Gas Diffusion Plant Site Specific Advisory Board (PGDPSSAB). 2002. "Recommendation 02-1: Resolution 02-1 Rejection of DOE Cleanup Plan." Available through the Oak Ridge SSAB library, Oak Ridge, TN; copy also in author's files.

———. 2003. "Letter to Jessie Roberson about Recent SSAB Resignations." Available through the Oak Ridge SSAB library, Oak Ridge, TN; copy also in author's files.

Renn, Ortwin, Thomas Webler, and Peter Wiedemann. 1995. *Fairness and Competence in Citizen Participation: Evaluating Models for Environmental Discourse*. Dordrecht: Kluwer Academic.

Ringholz, R. 1989. *Uranium Frenzy: Boom and Bust on the Colorado Plateau*. New York: Norton.

Rocky Flats Citizens Advisory Board (RFCAB). 2003. "Comments and Recommendations on Proposed Modifications and Additions to Attachments to the Rocky Flats Cleanup Agreement." Available at http://www.rockyflatssc .org/rfcab_recommendations/2003/RFCAB_ Rec_2003_01.pdf (accessed January 3, 2008).

Sanger, David E. 2005. "Month of Talks Fails to Bolster Nuclear Treaty; No Gains on Proliferation." *New York Times*, May 28, A1.

Schneider, Keith. 1988a. "Operators Got Millions in Bonuses Despite Hazards at Atomic Plants." *New York Times*, October 26, A1.

———. 1988b. "U.S. for Decades Let Uranium Leak at Weapons Plant." *New York Times*, October 15, A1, A7.

———. 1989a. "Candor on Nuclear Peril." *New York Times*, October 14, A1.

———. 1989b. "Chronic Failures at Nuclear Plant Are Disclosed by the Energy Department." *New York Times*, October 6, A1.

———. 1989c. "DuPont Asserts It Fully Disclosed Reactor Problems." *New York Times*, October 3, A1.

———. 1989d. "Energy Dept. Says It Kept Mishaps at Nuclear Weapon Plant Secret." *New York Times*, October 4, A1.

———. 1989e. "Ex-Nuclear Aides Deny Being Told of Plant Mishaps." *New York Times*, October 5, A1.

———. 1989f. "Reactor Shutdown Could Impede Nuclear Deterrent, Officials Say." *New York Times*, October 9, A1.

———. 1989g. "Second Nuclear Plant Is Ordered Closed by Energy Dept." *New York Times*, October 11, A1.

———. 1993a. "A Longtime Pillar of Government Now Aids Those Hurt by Its Bombs." *New York Times*, June 9, A8.

———. 1993b. "A Valley of Death for the Navajo Uranium Miners." *New York Times*, May 3, A1.

Schwartz, Stephen I. 1995. "Four Trillion Dollars and Counting." *Bulletin of the Atomic Scientists* 51, no. 6:32–52.

———. 1998. *Atomic Audit: The Costs and Consequences of U.S. Nuclear Weapons since 1940*. Washington, DC: Brookings Institution Press.

———. 2002. "Nukes You Can Use." *Bulletin of the Atomic Scientists* 58, no. 3:18–19.

Sigal, Lorene. 2000. Interview with the author. July 10. Oak Ridge, TN.

"The Sites." 2001. *Bulletin of the Atomic Scientists* 57, no. 4:58–60.

Smith, Brice. 2003. "The 'Useable' Nuke Strikes Back." *Science and Democratic Action* 11, no. 4:1–7. Available at www.ieer.org (accessed January 2007).

Speed, Roger, and Michael May. 2005. "Dangerous Doctrine." *Bulletin of the Atomic Scientists* 61, no. 2:38–49.

Stieber, Tamar. 1995. "Uranium Cleanup Bombs at DOE." *Nation*, October 23, 460–64.

Susskind, Lawrence, Paul F. Levy, and Jennifer Thomas-Larmer. 2000. *Negotiating Environmental Agreements: How to Avoid Escalating Confrontation, Needless Costs, and Unnecessary Litigation*. Covelo, CA: Island.

Wald, Mathew. 1997. "U.S. Atomic Tests in 50's Exposed Millions to Risk." *New York Times*, July 29, A8.

Waldman, Meredith. 1997. "NCI Apologizes for Fallout Study Delay." *Nature* 389, no. 534 (October 9): 534.

Webler, Thomas. 1995. "'Right' Discourse in Citizen Participation: An Evaluative Yardstick." *Fairness and Competence in Citizen Participation: Evaluating Models for Environmental Discourse,* ed. Ortwin Renn, Thomas Webler, and Peter Wiedemann, 35–86. Dordrecht: Kluwer Academic.

Webler, Thomas, and Ortwin Renn. 1995. "A Brief Primer on Participation." In *Fairness and Competence in Citizen Participation: Evaluating Models for Environmental Discourse,* ed. Ortwin Renn, Thomas Webler, and Peter Wiedemann, 17–34. Dordrecht: Kluwer Academic.

Weisgall, Jonathan, and Mike Moore. 1994. "The Able-Baker-Where's Charlie Follies." *Bulletin of the Atomic Scientists* 50, no. 3:24–34.

Williams, Walter Lee, Jr. 2002. *Determining Our Environments: The Role of Department of Energy Citizen Advisory Boards*. London: Praeger.

Yih, Katherine, Albert Donnay, Annalee Yassi, A. J. Ruttenber, and Scott Saleska. 1995. "Uranium Mining and Milling for Military Processes." In *Nuclear Wastelands: A Global Guide to*

Nuclear Weapons Production and Its Health and Environmental Effects, ed. Arjun Makhijani, Howard Hu, and Katherine Yih, 105–68. Cambridge, MA: MIT Press.

Zuckerbrod, Nancy. 2004. "Nuclear Safety Rules Targeted; Contractors Would Write Plant Standards." *Cincinnati Enquirer,* January 29, A1.

Chapter 7

Grassroots Environmental Opposition to Chemical Weapons Incineration in Central Kentucky

A Success Story

David Zurick

On January 13, 1993, the United States signed the International Comprehensive Ban on Chemical Weapons, obligating the country to dispose by 2004 of an estimated twenty-seven thousand tons of blister agent (mustard gas) and nerve agents (GB-Sarin, VX, and GA). The U.S. Army had amassed these chemical weapons during the period 1943–1969 as a "retaliatory stockpile," storing them in rockets, tanks, projectiles, and bulk containers inside earthen bunkers called "igloos," maintained at eight army depots around the United States and on Johnston Atoll in the South Pacific. The largest such facility, the Tooele Army Depot in Utah, holds 42 percent of the chemical stockpile, enough nerve agent to kill every creature on earth many times over (see figure 7.1). The smallest stockpile (1.6 percent of total) is stored at the Bluegrass Army Depot located in central Kentucky, approximately thirty miles south of Lexington (see photograph 7.1 and figure 7.2).

Ironically, given its relatively minor contribution to the national chemical weapons stockpile overall, the Bluegrass Army Depot is at the center of the nation's public debate over the disposition of the U.S. chemical stockpile. Because the rockets holding the chemical weapons are leaking, international treaties require their elimination, and the U.S. Congress has mandated their destruction. Clearly, some action is necessary, but the disposition of the deadly chemical weapons

is controversial owing to the potential environmental and public health risks. As early as 1982, the army considered incineration to be the best technology for destroying the chemical stockpile and, in 1986, formally proposed the adoption of this method in the Chemical Stockpile Disposal Program (Public Law 99-145; Department of Defense Authorization Act, 1986). Prior to this, during the 1950s and 1960s, the army got rid of its obsolete chemical agents by open-pit burning or by ocean dumping. The ocean-dumping program was code-named OPERATION CHASE, CHASE being the U.S. Navy's acronym for "Cut Holes and Sink 'Em." For obvious reasons, the public, once informed of these practices, demanded a halt to them. The stated intention of the army's proposal for the on-site incineration of chemical weapons is to provide higher standards of public and environmental safety (Silton 1993).

When the idea of building such an incinerator at the Bluegrass Army Depot was first made public in 1984, however, it met with strong and unforeseen opposition by local citizens, who expressed concerns about the risks posed to the local population and the natural environment. In a series of public assemblies, called "scoping meetings," held in communities near the depot during a period spanning almost two decades, the army maintained the safety of its incineration program, while members of the public voiced opposition to it. Two leading citizens' groups

Umatilla, OR
(11.6%)

Tooele, UT
(42.3%)

Pueblo, CO
(9.9%)

Newport, IN
(3.9%)

Aberdeen, MD
(5%)

Richmond, KY
(16%)

Pine Bluff, AR
(12%)

Anniston, AL
(7.1%)

Hawaii

Kalama Island (Johnston Atoll)
(6.6%)

Figure 7.1. Chemical weapons storage sites in the United States.

Photograph 7.1. Underground igloos store chemical weapons awaiting safe disposal at the Bluegrass Army Depot in central Kentucky. (Courtesy *Lexington Herald-Leader*. Photograph by David Stephenson.)

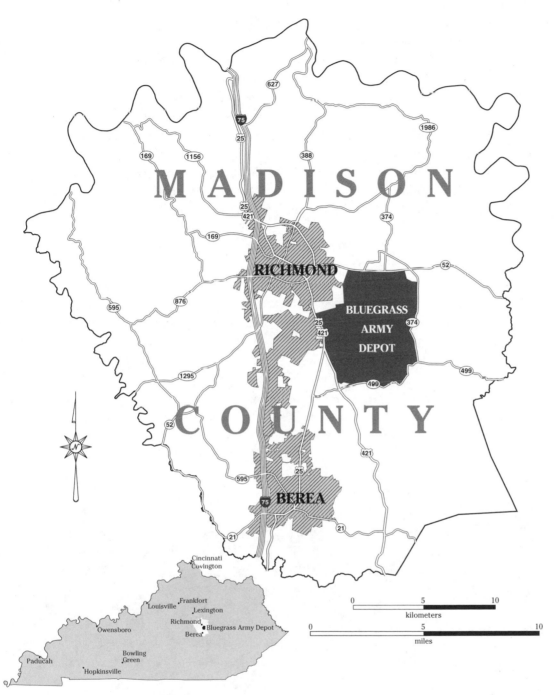

Figure 7.2. Bluegrass Army Depot.

emerged in the late 1980s—the Concerned Citizens of Madison County and Common Ground, which together spearheaded a grass-roots environmental movement in Kentucky and, later, across the United States against the army's chemical weapons incineration plan. These central Kentucky community groups, along with citizens' opposition groups at other chemical weapons stockpile sites, coalesced in 1990 into the Chemical Weapons Working Group (CWWG), which is led by the Berea-based Kentucky Environmental Foundation

(KEF). This coalition, formed initially to focus on the nerve gas issue, now supports a range of community-based environmental campaigns in the United States and abroad.

For two decades, the citizens of Madison County have steadfastly maintained their opposition to the Pentagon's incineration plans at the Bluegrass Army Depot (see photograph 7.2). Initially advocating removal of the ordnance to other less-populated sites for incineration, the movement later spearheaded an alternative plan for weapons disposal that called for their neutralization and biodegradation. On November 20, 2002, the Bluegrass-based environmental movement won its battle. The army formally announced that it would abandon its plan to incinerate chemical weapons at the Bluegrass Army Depot in favor of neutralization technologies long advocated by the citizen activists. This victory is a hopeful story, with immense benefit for the communities of central Kentucky and for other localities where chemical stockpiles

exist. And, because it successfully engages a relatively small group of determined citizens against the monolithic powers of the Pentagon, the environmental movement has far-reaching symbolic consequences as well. It suggests ways in which citizens can participate meaningfully in the discourse on national environmental policy—to practical ends. In this chapter, I outline a brief history of the Kentucky-based grassroots effort to oppose chemical weapons incineration, discussing why it was successful when so many environmental movements fail.

Nerve Gas Incineration as an Environmental and Public Health Hazard

From the onset, local citizens' opposition against nerve gas incineration at the Bluegrass Army Depot has focused on the environmental and human health hazards of

Photograph 7.2. Kentucky citizens' demonstration against incineration. (Courtesy Kentucky Environmental Foundation/Chemical Weapons Working Group. Photograph by Tim Hensley.)

David Zurick

burning chemical agents in the manner proposed by the army. This controversy centers on the confrontation between a government bureaucracy committed to a course of action that will bring environmental risks to a community and local citizens who seek a voice in such decisions. The safe disposal of lethal chemical weapons was not a primary concern to the army in their early manufacture, mainly because it was assumed that they would be used and not stored for long periods (CBS, *60 Minutes*, January 5, 1992). Once it became apparent, however, that the national chemical stockpile was aging and posed dangers from leakage or explosion, attention was given by the Pentagon to its destruction. The disposal of chemical weapons by open-pit burning or ocean dumping, as practiced in the 1950s and 1960s, was also proposed initially as the easiest method of disposal for the twenty-seven thousand tons of outstanding chemical ordnance. The idea was abandoned in 1970, however, because of widespread public and scientific protest against the policy in the United States and abroad.

In the early 1970s, on the basis of the recommendation of the National Academy of Sciences, the U.S. Department of Defense began looking into incineration as a preferred method of chemical weapons destruction. The technology requires that chemical munitions (e.g., M55 rockets, which contain agent VX or GB) be dismantled and drained of chemical agents and that the extracted materials be fed separately into a high-temperature incinerator complex (see photograph 7.3). Nerve agents, metal parts, explosives, and packing materials would be burned in separate, specially designed incinerators. The safety of the incinerator technology, especially the risk to the public and the natural environment of exposure to chemical agents from factory emissions or from a catastrophic meltdown, lies at the heart of the public's concerns. Concerns also focus on the potential release of chemicals during the handling and transportation

Photograph 7.3. Ammunition inspector Elmer Rogers, shown in 1999, stood among pallets containing more than two thousand M-55 rockets with nerve agent payloads. (Courtesy *Lexington Herald-Leader.* Photograph by David Stephenson.)

Photograph 7.4. The active involvement of local environment groups led to the development of a strategy for the safe disposal of thousands of M55 rockets with a payload of VX agent stored at the Bluegrass Army Depot in Central Kentucky. (Courtesy *Lexington Herald-Leader.* Photograph by David Stephenson.)

phases of munitions destruction and on the fact that a great deal of stress is placed on the hardware by the extremely high temperatures of the incinerators (1,600–2,700°F), which may, in the long run, compromise the ability of the incinerators to meet high efficiency and safety standards (Davies 1998; see also photograph 7.4).

These concerns are based in part on problems that plague the army's first chemical weapons incinerator, built on Johnston Atoll in the South Pacific in 1985. As a prototype of the Bluegrass Army Depot, it has proved to be a poor model. The Johnston Atoll Chemical Agent Disposal System has functioned less than 50 percent of the time and has suffered countless problems, including three live nerve agent releases, explosions, furnace fires, and equipment failures (Johnston, Stringer, Santillo, and Costner 1996). A number of by-products of the incineration

process also pose known dangers to human health and the natural environment, including emissions of dioxin, benzene chromium, mercury, and vinyl chloride. Environmental Protection Agency documents cite exposure to these emissions, and their biological concentration in the food chain, as, in fact, the major environmental health threat posed by the incineration technology.

In the event of a release of nerve and blister agents, immediate and known health risks ensue for human populations. GB, also known as Sarin, vaporizes instantly and, when inhaled, causes death. VX exposure through inhalation or skin contact results in nervous system disorders, including convulsions, coma, and death owing to paralysis. Blister agents, commonly known as mustard gas, cause eye injury and skin burn as well as systemic effects such as intestinal disorders ("Chemical Stockpile Disposal Program" 1987). The dispersal of these agents through the air will affect nonhuman life as well, and their entry into soil and water regimes may contaminate local natural resources for many years.

As reports emerged about the army's incineration program and its potential impacts on public and environmental health, it became clear to military leaders that local communities were not adequately equipped to respond to the kinds of emergency situations that would result from accidents occurring during the incineration process. In 1988, therefore, the army established the $700 million Chemical Stockpile Emergency Preparedness Program (known as CSEPP), which was designed to protect citizens from chemical accidents at the incinerators. CSEPP offices, established in communities near the proposed incineration sites, were to serve as clearinghouses of information and to provide support for such actions as disaster evacuation and the distribution to households of protective clothing, gas masks, duct tape and plastic, and other forms of in-place protection (Smithson 1994). The initiation of the CSEPP was viewed with skepticism by incinerator opponents, however, mainly because the public education arm of the program, through its various outreach campaigns, is believed to simply justify the Pentagon's plans to incinerate chemical weapons rather than to promote a balanced perspective on the issue and a meaningful platform for community dialogue and protection.

The credibility problems of the CSEPP arise, in part, from the army's long history of being less than candid to citizens living near the Bluegrass Army Depot about its chemical weapons operations. The depot was built in 1942 on fifteen thousand acres of farmland that the army took from local landowners in central Madison County, after razing numerous standing homes and country estates. At the beginning, the Bluegrass Army Depot maintained good relations with neighboring communities, providing jobs during World War II and offering public services such as fire protection. That relationship suffered, however, when nerve gas and blister agent shipments first arrived unannounced at the site in 1962, the shipments continuing into the mid-1960s. Open-pit burns of munitions stockpiles occurred at the depot during the 1960s without the knowledge of the local communities. The serious erosion of public trust began in the 1970s, when local citizens learned about the open-pit burns, and increased notably after the "Smoke Pot Incident" of 1979.

On August 16, 1979, a cloud of dark fumes emanated from open-pit burns of ordnance at the depot and drifted west to Interstate 75. Forty-five persons were hospitalized for burning eyes and respiratory problems. The army first denied responsibility for the incident, but, when later presented with evidence to the contrary, it admitted that the depot was the source of the fumes. There was no reported release of nerve agents in the incident, but the army's initial denial of involvement deepened the level of public distrust. Since then, on separate occasions, cattle and deer were discovered dead on the depot property, their blood samples testing positive for nerve agent. In these cases as well, the army initially denied culpability and then later recanted its story. Numerous reports from chemical stockpiles elsewhere in the country include nerve gas leaks on much larger scales, with similar army denials and recantations. The pattern of disavowals and then forced acceptance of responsibility over a period of two decades created a heightened sense of distrust, which extended to the army's CSEPP.

The Geographic Setting

The Bluegrass Army Depot is located in central Madison County on the eastern edge of the Bluegrass region. It is one of Kentucky's fastest-growing counties with 72,500 residents. The communities of Richmond (population 27,000) and Berea (population 10,000) are located within eight miles of the proposed incineration site, while the city of Lexington (metropolitan population 450,000) is situated twenty-seven miles north of the depot. In addition to these three major population centers and the dispersed rural population of Madison County, the depot's immediate environs include a major university (Eastern Kentucky University, with 15,000 students), a college (Berea College with 1,500 students), and several elementary and high schools. Furthermore, Interstate 75, one of the nation's busiest transportation arteries, is located only a few miles west of the depot.

A major factor in the initial rise of citizens' environmental activism against the army's plan to incinerate chemical ordnance at the Bluegrass Army Depot is the depot's close proximity to major population centers, which accentuates the risk to the public. This local concern, embedded in the spatial problems of siting the incinerator, prompted initial efforts to transport the ordnance to Utah or to other incineration sites that are more geographically remote. This so-called not-in-my-backyard movement, which was centered on local concerns and risks, gradually gave way to a more generalized opposition to the army's disregard for the geographic and environmental rights of all affected communities (Miller 1993). It was this transition from specific to general concerns that eventually held promise for a national outreach of the Bluegrass environmental movement, such that communities located at other chemical weapons incineration sites might benefit. Along its geographic journey from the Bluegrass to the nation and beyond, the grassroots environmental movement remained focused on the practical matters of technology and science, of public education, and of diplomacy in order to garner the political support considered necessary to achieve its aims.

Grassroots Mobilization

It is noteworthy that the mobilization of the grassroots environmental campaign against the Bluegrass Army Depot chemical weapons incineration program began, not with outside radicals, but with the social aristocracy of Richmond, who joined forces with political activists in Berea. Richmond and Berea are the two communities in Madison County located closest to the depot, but they share very different origins. Richmond's history is bound up with a conservative, landed gentry who, since the town's beginnings in 1775, have controlled its politics and economics. Many of the grassroots environmental activists in Richmond descend from the community's founding families, whose land was confiscated by the army for purposes of the depot in 1942. That initial land transaction, and the subsequent management by the army of the depot land, violated deeply held convictions among some of the social aristocracy of Richmond, whose sense of self-identity comes from their attachment to the land and to a sense of place (Ellis, Everman, and Sears 1985). In effect, the army disenfranchised a generation of rural social elite from its hereditary landed status in Richmond society.

Berea, meanwhile, is a more liberal community whose contribution to the grassroots movement includes experienced environmental campaigners and peace and justice activists. Berea, established in 1855, was based on the principles of social justice and abolition and has a long history of politically aware and environmentally active citizens. Historically at odds over the issue of slavery, the two modern communities of Richmond and Berea, one conservative and the other liberal, joined hands in the contemporary struggle against chemical weapons incineration in their shared backyard.

The Richmond activists, led by influential community leaders, formed the Concerned Citizens of Madison County in 1984, and the Berea activists, led by a Vietnam War veteran, formed Common Ground in 1987. These two lead groups in the local grassroots movement represent a broad constituency, including members of the founding families, the business communities, the political leadership, churches, and a wide range of self-described "solid citizens." The citizens were called to action in 1984 by the army's first public briefing about its incineration plan, at which it became apparent to members of both groups that their communities were at risk. Concerned Citizens formed in Richmond specifically to provide public input into the decisionmaking process and to work conservatively from within the system by engaging local elected officials on the issue. Common Ground, meanwhile, proposed to engage the army in a strategy that included gathering testimony from scientists and other experts to refute the army's claims about safety and risks, educating the public and organizing the opposition to incineration, lobbying key political officials, and sponsoring litigation when necessary to force a new trajectory of decisionmaking. While Concerned Citizens remained locally focused, Common Ground adopted a broader national perspective.

In 1990, Common Ground became the KEF and attained nonprofit status. The organization's primary purpose is to disseminate information and educate the public on environmental issues, with the nerve gas issue as its focus. The KEF determined early on that it could not achieve its goal of stopping the incineration of chemical weapons at the Bluegrass Army Depot without forging a broader alliance with community groups based at the other proposed incineration sites in the United States. This decision extended the tactics of the grassroots movement beyond simply local activism and moved the debate into national view.

In 1991, the KEF-led CWWG sponsored the first international meeting of citizens opposed to the army's incineration plan. The so-called Citizens' Summit, held at a Holiday Inn in Richmond, included representatives from all eight continental U.S. sites as well as delegates from Hawaii representing the interests of Pacific Islanders opposed to the army's Johnston Atoll facility and delegates from Russia. A consensus document from that meeting, entitled *The International Citizens' Accord on Chemical Weapons Disposal* (1991), includes the following points:

- All plans to use incineration for chemical weapons destruction should be halted.
- The Department of Defense should expand its investigation into alternative technologies.
- There should be greater citizen involvement in all decisionmaking processes.
- Environmentally unsound technologies for the disposal of chemical weapons must not be exported.

The document makes it clear that the intentions of the movement are not to deny the need to act on the aging and unsafe stored ordnance but, rather, to force the army to critically examine and consider the alternative methods of disposal that promise greater safety for humans and the natural environment. Moreover, the document reveals the deep reluctance of the community activists to allow the transportation of chemical weapons from one storage site to another, an idea initially proposed by some in the Bluegrass and in the army as a possible solution to the local problem of the Bluegrass chemical ordnance. The CWWG felt that such transportation options simply shunted the danger from one community to another as well as exposing to it all the people located along transit routes. With the Citizens' Summit, the environmental movement moved away from a determinedly local perspective on the problem (not in my backyard) to a national and international one (not in anyone's backyard) and embraced a consensus position on the development of safe disposal technologies. In effect, it was no longer an "anti-incineration" movement, having become a "pro–safe disposal" movement, a watershed change that, ultimately, set the stage for its success more than ten years later.

The incinerator opposition movement led by the KEF has from its early days enjoyed the support of a wide-ranging assemblage of political officials and organizations. They include several U.S. congressmen, a former state governor, and community organizations such as chambers of commerce and local workers' unions as well as the local chapters of several national environmental groups. This diverse base of support has helped considerably in gaining momentum among local communities in central Kentucky and in influencing wider political circles. With the inception of the CWWG, this mobilization effort extended nationwide to include the Anniston Army Depot in Alabama, the Pine Bluff Arsenal in Arkansas, the Pueblo Depot in Colorado, the Newport Army Amunition Plant in Indiana, the Aberdeen Proving Ground in Maryland, the Umatilla Depot in Oregon, and the Tooele Army Depot in Utah. All these sites (referenced in table 7.1) have organized community opposition movements, and in a few of them—the so-called army towns of Anniston and Tooele—counteropposition movements (i.e., proincineration movements) have formed as well in support of the army. Over a period of a decade, the mobilization efforts of the KEF in these dispersed communities have produced a national coalition in opposition to incineration that cuts across geographic, racial, ethnic, and class lines. This coalition extends internationally to include grassroots activists from Russia (who are concerned about that country's fifty thousand tons of chemical weapons stockpile) and from the Pacific (who oppose incineration on Johnston Atoll [also known as Kalama Island]).

From the onset, the army challenge to the legitimacy of the grassroots environmental movement was based on the technology of its incineration proposal. The very nature of the deadly chemical weapons, their storage devices, their deterioration rates, and the alternative means for their destruction all involve highly complex technical operations, which, the army argued, is beyond the understanding of local citizens. This rationale promoted a kind of "technical adversarialism," which limited public involvement in the decisions made about chemical weapons destruction in Kentucky and elsewhere in the country (Futrell 1999). In public hearings held in the Bluegrass communities during the 1980s, the army maintained its "expert knowledge" as the primary justification for its environmental policies regarding chemical weapons destruction. Meaningful public participation was diminished by the army's dismissive attitude toward citizens' concerns regarding the safety and risks of the technology, and residents simply were told to trust the experts (Hindman 1989). At these hearings, army personnel often appeared unconcerned about the safety issues raised by citizens and were perceived by many in the community to be simply going through the motions of a public hearing, as mandated by federal law. As the grassroots environmental movement took shape in the mid-1980s, it became clear from the public hearings that a primary need was for alternative expert witnesses on the technology issues.

Grassroots Tactics

Like most grassroots environmental movements, the central Kentucky opposition to the incineration of chemical weapons at the Bluegrass Army Depot has relied on a small cadre of devoted individuals who have given much of their time and energy to the controversy. In the words of one participant: "What

Table 7.1. Chemical Stockpiles and Citizens' Movements

State	Site	% of National Stockpile	Local Population	Grassroots Movements
Alabama	Anniston Army Depot	7.1	27,000	Families Concerned about Nerve Gas Incineration; Burn Busters; Serving Alabama's Future Environment
Arkansas	Pine Bluff Arsenal	12	60,000	Pine Bluff for Safe Disposal; Arkansas Fairness Council
Colorado	Pueblo Depot	9.9	100,000	Sangre de Cristo Group of the Rocky Mountain Sierra Club; Citizens for Safe Weapons Disposal
Indiana	Newport Army Amunition Plant		16,773	Newport Study Group; Citizens against Incinerating at Newport
Maryland	Aberdeen Proving Ground	5	Baltimore	Concerned Citizens for Maryland's Environment; Coalition for Safe Disposal; Aberdeen Proving Ground Superfund Citizen's Coalition
Oregon	Umatilla Depot	11.6	10,000	Citizens for Environmental Quality
Utah	Tooele Army Depot	42.3	Salt Lake City	Utah Sierra Club; West Desert HEAL; Families against Incinerator Risk

we learned early on was that we could get hundreds of people to a meeting, but if you needed people to lick stamps next Thursday night you were back down to about 12 or so" (Futrell 1999, 128). The movement has benefited tremendously from the charismatic and financial leadership of the KEF, which has a small paid staff as well as interns and volunteers. As we will see below, the institutionalization of the environmental movement through the formation of the KEF in 1990 has been a key factor in the ongoing commitment of the grassroots effort and its eventual success.

The KEF has employed several strategies in its grassroots campaign. These are worth noting, not only because they help explain the success of the Bluegrass-based movement, but also because they provide a useful framework for possible adoption by other grassroots environmental movements. These strategies include the following: Form a broad-based coalition representing a wide cross section of local society. Provide an alternative source of expert information. Adopt a proactive rather than a negative stance on the issues. Navigate the political process from local to national scales of power. Institutionalize the movement to increase influence and gain funding for sustained grassroots-based activity.

Broad-Based Coalition

The grassroots environmental movement in central Kentucky contains a wide cross section of local society, including members of the region's founding families, businesspersons and business organizations, politicians, health-care givers, educators, farmers, and householders. The early organizers of the opposition movement considered it important to show the army in the public hearings that the local opposition to the incinerator plan was widespread and deeply entrenched within the community and not simply restricted to a small band of vocal radical activists. The mobilization of this broad cross section of society was facilitated by the requirements of the National Environmental Policy Act and by the environmental impact statement process, which compelled the army to hold the public hearings among the central Kentucky communities. Without such federal laws, which essentially sparked the public debate by providing a forum for citizen participation and a legitimate channel for public input for the more conservative members of local society, it would have been difficult to garner the support of such a diverse and broadly construed constituency. Furthermore, legitimate citizen action for many early opponents included lobbying local politicians. Again, the broad-based coalition of citizens was important in attracting the sustained interest of politicians, who came to see their political futures as staked to their role in the public environmental debate about chemical weapons incineration. Finally, the broad coalition helped ensure meaningful media coverage by local news organizations, which was important in the public education outreach campaign of the movement.

The conservative Richmond-based group (Concerned Citizens of Madison County) and the liberal Berea group (Common Ground, later the KEF) appealed to very different constituencies. This worked to the benefit of the movement because it provided the means for reaching a wide spectrum of people. In effect, the conservative outlook of the Concerned Citizens of Madison County brought people and organizations into the movement that otherwise would not have gotten involved. The Berea group, meanwhile, elicited the support of a liberal community deeply engaged in issues of social and environmental justice and with a significant community-organizing capability already in hand. In sum, the specific benefits of such a broad coalition to the

movement include the following: greater success in lobbying local politicians; heightened interest in and attendance at public hearings (garnering audiences that ranged in size from three hundred to two thousand citizens); more comprehensive news coverage of the controversy by diverse media; and, perhaps most important, a growing recognition among army officials that the central Kentucky grassroots movement had considerable depth and would prove to be a formidable adversary.

Alternative Source of Expert Information

It was clear from the onset that any successful community campaign to contest the army's decision to incinerate chemical weapons in central Kentucky had to include a science and technology component. The public hearings held in the 1980s demonstrated the community concerns about technological risks, which the army countered with expert testimony obtained from army engineers and scientists under contract with the army (KEF et al. 1991). This testimony maintained the overall safety of incineration, downplaying the operating malfunctions experienced at incineration facilities on Kalama Island and in Tooele that released nerve agents into public spaces as being essentially human, not technological, problems. The army's expert testimony appeased some members of the public, who took assurance in it and in related documents, such as those published by the National Resource Council (NRC 1984). Others in the public were skeptical, however, of the army's scientific testimony, arguing that it failed to allay their concerns about incineration. This skepticism prompted action on the part of both the army and the local citizens' movement.

To allay fears about the incineration technology, the army produced additional scientific documents commenting on the safety of the technology and organized fact-finding missions for incinerator opponents to its prototype facility in Tooele. The army reasoned that, if citizens could see for themselves the incineration technology and the safeguards in place at the Utah site, their opposition to a similar facility in Kentucky would abate. The first such Utah site tour included an inspection of the technology, a survey of the depot grounds, and meetings with the director of operations, who commented that "a person is safer in the incinerator plant than he is at home," despite the fact that the plant had been shut down "a number of times" because of chemical contamination (*Richmond Register,* August 16, 1984). Contrary to the army's intention, the delegation returned to Kentucky with an even deeper conviction that incineration in densely populated Madison County was wrong. In accordance with the safety concerns of the public, the army also stepped up its campaign of public education, including numerous outreach efforts organized under the umbrella of its CSEPP.

The KEF, meanwhile, sought to bolster the grassroots movement by commissioning scientific expertise to counter the disputed claims of the army. In effect, the decision was made for KEF to become a public clearinghouse for information about alternative technology. Initially, the focus of this effort was on collecting documentation about environmental toxicity and soliciting expert opinions about the safety of the proposed chemical weapons incineration technology. The KEF turned to national organizations such as the Citizens Clearinghouse for Hazardous Waste, the National Toxics Campaign, and Greenpeace for technical information and assistance. Internationally renowned experts on incineration technologies and toxicity were invited to Madison County for public hearings. Whistleblowing engineers from the army were brought in with testimony that refuted many of the army's claims about safety and risks. In effect, the army was put on the defensive as KEF marshaled convincing scientific evidence about the dangers posed by incineration technology to the environment and

public health, thus anchoring its opposition, not in emotional appeals, but in the scientific/ technical paradigm employed by the army.

A watershed occurred in the KEF information campaign when it turned away from an exclusive anti-incineration platform and focused instead on alternative technologies for safe ordnance disposal. Such alternatives had been proposed much earlier in the controversy but were given little initial attention in light of the immediate concerns about the incineration plans. However, as we shall see below, once the grassroots movement shifted its stance from anti-incineration to proneutralization, advocating alternative disposal technologies, and brought expert information to bear, the influence of and public trust in the movement increased at all levels, from the community to the U.S. Congress. KEF, working with the national coalition of the CWWG, lobbied successfully to get the National Academy of Sciences and the National Research Council to investigate alternative disposal technologies, including chemical neutralization and biodegradation technologies utilizing steam gassification and enzyme digestion (Lambright 1998). In 1993, the environmental movement began a campaign to educate the local public and members of Congress about these alternatives.

Proactive Stance on Alternative Disposal Technologies

In its early years, the Bluegrass environmental movement was centered on opposing incineration. As the movement grew in sophistication and influence, it shifted tactics to adopt a more proactive stance in support of the development of alternative disposal technology. This was prompted by an understanding that the deteriorating chemical munitions had to be destroyed, not an acceptance of the army's premise that on-site incineration was the best means of doing so. With this shift away from anti-incineration to pro–alternative technology came a renewed sense of purpose for the movement and a effective strategy for meeting the public's need—which always has been to safely destroy the chemical stockpile. This shift in focus and renewed support came as several events unfolded locally and nationwide, in part as a result of the efforts of the Bluegrass citizens' movement.

The 1990s ushered in a heightened level of interest among members of Congress in the army's plan to incinerate chemical weapons around the country. This attention was due to the acknowledged technical malfunctions at the army's prototype incinerator on Kalama Island, to schedule delays and budgetary overruns (in 1994 the estimate for incineration technology was $1.7 billion; by 1998 this had risen to $16 billion), to the growing local and state opposition to incineration, and, importantly, to legislation introduced at state and national levels that required the army to reevaluate its plan for chemical weapons disposal in Kentucky as well as Indiana and Maryland—the three national sites with the smallest percentages of the U.S. chemical stockpile.

It was clear by the mid-1990s that alternatives such as neutralization and biodegradation were viable scientific, engineering, and political options for the elimination of the chemical stockpile. At this time, the KEF and the CWWG became key players in organizing local and national campaigns to gain their acceptance among communities and politicians and, eventually, by the army. They organized conferences, solicited expert opinions, lobbied for congressional action, wrote legislation, and initiated media campaigns to inform the public in all the communities located near the chemical weapons stockpiles. Ultimately, these efforts took the grassroots movement into the halls of Congress.

Navigating the Political Process

From the early days of the incineration opposition movement, the citizens' effort had the support of some locally elected officials, many of whom initially advocated transporting the chemical ordnance out of central Kentucky to some other disposal site. As the movement gained influence, however, it reached a wider political audience, including state governors and U.S. congressmen and senators. In the 1980s, Kentucky representatives to Congress brought national attention to the chemical stockpile program. The signing of the Bilateral Destruction Agreement by the United States and the Soviet Union in 1990 spurred additional congressional interest. The Kentucky congressman Larry Hopkins introduced a bill in 1992 that required the army to study all possible disposal alternatives, soliciting in the process the congressional testimony of the Bluegrass-based grassroots activists along with that of army officials. At the state level, bills were introduced to the Kentucky legislature in 1992, again with testimony by grassroots activists, that required the army to prove that incineration was the safest of all available disposal technologies before the state would grant the necessary environmental permits. Congressional hearings on the controversy proceeded through the 1990s, amid a major KEF-sponsored campaign that resulted in more than twenty-five letters being sent to members of the House and Senate Armed Services and Defense Appropriations committees (Futrell 1999).

A consolidation of political support occurred in 1996, when Senator Mitch McConnell introduced legislation to halt the incineration program and to increase congressional oversight in the disposition of chemical stockpiles. This legislation set the stage for a bonafide alternative technology program and greatly increased the influence of the grassroots movement within federal agencies and among affected communities across the United States. The Kentucky-based citizens' movement was engaged at all stages in the political process and, in fact, was largely responsible for it.

David Zurick

The grassroots effort was able to successfully navigate the complex bureaucratic and administrative labyrinth of political support by approaching it simultaneously on several levels. Local political officials were directly lobbied, their positions were made clear in local news media, and their electoral support was tied to immediate constituency demands. State congressmen were persuaded to view the controversy as being a problem in their own political backyards, hence requiring their personal and immediate attention. U.S. congressmen whose political base was in Kentucky also had political reasons to support the movement once it became clear that the interests of their constituencies were at stake. Political expedience was served at the national level when it became apparent that the local concerns dovetailed nicely with foreign policy mandates for timely disposal of chemical stockpiles. Finally, the economic argument behind the development of alternative disposal facilities resonated at all levels of the political hierarchy, where jobs and economic development would ensue among their constituencies as an outcome of the deliberations about disposal technologies. In the case of the Bluegrass Army Depot, more than $1.5 billion will enter the local economy over a period of twelve years as a result of the adoption of neutralization technology.

Institutionalizing the Grassroots Movement

The KEF was formed as a nonprofit organization in 1990 with the goal of "improving public access to information and fostering cooperation between government and citizens" ("KEF Mission Statement" 1992, 1). Organizationally, the KEF provided a stable platform for public outreach campaigns centered on the incineration proposal at the Bluegrass Army Depot. The nonprofit status of the KEF restricted its direct political activity, but Concerned Citizens of Madison County and Common Ground still existed as the action branch of the grassroots movement. KEF, meanwhile, concentrated on educational efforts directed at the general public and at state and federal officials. It also entered the fund-raising arena, obtaining grants to finance both its educational efforts and the activist work of its coalition members.

Recognizing the need to create a unified opposition to incineration composed of all the stockpile communities, the KEF in 1991 spearheaded the formation of the CWWG, which served to bring together community groups from the continental U.S. sites as well as groups from the Pacific and Russia. The CWWG recognized early on the national and even international dimensions of the chemical stockpile controversy. The formation of these organizations has resulted in increased credibility for the grassroots movement among the various political offices, in a consensus-based mode of operation that creates a united front among the diverse opposition communities, in more assured access to funding, and in the establishment of a full-time dedicated staff to coordinate the public education campaigns, political lobbying efforts, and other activist-related strategies of the movement. The institutional personality of the grassroots movement reflects the maturation of both its membership and the nature of the issues it confronts. Ultimately, it is this institutional formation that provides the robust and effective organizational capacity of sustained environmental action.

Conclusion

The success of the grassroots environmental movement opposed to incineration at the Bluegrass Army Depot is the result of two decades of steadfast work, first by local citizens from Kentucky, then by people from across the nation and around the world. From its origins as a small group of concerned citizens to its development as a national coalition, the movement has proceeded with the same basic goal in mind—to involve citizens in decision-making that affects the quality of their lives, their communities, and the local environment. In the case of opposition to the incineration of chemical weapons, the grassroots movement has had to negotiate a complex bureaucracy that extends from local to federal institutions, including the U.S. Congress and the Department of Defense. It learned to do so by insisting all along on a kind of participatory collaboration that allowed the voices of people to be heard in meaningful ways among an entrenched and self-serving bureaucracy. This effort has by necessity required public education campaigns, science and knowledge building, and political lobbying. In the process, the adversarial relationship between the army and the opposition groups gave way to a more cooperative engagement to solve the incinerator controversy in central Kentucky. Equally important, the institutional legacy of the movement—the lessons learned and strategies devised—has far-reaching consequences; it provides a model for a kind of environmental democracy that will support grassroots efforts elsewhere and into the future.

Postscript

Since 2002, when the army first announced that it would abandon incineration of chemical weapons at the Bluegrass Army Depot in favor of neutralization, the Pentagon has stepped back several times from its financial commitment to the incineration project. In these cases, funding was halted and returned only after the intense lobbying efforts of congressional leaders working in consort with the Berea-based KEF and the CWWG. The constant vigilance of the citizens' environmental group has proved necessary to ensure that its initial victory in getting the army to adopt neutralization technologies for the disposal of chemical weapons is, in fact, sustained in practice.

References

"Chemical Stockpile Disposal Program: Report of the Kentucky Study Group." 1987. Unpublished report. Aberdeen, MD: Aberdeen Proving Ground.

"Chemical Stockpile Disposal Program Draft Environmental Impact Statement." 1986. Unpublished report. Aberdeen, MD: Aberdeen Proving Ground.

Davies, C. G. 1998. "Standing up to the Military-Industrial Complex: Public Policy, Power and the Destruction of Chemical Weapons." Archives, Kentucky Environmental Foundation, Berea. Typescript.

Ellis, W., H. E. Everman, and R. Sears. 1985. *Madison County: 200 Years in Retrospect.* Richmond, KY: Madison County Historical Society.

Futrell, Robert. 1999. "Struggling for Democracy: Environmental Politics of Chemical Weapons Disposal." Ph.D. diss., University of Kansas, Department of Sociology.

Hindman, D. 1989. "Public Input to the Chemical Stockpile Disposal Program NEPA Process." *Environmental Professional* 11:291–96.

The International Citizens' Accord on Chemical Weapons Disposal. 1991. Berea: Kentucky Environmental Foundation.

Johnston, P., R. Stringer, D. Santillo, and P. Costner. 1996. *Development of Technologies for Chemical Demilitarisation.* Berea, KY: Kentucky Environmental Foundation.

"KEF Mission Statement." 1992. Berea: Kentucky Environmental Foundation.

Kentucky Environmental Foundation (KEF), Common Ground, and Concerned Citizens of Madison County, comps. 1991. *The Citizen's Viewpoint: Citizen Perspectives on the Army's Plan to Build a Nerve Agent Incinerator in Madison County, Kentucky.* Berea, KY: Kentucky Environmental Foundation.

Lambright, W. H. 1998. "Searching for Safer Technology: Army-Community Conflict in Meeting the Chemical Weapons Convention." Draft Document. Syracuse University, Maxwell School of Citizenship and Public Affairs, National Security Studies.

Miller, Roger. 1993. "The Social Spatiality of

NIMBY: A Case Study of Citizen Opposition to an Army Nerve Gas Incinerator in Central Kentucky." Master's thesis, University of Kentucky, Department of Geography.

National Resource Council (NRC). 1984. *Disposal of Chemical Munitions and Agents*. Washington, DC: National Academy Press.

Silton, T. 1993. "Out of the Frying Pan . . . : Chemical Weapons Incineration in the United States." *Ecologist* 23, no. 1:18–24.

Smithson, A. 1994. *The U.S. Chemical Weapons Destruction Program: Views, Analysis, and Recommendations*. Washington, DC: Henry L. Stimson Center.

Part III

Seeking to Preserve
Rural and Urban Landscapes

The Role of Local Groups in the Protection of Urban Farming and Farmland in Tokyo

Noritaka Yagasaki and Yasuko Nakamura

In the age of globalization, when we are able to enjoy a variety of food from the world over, there is a growing interest among consumers in the safety of food and in the sustainability of the food supply. This is so, partly because of the globalizing threat to food production from such things as BSE (bovine spongiform encephalopathy, or mad cow disease), bird flu, agricultural chemicals, and water pollution, and partly because of the limited food supply, a result of the ever-increasing world population. Urban consumers in Japan concerned about safe food are beginning to realize the importance of local production, and they want to see the crop-production process for themselves. There are many individuals, local groups, and institutions in Japan concerned about the production and consumption of agricultural products. The number of grassroots movements reflects the growing interest in the supply of food in urban areas, and grassroots groups contribute significantly to the preservation of urban farmland and the urban environment generally. This chapter intends to depict the nature of the grassroots movement to preserve urban farming and farmland in Tokyo by examining and evaluating farmers' role in it.

Farmland as a Green Island in the Metropolis

Tokyo is one of the most highly urbanized parts of the world. While urban land use dominates this world city, scattered farmland also characterizes Tokyo's urban landscape. It may be surprising to learn that farming is undertaken in such a large, economically advanced city. But there are, in fact, progressive farmers who play an important role in promoting urban farming, supplying locally grown vegetables to local residents. And urbanites also participate in recreational farming. This chapter analyzes the sustainability of urban farming by focusing on farmers' grassroots movements.

Japan is a mountainous country where arable land is restricted to narrow coastal plains and small intermountain basins (Yagasaki 2000). In such flatland, various human activities have been concentrated, and urban functions and farming activities have competed with each other. As the population grew, efforts were continually being made to reclaim land on the mountain slopes as well as in the coastal wetlands. Intensive-farming methods were developed in order to utilize the limited arable land and to maximize the harvest. The amount of farmland in Japan, and, with it, the number of farms and farmers, has, however, been decreasing. The farming population constituted only 4.5 percent of the total economically active population in 2000. The Japanese depend more and more on imported foods. At the same time, people are increasingly concerned about sustainable food production, healthy food, and environmental conservation.

Farmland-protection movements appear to be important in many aspects of Japanese daily life. Various attempts are being made to protect and even promote urban farming. Urban farmland is considered important as it is the basis of food production and also sustains green space in the urban landscape. It plays an important role in providing open space when such disasters as earthquakes and fires strike, and it also contributes to flood prevention. There is, therefore, a growing interest in the protection of urban farming and farmland, an interest generated in important ways by grassroots movements.

Tokyo differs from other world cities in that farming is a sustained economic activity and farmland is an important landscape element. Tokyo may look disorganized to Western observers, partly because of the irregular road patterns and narrow streets, and partly because of the mixture of buildings constructed at different times and in different styles. The chaos of the urban landscape is exacerbated by the existence of scattered farmland surrounded by land put to urban uses. To un-

derstand how such a land-use pattern developed, a review of the process of urbanization in Tokyo is helpful.

Tokyo's urban settlement was, in the 1940s, surrounded by farmland, for the flatland around the city was important to urban residents as the source of its food supply. As urbanization proceeded and the urban population grew, residential areas sprang up along major suburban railroads. Bus routes connected one suburban station to another, and along these bus routes more residential areas appeared. Since these residential developments were built on what had been farmland, and since individual farmers owned only limited amounts of land, the scale of development was small, tiny houses and apartment building being constructed in small subdivisions for urban workers who commuted to downtown Tokyo. These early subdivisions looked like islands in the farmland (see photograph 8.1).

Photograph 8.1. Aerial photograph of a part of Nerima Ward in 1948. (Courtesy U.S. Army.)

Noritaka Yagasaki and Yasuko Nakamura

When urbanization proceeded further during the high-economic-growth period beginning in the mid-1950s, farmland was increasingly converted to residential use. Under the Town Planning and Zoning Act of 1968, urbanization-promotion zones were designated where high tax rates accelerated the process of conversion from rural to urban land use. Consequently, only small patches of farmland now remain as islands in the urban landscape (see photograph 8.2). The existence of urban farming and farmland is one of the characteristics of Tokyo, in sharp contrast to the urban landscape of such large American cities as New York and Los Angeles. Tokyo also differs from cities in Europe, where urban settlements were built to be distinct from the surrounding rural areas.

Urban farming is clearly observed in Tokyo's twenty-three wards, which contain two-thirds of the city's population of about 12 million. The density of population in these wards is nearly thirteen thousand persons to the square kilometer, while the Japanese average is some 340 persons to a square kilometer. Besides the central twenty-three wards, cities in the western suburbs, mountainous areas farther west, and islands in the Pacific Ocean are also considered part of metropolitan Tokyo.

How much farmland remains in the central section of Tokyo? Figure 8.1 shows the area of farmland in the central section of Tokyo. The Japanese government is located in Chiyoda Ward, and the Tokyo metropolitan government is headquartered in Shinjuku Ward. While the core wards—Chiyoda, Shinjuku, Chuo, Minato, Shibuya, Bunkyo, and Toshima—have no farmland, those in the periphery have substantial amounts. Particular attention is paid in this chapter to Nerima Ward, where a large area of farmland is sustained and various attempts are being made by progressive farmers to protect farmland and to promote urban farming.

Photograph 8.2. Aerial photograph of a part of Nerima Ward in 1992. (Courtesy Geological Survey, Government of Japan.)

Protection of Urban Farming and Farmland in Tokyo

Progressive Farmers and Urban Farmland

Urban farmers are faced with pressures such as zoning and taxes. Farming activities have been controlled by the Town Planning and Zoning Act as well as the Agricultural Land Act (1952) and other acts related to farmland. The Productive Green Land Act, originally enacted in 1974, was revised in 1992 to designate two categories of urban farmland: farmland to be converted to residential use, or urbanization-promotion farmland, and farmland to be conserved, or productive green land. If a parcel has an area of five hundred square meters or more, it could be designated as productive green land, obligating it to remain under cultivation for thirty years. The property tax for 0.1 hectare (0.25 acre) of productive green land is ¥3,090; that for urbanization-promotion farmland is ¥1 million. The payment of the inheritance tax on productive green land can be postponed until the death

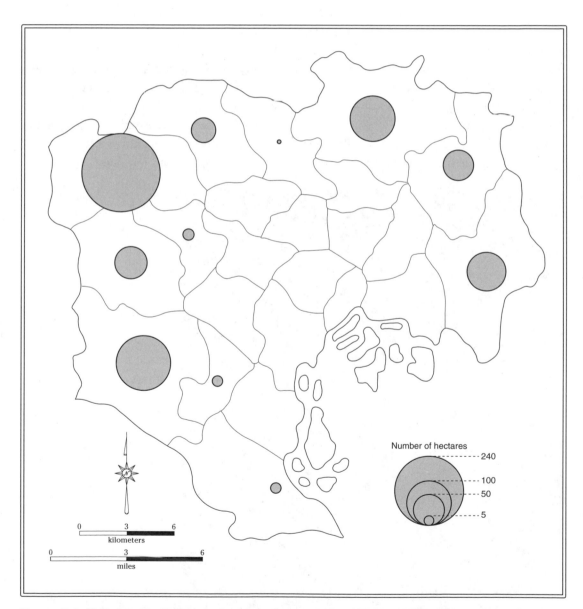

Figure 8.1. Farmland in the central twenty-three wards of Tokyo, 2002.

Noritaka Yagasaki and Yasuko Nakamura

of the heir if he or she maintained the property as productive farmland.

Nerima Ward had 287.6 hectares of farmland in 2005. Registered as productive green land are 217.87 hectares in 717 parcels, accounting for 70 percent of the total farmland in the ward. These parcels are small and scattered. The average size of a parcel is 0.3 hectare, or 0.76 acre. This farmland is cultivated by 571 farm households. Average farmland per household is 0.50 hectare (1.2 acres). While 52 households (9 percent) own 1 hectare (2.5 acres) or more, 231 households (40 percent) own less than 0.3 hectare (0.7 acre). Only 15 of those 571 households (or 2.6 percent) engage in farming full-time, that is, derive 80–100 percent of their income from farming. The great majority depend on nonfarming income, farming income accounting for less than 50 percent of their total income. In such instances, income is supplemented by such real estate ventures as managing apartments, rented houses, and parking lots.

A great diversity of farm products is grown by urban farmers in Nerima Ward. A variety of vegetables and fruits are harvested all year round. Nearly 70 percent of the total farmland is planted in vegetables. The ten leading vegetables are cabbage, broccoli, Japanese radish, potato, spinach, sweet cone, green soybean, taro, cauliflower, and Welsh onion. Overall, more than thirty kinds of vegetables are grown in the ward. In addition, fruit trees (thirteen kinds), nuts, and grapes are grown, while flowers, nursery plants, and turf are also planted. All these products are locally marketed and consumed. Despite the fact that there exist various restrictions and pressures on farming activities, easy access to consumers is an advantage for urban farmers.

Some farmers have become actively involved in promoting urban farming, seeking new possibilities for both farmers and local residents. The so-called agriculture-experience farms started by leaders in the farming community represent one such venture, playing an important role in promoting urban farming and the environmental protection movement in Nerima Ward. The case history of one such farmer, Mr. K, clearly illustrates the development of this particular grassroots movement.

When Mr. K began farming in 1980, agriculture in Nerima Ward specialized in producing such vegetables as cabbage, broccoli, and cauliflower for the central produce markets of Tokyo. During that time of rapid urbanization and soaring land prices, urban farmers were often criticized for cultivating valueless products on the high-value land and were suspected of simply enjoying watching their property values rise. Mr. K felt that urban farming could not survive without the support of those local residents who fully understood the importance of agriculture in urban areas. He also recognized how problematic it was that urban farmers made no attempt to justify the existence of urban farming or otherwise maintain a good relationship with the nonfarming urban community.

Mr. K started a vegetable stand on his farm in the early 1980s. This gave him the opportunity of gauging the needs of local residents. Receiving a positive response from his customers, he began to modify his farming practices by producing more for local sales and less for the central produce markets. The variety of vegetables needing to be grown in response to local demand required a more labor-intensive kind of farming.

In the late 1980s, when Mr. K's daughter attended the local primary school, her fellow students, accompanied by their parents, took an educational field trip to his farm. The students responded well to the new experience, seeming interested in his explanations of plants, soil, and pests. Although at the time farmers thought that farmland should not be open to nonfarmers, Mr. K came to understand the importance of farming and farmland to the nonfarming urban population. The field trip led to the idea of starting agriculture-experience farms in Nerima Ward. Another progressive farmer in the neighborhood came to participate in this new promotional tactic.

Mr. K started the Private School for Green and Agriculture [Nou to midori no taikenjuku], the first agriculture-experience farm (see photograph 8.3), in 1996. It differs from ordinary community gardens in that it is owned and operated by a farmer as a part of his farming enterprise. The way it works is that individuals rent small plots from, and grow and harvest vegetables under the guidance of, the farm owner, leases running for eleven months, from March through January. Short courses in vegetable growing are regularly provided so that those participants without farming experience will be able to grow good-quality vegetables (these short courses are discussed further below). Participants in the program range in age from the twenties to the eighties and pursue a variety of occupational paths. Many are concerned with producing vegetables without the use of chemicals. Participants can renew the lease on their plot for up to five years. Sponsoring farmers receive ¥29,000 from participants, and ¥12,000 from the ward office, per plot, the fees supplementing and stabilizing the farm household's income. Of the little more than one hectare of farmland owned by Mr. K, nearly half is devoted to the agriculture-experience farm plots.

The agriculture-experience farm provides new opportunities for both Mr. K and the participants in the program. Through the continuing interaction with his farming students, Mr. K is able to learn what the urban residents think and how they behave. Participants enjoy farming with the help of a professional farmer. The good-quality vegetables that they produce are also important to them.

Photograph 8.3. Private School for Green and Agriculture [Nou to midori no taikenjuku] operated by Mr. K in Nerima Ward. (Courtesy Yasuko Nakamura.)

Noritaka Yagasaki and Yasuko Nakamura

Over time, Mr. K has come to believe that his farm is creating a sense of community among the participants as they attend lectures, engage in farmwork, and hold harvest festivals and parties together. In an attempt to broaden participants' experiences, such other ventures as cooking classes, rice-growing classes in the paddy district, and trips to foreign countries to observe alternative farming methods are also being planned.

Mr. K also understands that urban farmers can also help farmers in rural areas connect with urban consumers. Having no direct contact with urban consumers, farmers in rural areas have difficulties gauging urban demand. Urban farmers who do have close contact with urban consumers can suggest appropriate farming methods and products. By operating his agriculture-experience farm, Mr. K has created links with those farming in other regions.

Mr. K has been successful in encouraging his fellow farmers to expand the project with the collaboration of the Nerima ward office. The number of agriculture-experience farms has gradually increased as one new farm has been opened every year. As of 2006, there were eleven agriculture-experience farms operating in the ward and more on the way. This system has attracted public attention and contributed to the understanding of farming and farmland in urban areas of Tokyo (Shiraishi 2001).

The Structure of the Grassroots Movement

Although progressive farmers, in collaboration with local residents, have been playing an important role in sustaining farming in Nerima Ward, their efforts need to be evaluated in the larger framework of interrelated economic, political, and geographic factors. Figure 8.2 shows our analytic framework. It consists of three levels: the administrative,

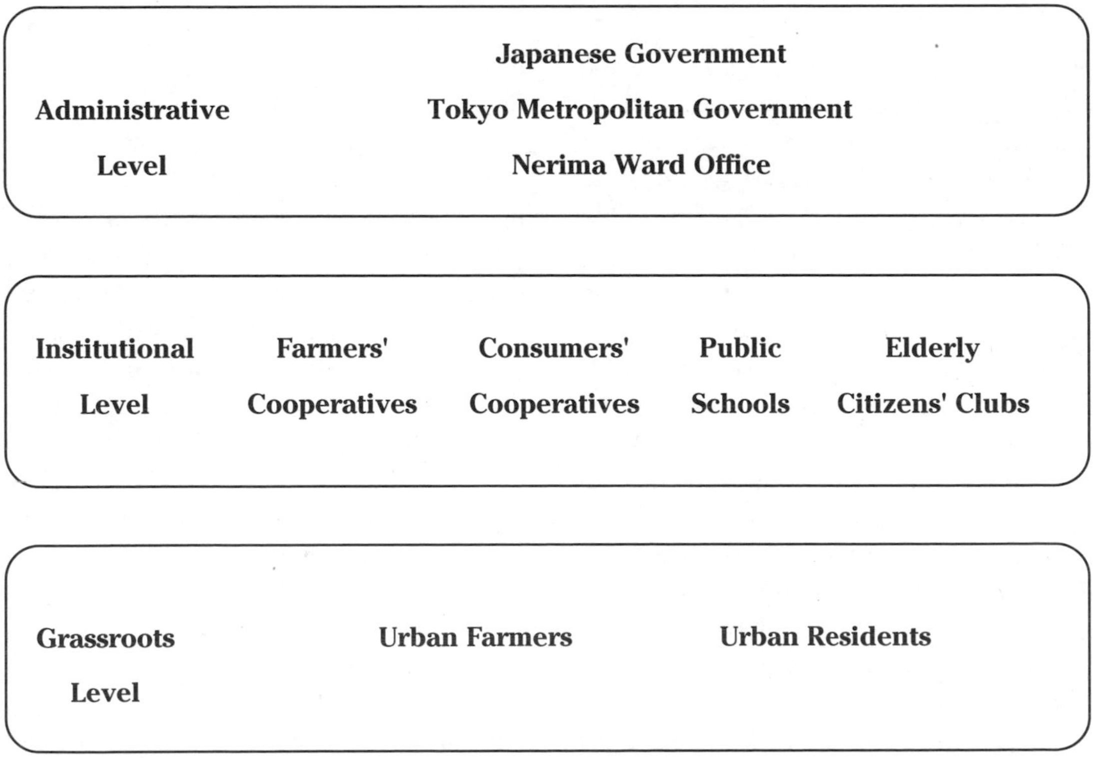

Administrative Level	Japanese Government
	Tokyo Metropolitan Government
	Nerima Ward Office

| Institutional Level | Farmers' Cooperatives | Consumers' Cooperatives | Public Schools | Elderly Citizens' Clubs |

| Grassroots Level | Urban Farmers | Urban Residents |

Figure 8.2. Three levels for sustaining urban farming and farmland.

the institutional, and the grassroots. At the administrative level, the Japanese government controls farming activities, land use, and food production via legislation. Within the framework of the national government, provincial governments, such as the Tokyo metropolitan government, make their plans to promote urban farming. Municipal administrations, including Nerima ward office, carry out actual projects in cooperation with local organizations, farmers, and residents. National and local policies play a crucial role in the preservation of urban farmland. The institutional level is composed of all groups, such as farmers' cooperatives, consumers' cooperatives, schools, and senior citizens' clubs, involved in farming and farmland preservation. The grassroots level consists of farmers and urban residents actually engaged in farming.

The Nerima ward office has been supportive of the Industrial Promotion Department's Urban Farming Division's efforts to promote urban farming. In addition to various attempts to facilitate urban farming and to expand outlets for fresh produce, it also carries out educational programs designed to promote an understanding of urban farming among local residents and to increase opportunities for contact between farmers and residents.

The Agricultural Fair, which started in 1947, is held annually in November. The Flower and Nursery Show has been held annually in April since 1967. In addition to these traditional events, the Vegetable Walk Rally is held twice a year, in June and October. This new event, introduced in 1990, is intended to facilitate communication between local farmers and residents and to promote the understanding of local farming. Participants—who range from schoolchildren to the elderly—walk from one farm to another, harvesting vegetables, touching farm machinery, and generally getting the feel of urban farming. The fee is ¥500 per party, and, in 2002, some 240 people in

more than 70 parties participated in both June and October.

The Experience of Harvesting the Nerima Radish is another event held every November. Nerima was once famous for its radishes. Cultivation of the radish began there during the Edo period and continued, as population increased, until the 1920s. Due to disease problems, urban encroachment on farmland, and the Westernization of Japanese eating habits, the production of the Nerima radish declined. However, the Nerima ward office is attempting to revitalize the production of the Nerima radish as a key to promoting urban farming and agricultural understanding.

Large agricultural forums are occasionally held in Nerima Ward. One on the topic of Nerima's agriculture was held in early March 2003, with some two hundred residents attending. A keynote speech was delivered by an actress who is active in farming and the environmental movement, and a panel discussion on local agriculture was held. The event was sponsored by the Nerima ward office and a local agricultural cooperative with financial assistance from the Tokyo metropolitan government (Nerima-ku and Nerimaku Nogyo Iinkai 2003).

In 1991, the Nerima ward office began a program that, with the help of local farmers, provides fresh vegetables for school lunches. Seventeen primary schools and five junior high schools now receive seventeen different kinds of locally grown vegetables, including cabbage, onions, and sweet corn. A new program begun in 2002 converts leftovers from school lunches to compost for local farmers. The compost is called the "earth of Nerima."

Public schools also operate farms with the assistance of the Nerima ward office, which arranges for local farmers to provide the farmland free of charge. As of March 2003, twelve primary schools and three junior high schools operate such school farms. The farms are generally small, averaging 411.6 square meters. The first school farm appeared in

Noritaka Yagasaki and Yasuko Nakamura

1980, and the number of such farms gradually increased during the 1990s. When the Ministry of Education, Science, and Culture revised its course of study recently, a component involving practical field experience was introduced. School farms offered students the perfect opportunity to fulfill that requirement. The Nerima ward office also arranges for preschool children to have similar experiences, its Farms of Contact program introducing them to the harvesting of potatoes.

Senior citizens' clubs are also involved with urban farming. The Nerima ward office rents farmland from local farmers and provides the clubs with space for community gardening. As of March 2003, there were thirty-three such farms in Nerima Ward. Considering the aging of population in Tokyo, these farms are likely to become more important in the future.

The Participation of Urban Residents in the Movement

As we have seen, urban residents sustain urban farming both as consumers and as recreational farmers. In this section, we take a brief, closer look at their role in the grassroots urban-farming movement.

To begin with, the local consumption of their produce is the mainstay of urban farmers. Some 60 percent of farmers in Nerima Ward have sales facilities on their farms. For small-scale farmers, vegetable stands are their main sales outlet (see photograph 8.4). The directory map of farm produce stands prepared by the Nerima ward office for the convenience of consumers clearly shows that such stands are scattered across the ward. Farmers' marketing strategies are gradually shifting from shipping to the central produce markets to

Photograph 8.4. A vegetable stand using coin-operated lockers in Nerima Ward. (Courtesy Yasuko Nakamura.)

Photograph 8.5. An urban farm surrounded by homes in Nerima Ward. (Courtesy Noritaka Yagasaki.)

selling from their own vegetable stands. Only 23 percent of the farmers in Nerima Ward now depend on the central produce markets. Consequently, instead of specializing, farmers are now planting a greater variety of crops (see photograph 8.5).

Local farmers' cooperatives also promote the direct sale of local produce. JA Tokyo Aoba, an agricultural cooperative organized in 1997 with the merger of three local cooperatives, operates three stores (see photograph 8.6). These stores are where farmers and local consumers regularly meet as farmers take turns staffing the store. The farmers' photographs are posted so that consumers can actually see who grew which vegetables (see photograph 8.7). JA Tokyo Aoba actively promotes the stores as well as conducting educational programs such as lectures on urban farming at local public schools.

Urban residents also participate in farming by way of community gardens. Not only are community gardens yet another means of preserving the urban-farming tradition; their growing popularity suggests that recreational farming also fulfills some unmet need among urban residents. Indeed, community gardens are becoming popular throughout Japan.

Nerima Ward is a leader in community garden programs in Japan. Three types of community gardens are operated there: ward residents' farms, citizens' farms, and agriculture-experience farms.

Community gardens first began to appear in Nerima Ward in 1973 when an experimental program establishing so-called ward residents' farms was instituted. Under the terms of the program—which continues today—the Nerima ward office leases urbanization-promotion farmland from landown-

Noritaka Yagasaki and Yasuko Nakamura

Photograph 8.6. Farmers' shop named "Kogure-mura" operated by the agricultural cooperative of Nerima Ward. (Courtesy Yasuko Nakamura.)

Photograph 8.7. Photographs of farmers supplying produce at Kogure-mura. (Courtesy Yasuko Nakamura.)

ers free of charge and makes plots available to ward residents. Each plot is fifteen square meters, and monthly rent is currently ¥400 for a term of twenty-three months. There are today twenty-two hundred plots on twenty-three farms in the ward, for a total of 53,718 square meters. Residents renting the plots can use them as they like (see photograph 8.8), while farmers take advantage of the low tax rate applied to the farmland.

Responding to the growing interest in community gardens throughout Japan, the

Photograph 8.8. A ward resident's farm in Nerima Ward. (Courtesy Yasuko Nakamura.)

Photograph 8.9. A citizens' farm in Nerima Ward. (Courtesy Noritaka Yagasaki.)

Photograph 8.10. Agriculture-experience farm in Nerima Ward. (Courtesy Yasuko Nakamura.)

national government began to revise agricultural land policies on the basis of the Agricultural Land Act of 1952. In 1989, the Special Farmland Loan Act was enacted, facilitating municipal governments and farmers' cooperatives to lease farmland to nonfarmers for nonprofit cultivation. In 1990, the Community Garden Promotion Act was introduced to further facilitate the establishment—whether by municipal governments, farmers' cooperatives, or individual farmers—of community gardens on productive green land in urban areas.

The Nerima ward office, which had been long engaged in its own community garden programs, began the so-called citizens' farm program in 1992 under the Community Garden Promotion Act. There are nine citizens' farms in Nerima Ward, covering 29,353 square meters. The Nerima ward office pays ¥350 per square meter annually to the farms' landlords. There are 426 plots in all, each 30 square meters. Monthly rent for a plot is ¥1,200 for twenty-three months. Citizens' farms each have a clubhouse (with an office, a kitchen, restrooms, a lounge, and dressing rooms), storage facilities, and picnic tables, thus providing a comfortable place in which to spend spare time (see photograph 8.9). People can use their plots as they like, and a variety of vegetables are grown. Plots are

allotted by means of a lottery system, and demand is so great that plots are hard to come by.

One important difference between the ward residents' and citizens' farms and the agriculture-experience farms (discussed earlier) is that, on the latter, the recreational farmers work under the supervision of the farmer. Agriculture-experience farms regularly offer short courses in vegetable growing from planting through harvesting (see photograph 8.10). Farm owners supply the seeds and seedlings, fertilizer, and the farm implements, and their instruction emphasizes sustainable agriculture by teaching traditional farming methods. The result is that even novices can grow high-quality vegetables. Another important difference is that the administration and maintenance of the agriculture-experience farms cost local governments less, as farmers themselves maintain the farmland.

Maintaining Urban Farming

The preceding discussion shows that grassroots environmental movements cannot be understood in terms of a simple schema, in this case farmers versus the government, but must be contextualized, in this case placed in an urban framework determined by many interrelated factors (schematized in figure 8.3). The future of urban farming and farmland is even harder to determine, based as it will be on not necessarily predictable social and demographic changes.

Tokyo's population, including the farming population of Nerima Ward, is aging. Of the 1,493 people engaged in farming in Nerima Ward in 2002, 798 (53 percent) were sixty-one years of age or older, while only 82 farmers (5.5 percent) were thirty-five years of age or younger. The aging of the farming population, which is a common characteristic of Japanese agriculture, seems likely to accelerate the decline of urban farming in the near future as retirement means a loss of farmland. Thus, successors are important if urban farmland is to be maintained in Tokyo. In a survey conducted by the Nerima ward office in 2002, 247 of 681 farm households, or 36 percent, had successors. These successors are from younger generations who have already been engaged in family farming or who intend to take over their parents' farms in the future. Our interview survey reveals that younger generations around forty years of age, who had been employed in nonfarming sectors, are increasingly engaged in farming. In this sense, urban farming is far better off than farming in remote agricultural districts, where the continual decline of the farming population makes it difficult to sustain agricultural production, not to mention rural communities.

Perhaps one reason for the growth of urban farming (vs. the decline of rural farming) is the fact that many of the younger generation of urban farmers are college graduates who took up nonfarming occupations for a while and, thus, bring new ideas to traditional farming. One example is the personal computer club organized within the farmers' cooperative. This Nerima ward office–subsidized group attempts—via its Web site—to keep urban farm children on the farm and to lure back those who have strayed. Farmers have also established Web sites in an attempt to increase their visibility and their customer base.

Just as urban farming is changing, the so-

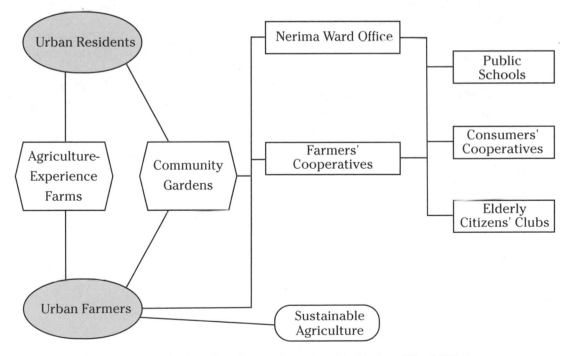

Figure 8.3. Organization of urban farming and farmland in Nerima Ward, Tokyo.

cial milieu in which it operates is changing as well. For one thing, owing to the current population decline in Japan, the price of land has begun to drop, and the housing market has opened up. As people have begun to return to downtown Tokyo, they have shown great interest in the idea of urban livability, their concern for how they live and what they eat being translated into support for grassroots movements promoting the maintenance of green space and sustainable agriculture. Recognizing this trend, the Tokyo metropolitan government began in 1997 a certification program for organic farm products (Tabayashi and Kikuchi 2000; Goto 2003). Under the program, vegetables grown with less than half the usual amount of chemical fertilizers and pesticides are certified as a "specially grown product." Products so certified—more than twenty to date—can be marketed as organic. A variety of methods (e.g., the use of special netting and natural enemies to protect plants from pests and fertilizing with compost), all subsidized by the Nerima ward office, are employed to reduce chemical use.

The trend toward urban livability was confirmed by a 1999 survey conducted by the Tokyo metropolitan government in which 94 percent of respondents indicated a desire to preserve farming and farmland in the city. The main reasons cited were the local provision of fresh produce, the preservation of the natural environment, and the availability of green space to take refuge in during natural disasters. Tokyo Metropolitan Government introduced "Agriculture Promotion Plan of Tokyo" in 2001 (Tokyo-to 2001). As compared with the cases of peripheral farming regions that are facing aging of population, depopulation, and degradation of farmland, there will be a fair chance that urban farming is sustained in Tokyo with grassroots movements.

Key to the protection of urban farming and farmland are the notions of coexistence, col-laboration, and education. Rural land use and urban land use could easily coexist in the city. But for them to do so will require the (increased) collaboration of farmers, residents, local institutions, and the municipal government to educate people about the importance of local food production. If younger generations can be made conscious of the importance of healthy food and, thus, sustainable farming and environmental preservation, urban farming and farmland can look forward to a bright future in Tokyo.

References

Goto, Mitsuzo. 2003. *Toshi nouchi no shiminteki riyou* [The use of urban farmland for urbanites]. Tokyo: Nihon Keizai Hyoronsha.

Nerima-ku and Nerimaku Nogyo Iinkai. 2003. *Nerimaku no nogyo* [Agriculture in Nerima Ward]. Tokyo: Nerimaku.

Shiraishi, Yoshitaka. 2001. *Tokaino hyakusho desu* [I am an urban farmer]. Tokyo: Komonzu.

Tabayashi, Akira, and Toshio Kikuchi. 2000. *Jizokuteki nouson sisutemu no chiikiteki joken* [The regional basis for a sustainable rural system]. Tokyo: Norin Tokei Kyokai.

Tokyo-to. 2001. *Aratana kanousei wo kirihiraku Tokyo nogyo no chosen: Tokyo nogyo shinko puran* [The challenge for Tokyo's agriculture to open up new possibilities: A plan for promoting Tokyo's agriculture]. Tokyo: Tokyo-to.

Yagasaki, N. 2000. *Japan: Geographical Perspectives on an Island Nation*. Tokyo: Teikoku-shoin.

Chapter 9

From Horse Farms to Wal-Mart

The Citizens' Movement to Protect Farmland in the Central Bluegrass Region of Kentucky

Dan Carey and Pradyumna P. Karan

A Home Depot and a Super Wal-Mart would soon rest in the same spot where eighteen famous racehorses, dating back to the 1898 Kentucky Derby winner, Plaudit, now rest in peace on Hamburg Place Farm. The famed horse farm was to make way for 436,900 square feet of retail space and nearly 2,500 parking spots (see photograph 9.1). When this story was reported in the *Lexington Herald-Leader* (November 18, 2004), it was only one of the numerous reports of the loss of the unique Bluegrass landscape of grassy horse farms, tobacco barns, and white fences to shopping malls and subdivisions chronicled by the newspaper since the 1960s. The influx of chain stores and strip malls leads to a loss of true community identity. The new stores can be found in virtually every other community across the United States. In addition, the loss of locally owned businesses results in a reduced commitment to the interests and vitality of the local community.

The Lexington and Central Bluegrass region of Kentucky is riding a wave of economic growth and expansion that, for the foreseeable future, shows little signs of abating. With growing economic opportunities in the booming Japanese automobile, auto parts, electrical, and information sectors, Bluegrass cities and towns are experiencing a rapid influx of newcomers and a redistribution of existing residents. Over the past several years, Kentucky has added thousands of new residents, most of whom have been absorbed into the Bluegrass communities that lie in an area traditionally cherished for its wide-open horse farms, productive agriculture, and extraordinary scenic splendor.

As a result of these developments, many Bluegrass communities are facing a loss of farmland and open space and disputes involving land-use politics, environmental protection initiatives, economic-growth strategies, and residential-development plans. In the absence of an overall regional plan for the entire area, different communities, all with their own unique sense of history, place, and politics, have charted their own way to handle commercial and residential growth. People are drawn to the Bluegrass region by its booming

Photograph 9.1. Hamburg Place commercial and residential development, future site of Wal-mart and Lowe's super stores. (Courtesy Dan Carey.)

economy, moderate climate, educational and recreational opportunities, and cultural life. Corporations and businesses such as Toyota, IBM, Lexmark, and Hitachi are also interested in the region because of its central location in the densely settled Eastern United States within overnight trucking distance from New York, Chicago, and Atlanta.

Loss of Farmland

Although Kentucky is still predominantly an agricultural state, with approximately 54 percent of the state's 25.8 million acres used for farming and livestock operations (U.S. Department of Agriculture 2006), development pressure is increasing each year. As a proportion of Kentucky's total nonfederal land acres, developed acreage has significantly increased from 1982 to 1997 (see figure 9.1). In 1997, 7.2 percent was developed into urban areas and roads, compared to 4.7 percent in 1982.

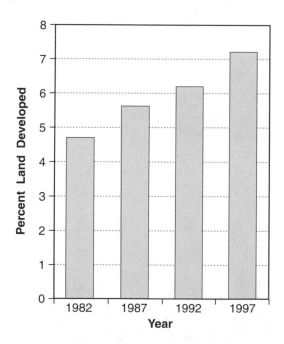

Figure 9.1. Percentage of nonfederal land developed in Kentucky, 1982–1997. *Source:* http://www.kyagr.com/marketing/farmland/index.htm.

Farmland under the heaviest development pressure is found near the metropolitan areas of Bowling Green, Louisville/Jefferson County, and Lexington/Fayette County. The Inner Bluegrass region has high-quality farmland under high development pressure (see figure 9.2).

Lexington/Fayette County, in the heart of the Bluegrass region, exemplifies, more than any of the surrounding counties, the intense development pressures in the region. During the period 1937–2001, the urban area of Fayette County grew more than tenfold, from 7.5 to 85 square miles (see figure 9.3).

In 1964, there were 978 farms in Fayette County, with 166,000 acres in farmland. In 2002, there were 738 farms, with 119,000 acres in farmland (U.S. Department of Agriculture 2006). Of the 47,000 acres of farmland lost during the thirty-eight-year period, more than half was lost during the period 1992–2002. In the face of the increasing encroachment of development into rural areas, the preservation of rural farmland in the county was deemed important for a number of reasons: agricultural-economic benefits and concerns; the development pressure on farmland; the high cost of infrastructure as development spreads from urban areas; the protection of cultural, environmental, and historic resources; the maintenance of the unique Bluegrass identity and "sense of place"; and the protection of a thriving tourism industry.

Of the 120 counties in Kentucky, Fayette County ranked second in total farm cash receipts, generating $354,260,000 in 2005. The county's agriculture supports more agricultural jobs than does that of any other county in Kentucky and provides a significant tax base.

As residential development moves away from urban areas, the cost of the supporting infrastructure exceeds the tax revenue. A study by the American Farmland Trust (AFT 1999), sponsored by the citizens' groups the Bluegrass Conservancy and the Land and Nature Trust, found that the cost of suburban

Dan Carey and Pradyumna P. Karan

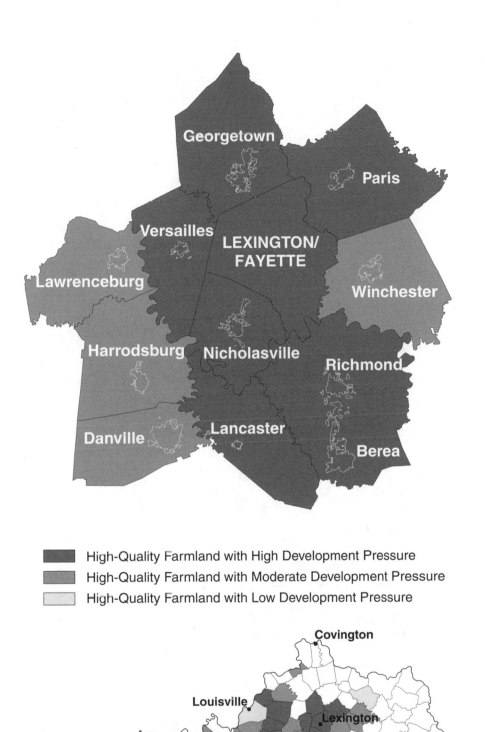

- High-Quality Farmland with High Development Pressure
- High-Quality Farmland with Moderate Development Pressure
- High-Quality Farmland with Low Development Pressure

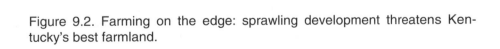

Figure 9.2. Farming on the edge: sprawling development threatens Kentucky's best farmland.

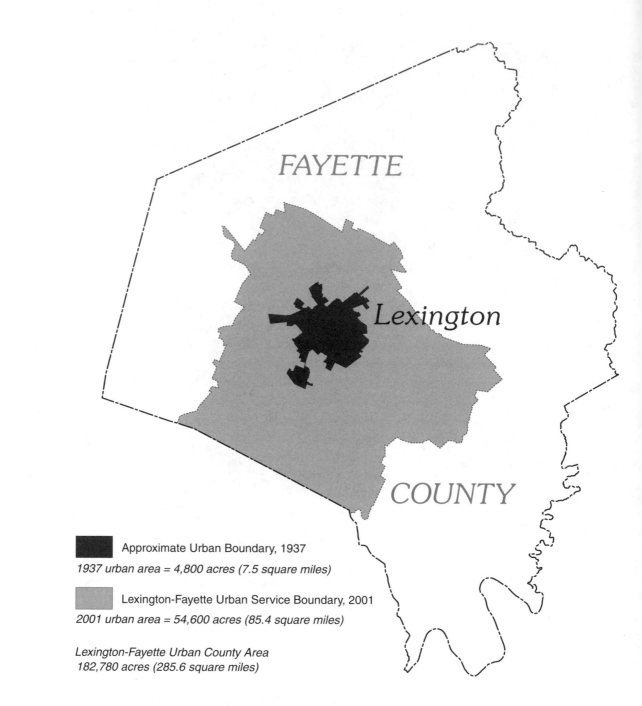

FAYETTE

Lexington

COUNTY

■ Approximate Urban Boundary, 1937
1937 urban area = 4,800 acres (7.5 square miles)

▨ Lexington-Fayette Urban Service Boundary, 2001
2001 urban area = 54,600 acres (85.4 square miles)

Lexington-Fayette Urban County Area
182,780 acres (285.6 square miles)

Figure 9.3. Fayette County, Kentucky. Growth of the urban area, 1937–2001. *Source:* Lexington-Fayette Urban County Government (2005).

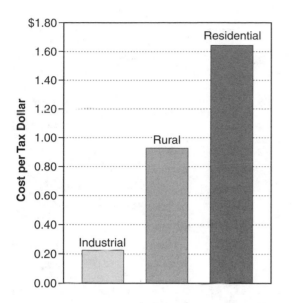

Figure 9.4. Cost of community services for Fayette County.

infrastructure was 164 percent of the tax revenue (see figure 9.4).

The loss of farmland and open space had a negative impact on cultural, environmental, and historic resources. The Bluegrass region, which includes Fayette and the surrounding counties, is world renowned for its rolling landscape, soils rich in phosphate, limestone springs, and identity as the thoroughbred capital of the world. The rural hamlets and horse farms harbor a unique cultural heritage of historic buildings and stone and plank fences. They provide an ongoing sense of place that is increasingly disappearing from American life.

Thousands of travelers come every year from all parts of the United States and the world to visit the Bluegrass. Fayette County is within one day's driving distance of 75 percent of the U.S. population. Visitors come to attend the horse auctions and races at historic Keeneland, to tour the Kentucky Horse Park (see photograph 9.2) and thoroughbred horse farms, and to enjoy the beauty of the rolling farmland. Tourism in Fayette County generated $600,000 and supported nearly fourteen thousand jobs in 2002. The Kentucky Horse Park at Lexington will host the 2010 World Equestrian Games, which will have an anticipated economic impact of $150 million.

Over the past forty-five years, the Bluegrass region has undergone rapid suburbanization. Urban sprawl claims more prime agricultural land every day. In the Bluegrass, there are now fewer extensive sweeps of rolling green grass and plank fences. The undulating countryside is splattered with subdivisions around towns such as Richmond, Nicholasville, Versailles, Danville, Frankfort, and Lexington—each laid out in the same dreary asphalt curves, connected to one another only by highways. Streams are buried in concrete culverts. Farms still remain, but for-sale signs invite the builders to turn the land into rural subdivisions. In 2005, the World Monuments Fund designated more than 1 million acres of the Bluegrass region one of the one hundred most-endangered cultural sites in the world (World Monuments Fund 2005). In addition to the lost farmland, historic structures, and rock fences that give the region a unique sense of place, the region is faced with losing a nearly $1 billion tourist industry.

Photograph 9.2. Kentucky Horse Park. (Courtesy Dan Carey.)

Photograph 9.3. Sir Barton Way runs through a former horse-training track at Hamburg Place in this April 2004 photograph. By December 2006, both sides of the road were filled with stores, businesses, offices, and housing developments. Citizen activists were unable to check the urban sprawl. (Courtesy *Lexington Herald-Leader.* Photograph by Charles Bertram.)

A large number of Kentuckians are concerned about the transformation of the Bluegrass landscape (see photograph 9.3). The growth and building boom has come at considerable environmental cost—the energy-wasting design of the residential and commercial development and the pollution of surface and groundwater from septic tanks and industrial waste. Over 65 percent of all streams in Fayette County are not clean enough for fishing or swimming.

What can be done to stop the destruction of the cultural and natural landscape of the Bluegrass region? As the twenty-first century unfolds, this question is at the heart of a new kind of movement—a citizens' campaign to save the Bluegrass. In response to the need for growth-management planning and to preserve and enhance the unique character of the Bluegrass, a number of citizens' groups arose during the 1990s, including Bluegrass Tomorrow, the Bluegrass Conservancy, the Rural Land Management Board, and the Blue Grass Trust for Historic Preservation. Urban planners in the Bluegrass communities have noted the costs of sprawl. The subject has attracted the attention of civic clubs and local leaders. Efforts to preserve the landscape also included those by government officials, conservationists, and advocates of parks and playgrounds.

Dan Carey and Pradyumna P. Karan

In many Bluegrass communities, the advance of development has led to the formulation of local groups, often led by women, to "save the farm" or "save the open space." The activists have made two kinds of arguments. One is the *conservation* argument. The other is the cultural and natural *amenity* argument.

Although many advocates of preservation sought to be comprehensive, the two lines of argument had different roots and often appealed to different groups of people. Because subdivisions were built on horse farms, horse industry advocates often argued that Kentucky was risking the loss of a distinctive landscape. Although the growth of suburbs destroyed thousands of acres unmatched for their productivity, the building of houses on prime farmland persisted. The issue, however, is usually a matter of culture, not just agricultural production. The outcry against the suburbanization of the Bluegrass countryside is partly a way to express anxiety about the social and environmental consequences of the landscape change as communities continue to swallow up the country. Advocates of environmental preservation have argued that the conservation of prime agricultural land in fast-growing areas of the Bluegrass would ensure that urban and suburban people retain a sense of place. Preservation was also necessary to maintain the "ecological balance"— that is, to preserve the complex community of living things that sustain human society.

The aesthetic argument is more popular and influential. To many people, subdivisions are simply ugly. They are places with no character. Underlying the aesthetic argument is the desire of the people to enjoy open space and the beauty of nature. More than a century ago, the leaders of urban America had begun to worry that the growth of cities threatened to alienate people from nature, and many sought to preserve a bit of green in the metropolis by building landscaped parks. While in the 1990s the appreciation of nature was not new, it was certainly more widespread than ever before. For many people in the Bluegrass, it was increasingly important to have everyday opportunities to appreciate nature in the rolling, green, grassy landscape of the region. By working both independently and as symbiotic partners with local government, citizens' groups have played a critical role in bringing public and government attention to the preservation needs of the region.

Citizens' Groups for the Protection, Preservation, and Enhancement of the Bluegrass

Citizens' groups have conducted educational programs, developed planning tools, and provided support to a variety of agencies, planning commissions, developers, businesses, and private citizens. Not only have these environmental groups in the Bluegrass region achieved results through their own programs, but they have also been instrumental in the establishment of local, state, and federal government programs for farmland and historic preservation. In this section, we discuss the activities of some of the major citizens' groups in the Bluegrass region.

Bluegrass Tomorrow

Perhaps the largest citizens' group is Bluegrass Tomorrow,[1] a regional planning coalition of business, farming, development, and preservation interests dedicated to promoting coordinated growth and preservation planning for the seven-county Central Bluegrass region of Kentucky. Its goal is to maintain the high quality of life and economic vitality that distinguishes the Bluegrass region. Its vision is that central Kentucky towns remain separate and distinct, people friendly, and full of architectural character; that the best farmlands remain secure and productive; that the beauty of the landscape remains evident from the roadways; that governments make informed, responsible decisions about growth and change; and that citizens are sensitive to the unique, fragile environment. In other words, it is working toward a strong, sustainable economic future. Bluegrass Tomorrow provides planning support and educational resources to planning commissions, private developers, civic groups, governments, individuals, utilities, and businesses in the region. Educational programs—through conferences, workshops, lectures, and roundtables—look at issues and trends facing the region, the range of local and regional solutions to problems, other regional, state, and national efforts, and unique research into areas of special concern.

Bluegrass Tomorrow provides advice and counsel to local groups on neighborhood planning; scenic, efficient, and safe road corridor planning; downtown-redevelopment strategies and design planning; agricultural-preservation strategies; transportation planning; commercial-development strategies; and green space planning. It publishes its annual *Vision Report* examining the year's activities.[2]

The Bluegrass Conservancy

The Bluegrass Conservancy began in 1995 as an outgrowth of Bluegrass Tomorrow.[3] The conservancy was a strong advocate for the creation of the Purchase of Development Rights (PDR) program in Fayette County (discussed below). It is a private, nonprofit regional land trust committed to the conservation and preservation of the unique rural and cultural resources of the Bluegrass region. Its mission is to promote the conservation of Bluegrass farmland. The primary tool of the conservancy is the conservation easement.

In addition to serving as a grantee for conservation easements and overseeing maintenance of the agreements, the conservancy conducts a wide variety of educational activities with schools, community groups, and others to promote the farmland-preservation effort. The conservancy provides, on request,

planning review of zone changes that would adversely affect farmland-preservation efforts.

The conservancy publishes a thirteen-page brochure providing images of the region, a discussion of the perils to the region's character posed by development, and an explanation of the conservation-easement-preservation program. Other publications include the *Bluegrass Conservancy Guide to Donating a Conservation Easement*, its *Fact Sheet on Conservation Easements*, its *Fact Sheet on Land Trusts, 11 Easy Things You Can Do to Save Farmland*, and its *Fact Sheet on Fayette County Purchase of Development Rights Program*. A quarterly newsletter—the *Bluegrass Conservancy News*—keeps members and friends up-to-date on the organization's activities.[4]

The Bluegrass Conservancy anticipates an annual budget of $250,000 administered by three full- and two part-time employees. It is funded and maintained by donations from public foundations and grants, private citizens, and corporate groups.

The Rural Land Management Board

The Rural Land Management Board administers the PDR program in Fayette County. It is a nonprofit, nonstock corporation with thirteen rotating members representing farm, conservation, development, business, and tourism organizations.[5] The board's duties are to solicit, acquire, and hold conservation easements, prioritize acquisitions, expend funds for acquisitions, disseminate public information and conduct forums, solicit contributions, establish administrative procedures, and release/terminate (perpetuity agreements). The board's goal is to provide funding of $140 million over the next twenty years for the purchase of development rights. Funding for the PDR program comes from three sources. The Lexington-Fayette urban county government provides about $2 million per year through general funds and bond issues. The program received $15 million in state match-

Dan Carey and Pradyumna P. Karan

ing funds and $10 million in matching funds from Kentucky's Purchase of Agricultural Conservation Easement (PACE) program. The federal government also provides matching funds through the federal Farmland and Ranch Protection Program.

The Blue Grass Trust for Historic Preservation

The Blue Grass Trust for Historic Preservation is a nonprofit advocate for historic preservation that strives to protect, revitalize, and promote the special historic places in Central Kentucky, thus enhancing the quality of life for future generations.[6] The trust is guided by three tenets: education, service, and advocacy. The trust believes that historic preservation is a partnership that cannot be accomplished without the support of the community. To that end, it strives to provide help with and education about preservation issues to concerned citizens through advocacy and other efforts. The trust offers technical assistance on a variety of preservation issues, through restoration seminars, workshops, and working with individual requests for information.

One of the ways in which the trust recognizes and promotes preservation is with its plaque program. For many individuals, the plaques that adorn houses all over the Bluegrass are the most identifiable aspect of the Blue Grass Trust. The program was created in the early 1970s to recognize buildings of architectural and historical significance by marking them with a recognizable symbol. The plaque program serves Fayette, Bourbon, Clark, Jessamine, Madison, Scott, and Woodford counties, and, over the years, hundreds of houses have been distinguished by this familiar symbol.

The Kentucky Chapter of the Nature Conservancy

The Kentucky chapter of the Nature Conservancy plays an active role in the preservation of valuable natural areas in the Central Bluegrass.[7] Its mission is to preserve the plants, animals, and natural communities that represent the diversity of life in the region by protecting the lands and waters they need to survive. Its preservation strategy is based on a science-based planning process called "conservation by design" that helps identify the highest-priority places—landscapes and seascapes that, if conserved, promise to ensure biodiversity over the long term.

The Nature Conservancy pursues non-confrontational, pragmatic, market-based solutions to conservation challenges. This makes it essential that it work collaboratively with partners—communities, businesses, government agencies, multilateral institutions, individuals, and other nonprofit organizations. The Kentucky chapter utilizes the Conservation Buyer Program—working with private purchasers to protect land that buffers biologically sensitive areas. The conservancy purchases the property and then resells it with conservation-easement restrictions. These properties typically do not harbor rare or endangered species but do offer a shield from such things as development, damaging agricultural practices, and pollution.

In the Central Bluegrass region, the conservancy is working cooperatively with state and nonprofit organizations for the preservation of the Kentucky River Palisades ecosystem, encompassing about one hundred river miles. The area includes deep gorges, wet-weather springs, caves, limestone outcrops, and variations in slope exposure that provide extremely diverse plant communities. This area harbors four species of endangered bats as well as several rare and endangered plant species. It provides a major migration route for birds, a travel corridor for wildlife, and foraging areas for bats. Limestone cliffs and wooded uplands along the river harbor the only natural vegetation remaining in the Inner Bluegrass region.

Town Branch Trail

Town Branch Trail is a nonprofit organization that began in 2000 and is dedicated to the revitalization of the Town Branch Creek in Lexington.[8] The vision for the eight-mile Town Branch Trail includes revitalizing older neighborhoods and commercial areas along the proposed trajectory of the trail and encouraging the adaptive reuse of older buildings and a mix of development types, including office, retail, live-work, residential, and neighborhood support services. Residents of these areas could use the trail to commute into the center of downtown and also have access to daily needs within easy walking distance. The Town Branch Trail will also connect to other trails and bikeways and to major roadways. The location and linkages of the trail mean that it will be well used, creating a safe, pleasant, and interesting facility. Workers from various downtown businesses will mingle with conventiongoers and tourists as schoolchildren ride their bikes or elderly neighbors go for a gentle stroll. Use will occur during normal business hours, after work, and on weekends, bringing energy to many areas near downtown and beyond.

Four committees administer the programs of the organization: Trail Planning, Public Education and Marketing, Environmental Education, and Mapping and Multimedia. *Town Branch Trail,* the semi-annual newsletter, keeps members and friends up-to-date on activities. The group has several publications: the *Town Branch Trail Guide* brochure, a narrative discussing the benefits of Town Branch Trail and giving descriptions of significant landmarks along the trail; *The Disappearance of Town Branch,* a history of the area; and the *Town Branch Trail Environmental Education Sign Project,* a sixteen-page brochure that discusses the biology and hydrology of and the human impacts on Town Branch.[9]

The Kentucky Rails to Trails Council

The Kentucky Rails to Trails Council (KRTC) was formed in 1994 and incorporated in 1995.[10] It is fully staffed by volunteers. The KRTC's mission is to enhance the quality of life in communities by developing a Kentucky rail-trail program. The KRTC works with local organizations to develop greenways and trails and serves as a source of information on project funding, design, and management. It also seeks to increase public awareness of the benefits of rail-trails and works closely with the national Rails to Trails Conservancy.

The KRTC has two to three hundred dues-paying members and a mailing list of sixteen hundred. On an annual budget of under $10,000, it provides information and education through speakers, presentations, booths, e-mails, newsletters, and its Web site. In 1999, the group served as a technical adviser to the Kentucky legislature's Special Task Force on Feasibility of Rail-Trails, leading to the passage of a bill establishing the state Office of Rail Trail Programs.

The KRTC functions as a land trust; it can purchase and/or hold "real property of strategic value in developing trails for use by the public to preserve its availability to public benefit."[11] Funding for preservation and trail-development projects comes from corporate and private donations, local governments, and the federal Intermodal Surface Transportation Efficiency Act.

The Bluegrass Rails to Trails Foundation

The C&O Railroad completed 109 miles of rail line connecting Lexington and Coalton, Kentucky, in 1881. It was not until 1985 that the C&O Railroad abandoned this line. In 1996, the KRTC put together a strategy for the development of the Lexington–Big Sandy Rail Trail.

In 1997, the Lexington Fayette urban county government purchased for $30,000

Dan Carey and Pradyumna P. Karan

the right-of-way to nearly one mile of privately owned rail bed between Man O'War Boulevard and I-75 in Fayette County. In the same year, the Bluegrass Rails to Trails Foundation (BRTF) was organized.[12] The BRTF was incorporated in 1998. It has been working toward trail development since that time. Meetings of the group are currently held quarterly.

The BRTF is a nonprofit foundation organized to serve as a catalyst to transform abandoned rail corridors in central Kentucky counties into useful recreational trails, all closed to motorized vehicles. Through its efforts, it hopes to secure support from city and county officials, local organizations, community leaders, members of the communities, landowners, and the state to build an approximately thirty-nine-mile continuous trail through Fayette, Clark, and Montgomery counties on abandoned rail corridors.

Local Government Efforts to Preserve Farmland

To address the problem of loss of farmland and open space and issues associated with population and economic growth, Fayette County began to take steps to ensure the upkeep of both natural and human environments. Citizens' groups were instrumental in working with local governments to establish and implement local, state, and federal preservation programs in the Bluegrass. The primary tools for preservation include the urban-service-area concept, the rural-land-management plan, the purchase of development rights through conservation easements, and donated easements. Conservation easements are purchased by government agencies, such as Lexington-Fayette County's PDR program. Donated easements, which provide substantial state and federal tax benefits, are administered by several citizens' groups, such as the Bluegrass Conservancy. Government

agencies also administer donated conservation-easement programs.

In 1958, Lexington became the first U.S. city to establish an urban service boundary, a concept that has been the cornerstone of community planning for the last fifty years. Outside the urban service area, intensive development is not allowed. Establishing such a boundary has helped manage the city's growth and contain urban sprawl. However, the urban service boundary has been continuously expanded to include land for development. In 1996, the urban-service-area boundary was significantly expanded, adding fifty-four hundred acres to meet the growth pressure.

In addition, the board of health's ten-acre rule, which regulates septic tanks, was included in the zoning ordinance and subdivision regulations of the county in the late 1960s. Prior to this regulation, subdivisions of one-half- to one-acre lot sizes were sprouting up across the rural area. By the 1990s, however, it became increasingly evident that the ten-acre rule was also resulting in a serious loss of land available for agricultural use (see photograph 9.4). During the period 1990–1998, forty-seven hundred acres of rural land were consumed by 429 residential lots. At the average density within the urban service area, that acreage would typically have resulted in 11,315 residential dwellings.

The Rural Land Management Plan

With the support, encouragement, and assistance of the citizens' preservation groups, the Rural Service Area Land Management Plan was adopted in 1999 to protect and preserve the rural resources of Fayette County. The plan included a land capability analysis for the rural service area of the county. The analysis examined land capability, soils, traffic, scenic areas, environmental areas, transportation, agricultural patterns, historic areas, rural settlements, aquifer protection, and future sewerability. Sewerability, the ability to

Photograph 9.4. Scattered rural estate development in southern Fayette County outside the urban service area, 2004. (Courtesy U.S. Department of Agriculture, Farm Services Administration, National Imagery Program.)

provide public sewers to an area, is a strong determinant in development and was examined in the plan. Sewerability categories were as follows: does not require major construction; requires moderate construction; and requires significant investment in new facilities. The first two categories represented about 680 acres, or 0.5 percent of the rural service area.

Land-use categories were determined for the entire rural service area: core agricultural and rural land, natural areas, rural activity centers, buffer areas, existing rural residential areas, and rural settlements. Special plan elements included environmentally sensitive areas, rural roadways, historic areas and rural settlements, special natural-protection areas, rural greenways, nonagricultural zoning in the rural service area, and potential development considerations. In core agricultural and rural land and natural areas, a forty-acre minimum lot size is required for residential development; existing smaller tracts were permitted to remain (see figure 9.5). These areas were designated for agricultural and supporting uses only. All lands in these categories would be eligible for the PDR program. The public acquisition of 1,000–2,000 acres of natural land for a major preserve was recommended in the plan. These two categories constitute 122,800 acres, or 95.5 percent, of the 128,300-acre rural service area; rural activity centers, buffer areas, existing rural residential, and rural settlements make up the remaining 5,500 acres (4.5 percent).

The PDR Program

Fayette County's PDR program is the first agricultural conservation-easement program by a local government in the state of Kentucky. In 2000, an ordinance was passed creating the PDR program and the Rural Land Management Board. In accordance with the urban

Dan Carey and Pradyumna P. Karan

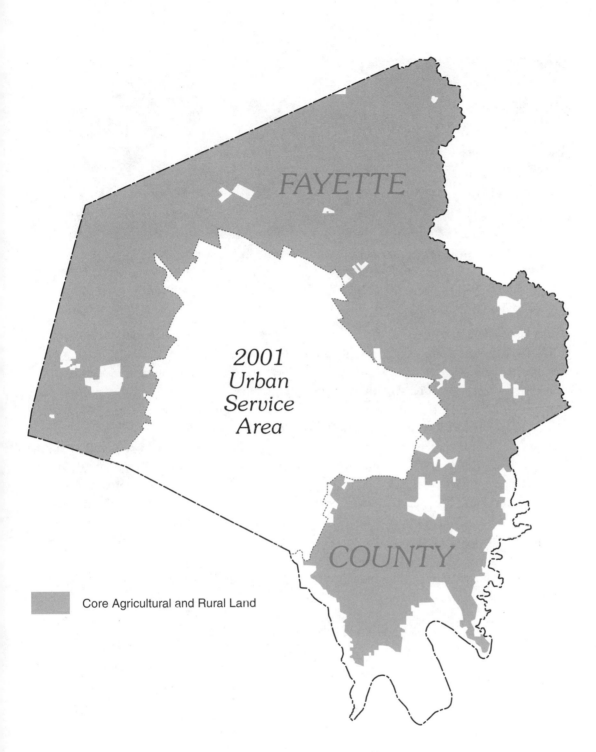

FAYETTE

2001
Urban
Service
Area

COUNTY

Core Agricultural and Rural Land

Figure 9.5. Fayette County, Kentucky, core agricultural and rural land, 2005.

Photograph 9.5. Preservation of the character of the land was a high priority in the design of the new four-lane Paris Pike. This required spacious green medians and, in some cases, the relocation and reconstruction of historic stone fences. (Courtesy Dan Carey.)

county government charter and the Rural Land Management Plan, the PDR program supports the integrity of the full urban services district. Concentrating growth in an urban center reduces the cost of public services to local government. The goals of the PDR program are to purchase conservation easements to protect fifty thousand acres of rural Fayette County by the year 2020, to protect the agricultural and horse economies by conserving large areas of farmland, to conserve natural, scenic, open space and historic and agricultural resources, and to protect the tourism economy by preserving the unique char-

acter and sense of place that attracts visitors from all over the world. The policy emphasis for natural areas is on the preservation and enhancement of the land in its natural state with a minimum of intrusions. The PDR effort is supported by transportation planning that makes a high priority of maintaining the character of the land (see photograph 9.5).

The value of development rights is based on the future development value of the land, over and above its intrinsic and agricultural value (see figure 9.6). The landowner sells the development rights in perpetuity. The major principles of the PDR program are that

What is the PDR, and how do you determine its value?

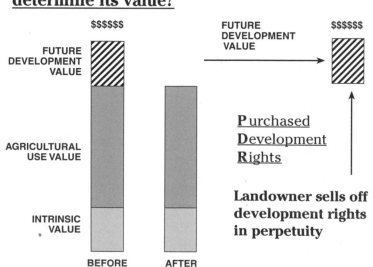

Figure 9.6. Determining the value of development rights.

Dan Carey and Pradyumna P. Karan

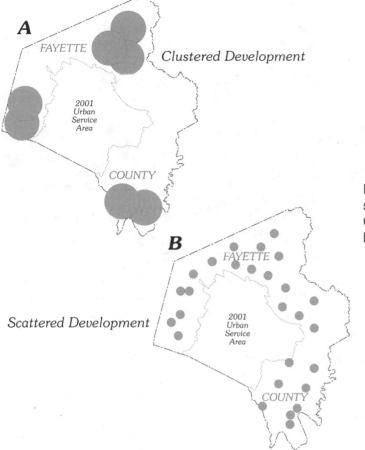

A

FAYETTE

Clustered Development

2001
Urban
Service
Area

COUNTY

B

Scattered Development

FAYETTE

2001
Urban
Service
Area

COUNTY

Figure 9.7. Clustered (*a*) and scattered (*b*) development. Conservation is the goal behind clustered development.

it is applied on a voluntary basis only; that all rural landowners with twenty or more acres, and subject to the forty-acre rule, are eligible; that priority order of acquisition is based on a fair and objective ranking system; that preservation is in perpetuity (with rare exceptions permitted); and that the program is not antigrowth. The PDR program encourages planned growth within and along the urban service area by establishing critical land masses in the rural service area. The assessment ranking is lower for properties adjacent to the urban service area. The program is pro-infill and is proagriculture and promotes rural preservation.

The enabling ordinance for the PDR program establishes definitions, PDR board memberships and duties, the Land Evaluation and Site Assessment (LESA) point system, negotiation and acquisition procedures, donation of easement procedures, release/transfer provisions, and coordination with the comprehensive plan. LESA priorities are based on the following factors: agricultural (farm size, road frontage, contiguous applications, soils, farm activities, and agricultural improvements), environmental/other (environmentally sensitive areas, greenway potential, natural-protection areas, links to parks, historic/cultural resources, and scenic resources, and reconsolidation of subdivided farms), and future urban factors lowering priority (proximity to an urban service area, high probability of sewerability, proximity to federal highways and interstate interchanges, with exceptions for "community icons" and other major protection priorities). The goal of the LESA process is to produce clustered development rather than scattered development (see figure 9.7).

Photograph 9.6. Farmland east of Lexington protected through the PDR program, 2005, Lexington–Fayette County, Kentucky, Rural Service Area Land Use Management Plan and Purchase of Development Rights. (Courtesy Rural Land Management Board, 2005.)

In November 2007, there were 181 farms totaling more than 20,771 acres preserved in perpetuity through the PDR program, indicating that the goal of 50,000 protected acres is attainable (see photograph 9.6 and figure 9.8). The categories of protected farms include 70 general agriculture, 96 equine, and 15 other (sod, wooded, etc.). The total includes 29 farms with donated conservation easements on 1,480 acres. The PDR program has had a positive regional impact on the agriculture, tourism, and horse industries and has forwarded land planning and preservation in the Bluegrass region by providing a successful model to other counties developing farmland-preservation programs.

Figure 9.8. Fayette County, Kentucky, PDR-protected farms, 2005.

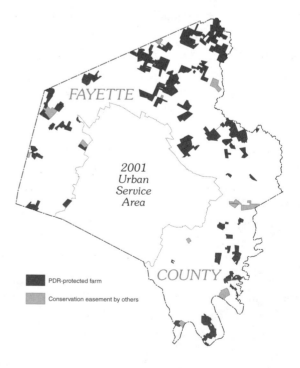

Dan Carey and Pradyumna P. Karan

Conservation Easements

A conservation easement is a voluntary legal agreement between a landowner and a conservation organization such as the Bluegrass Conservancy that permanently protects productive agricultural land, historic sites, scenic views, and natural resources. Conservation easements are flexible documents tailored to each property and the needs of individual landowners. They can cover an entire parcel or only portions of a property. Conservation easements are granted by landowners ("grantors"), who authorize a qualified conservation organization ("grantee") to monitor and enforce the restrictions set forth in the agreement.

After granting a conservation easement, landowners retain title to their property and can still use it for farming, restrict public access, use the land as collateral for a loan, or sell or transfer title to the property subject to the terms of the agreement. Future owners of the property are subject to the easement agreement. Easements are especially suitable for the Bluegrass region, where it is desirable to protect productive farmland essential to the equine industry and other general agricultural operations.

The value of a conservation easement is determined by comparing the value of the property before and after the easement restrictions are placed on the property. Landowners must substantiate the asserted value by providing an independent appraisal that meets federal standards. In general, more restrictive agreements and intense development pressure result in higher easement values. In order to qualify for the favorable tax benefits, donated conservation easements must be perpetual in duration.

Conservation easements permanently protect productive farmland while keeping the land in private ownership and on local tax rolls. Conservation easements are flexible and can be tailored to meet the needs of individual landowners and unique properties. Qualified conservation easements provide landowners with several potential tax benefits, including income, estate, and property tax reductions. By reducing estate taxes, conservation easements help landowners transfer their operations to the next generation.

Currently, the Bluegrass Conservancy holds twenty-four donated conservation easements protecting 3,980 acres in five counties.

The state of Kentucky also funds and administers a conservation-easement program. The Kentucky General Assembly in 1994 established the PACE Corporation and authorized the state to purchase agricultural conservation easements in order to ensure that lands currently in agricultural use will continue to remain available for agriculture and not be converted to other uses.[13]

All landowners who hold title to farmland that is being used, or is available for use, for agricultural production are eligible. Easement value is determined as follows: first, the fair market value of the land is determined according to its development potential; second, the value of the land is estimated if its use is restricted to agriculture. The easement value for any given acreage will fall somewhere between the two. Landowners can use these two figures to help them determine an asking price. The landowner and the PACE board will then negotiate the final conservation-easement value. All transactions are subject to negotiations and restrictions established by the board.

A two-hundred-acre farm appraised at a fair market development value of $2,500 per acre and a farmland-restricted value of $1,200 per acre would have a potential conservation-easement value of $1,300 per acre, or $260,000.

The features and benefits of the PACE program are that participation is strictly voluntary, that it enables landowners to realize a portion of the equity in their land without having to sell it, that proceeds from the sale of an easement can be used for any purpose, that it can contribute significantly to the diversification and expansion of the farm economy, that it can make land and farming more affordable to young farmers by effectively reducing the value of the land, and that it can aid in the transition from one generation of farmers to the next.

Statewide, the PACE Corporation has purchased agricultural conservation easements on 88 farms totaling 21,000 acres for $18 million (as of 2007). The easement costs have averaged $850 per acre. The farm size has averaged 240 acres. In addition, 34 easements on 4,600 acres have been donated to the program, bringing the total inventory to 122 farms containing 25,600 acres.

Since the program's inception, the department has received 816 applications from seventy-five counties statewide totaling more than 160,000 acres. A total of 667 applications are currently pending, for a total of more than 129,000 acres with an estimated easement value of more than $100 million.

The Farm and Ranch Lands Protection Program (FRPP) is a voluntary program that helps farmers and ranchers keep their land in agriculture.[14] The program provides matching funds to state, tribal, or local governments and nongovernment organizations with existing farm- and ranchland-protection programs to purchase conservation easements. The FRPP was reauthorized in the 2002 Farm Bill (the Farm Security and Rural Investment Act). The Natural Resources Conservation Service (NRCS) of the U.S. Department of Agriculture manages the program.

The USDA works through state, tribal, and local governments and nongovernment organizations to conduct the FRPP. These entities acquire conservation easements from landowners. Participating landowners agree not to convert their land to nonagricultural uses and to develop and implement a conservation plan for any highly erodible land. All highly erodible land enrolled must have a conservation plan based on the standards in the NRCS *Field Office Technical Guide* and approved by the local conservation district. Landowners retain the right to use the property for agriculture. To participate, a landowner submits an application to an entity—a state, tribal, or local government or a nongovernment organization—that has an existing farm- or ranchland-protection program. The NRCS state conservationist, with advice from the State Technical Committee, awards funds to qualified entities to purchase perpetual conservation easements.

To qualify for the FRPP, the land offered must be part or all of a farm or ranch and must contain prime, unique, or other productive soil or historical or archaeological resources; be included in a pending offer from a state, tribal, or local government or nongovernment organization's farmland-protection program; be privately owned; have any highly erodible land covered by a conservation plan; be large enough to sustain agricultural production; be accessible to markets for what the land produces; be surrounded by parcels of land that can support long-term agricultural production; and be owned by an individual or entity that does not exceed the adjusted gross income (AGI) limitation. The AGI provision of the 2002 Farm Bill affects eligibility for the FRPP and several other 2002 Farm Bill programs. Individuals or entities that have an average AGI exceeding $2.5 million for the three tax years immediately preceding the year the contract is approved are not eligible to receive program benefits or payments. However, an exemption is provided in cases where 75 percent of the AGI is derived from farming, ranching, or forestry operations. If the land cannot be converted to nonagricultural uses because of existing deed restrictions or other legal constraints, it is ineligible for the FRPP.

The FRPP share of the easement cost must not exceed 50 percent of the appraised fair market value of the conservation easement. As part of its share of the cost of purchasing a conservation easement, a state, tribal, or local government or nongovernment organization can include a charitable donation by the landowner of up to 25 percent of the appraised fair

Dan Carey and Pradyumna P. Karan

market value of the conservation easement. As a minimum, a cooperating entity must provide, in cash, 25 percent of the appraised fair market value or 50 percent of the purchase price of the conservation easement.

Through 2007, more than 533,000 acres have been protected in forty-nine states. For fiscal year 2007, $73 million purchased easements on fifty-four thousand acres in the United States. Through fiscal year 2007, $20 million in FRPP funds has been awarded to purchase easements on nearly twenty-six thousand acres of valuable agricultural land in Kentucky. Owing to increasing development pressure throughout the state, requests for FRPP funds have increased dramatically in recent years, far outpacing the program's funding capacity. In fiscal year 2003, partners in Kentucky requested $11.5 million in FRPP funds. That request represented 17.6 percent of the total amount of FRPP funds available nationwide that year. In recent years, allocations have been reduced. Kentucky was allocated $1,800,000 to purchase conservation easements in fiscal year 2007.

Donated Easements

The PDR program also solicits donated easements on the basis of the criteria of location, desirability for recreation, importance for the maintenance of ecosystems, and whether the land being donated represents open space for the public benefit or historically important land. The donation of a conservation easement is treated as a tax-deductible charitable contribution if it meets the "qualified conservation contribution" requirements of the Internal Revenue Service Code. Preservation easements are granted in perpetuity. Landowners in the PACE program can also donate a conservation easement, which is a tax-deductible, charitable gift, provided that the easement is perpetual and donated exclusively for conservation purposes. Conservation purposes include preserving open space

and farmland. If the conservation easement is calculated as $260,000 for a two-hundred-acre farm, the landowner is eligible to deduct an amount equal to 30 percent of his adjusted gross income each year for a total of six years or until the value of the gift is matched. For example, if the landowner has an annual federal adjusted gross income of $100,000, he can deduct $30,000 a year for the next six years, or $180,000 in all, if his income does not change. State income tax deductions for charitable gifts follow the same guidelines as do those for federal tax deductions. Thus, the potential state tax deduction is an additional $180,000.

Results of the Citizens' Groups' Efforts

Citizens' preservation groups have achieved results not only through their own programs. By working cooperatively with local officials and planners, they have also been instrumental in the establishment, implementation, and maintenance of local, state, and federal government programs for farmland and historic preservation.

Unfortunately, preservation successes have not been consistent across the Central Bluegrass region. The most intense development pressure in the region has been in the central, most-populous county of Fayette. The response of Fayette has been the development of programs and planning tools that could benefit the entire region. Adjacent Woodford County has adopted many of these tools. The smaller counties of Bourbon, Clark, Jessamine, Madison, and Scott, which could benefit from the experience of Fayette County, have been slower to respond, perhaps because of fewer resources. Fayette County, containing about 15 percent of the area of the region, has put more than twenty thousand acres of farmland under preservation. The remaining counties, constituting 85 percent of the region, have less than four thousand acres under preservation. If the surrounding counties are to preserve their cultural, agricultural, and historic resources and values, it will be imperative for citizens' groups to become more active. Citizen action will also be needed to maintain pressure on elected officials. Despite the success of the land-conservation programs and the increasing requests for participation, federal matching funding for the programs has decreased in recent years. Citizens' groups can play a large role in raising awareness of the importance of land-conservation programs.

For many Bluegrass residents, the integration of the natural and residential environments provides a key feature or amenity that makes Lexington and its surrounding towns a unique and high-quality place to live. Although some residents argue that the preservation of the natural environment threatens the economic vitality of the region by limiting the amount of residential and commercial growth, others hold that the natural environment and open spaces are economic resources in and of themselves. Consequently, as some residents argue, the denigration of natural spaces through residential and commercial development actually becomes an economic liability to the Bluegrass.

Regardless of one's particular stance toward farmland and open space, the Bluegrass region continues to grow. With its increased population, Lexington has become part of a national trend in which automobile ownership is growing at a faster pace than is the population. The extra vehicle miles threaten Lexington's ability to deal with traffic congestion and air-quality problems. Differing opinions concerning the way in which future development is handled lie at the center of local environmental groups' major concern.

In 2006, an additional seven thousand acres of land were scheduled to be added to the urban service area. In early 2007, the Lexington Urban County Planning Commission rejected the expansion and voted against any expansion of the urban-service-area bound-ary for the next five years. The concept of an urban service area seemed finally to have asserted its effectiveness to limit unplanned growth and preserve rural areas, largely due to the efforts of the local environmental and conservation groups.

Notes

1. See http://www.bluegrasstomorrow.org.
2. The *Vision Report* is available through Bluegrass Tomorrow's Web site (see n. 1 above).
3. See http://www.bluegrassconservancy.org.
4. All these publications are available through the conservancy's Web site (see n. 3 above).
5. See http://www.lfucg.com/pdr/index.aspx.
6. See http://www.bluegrasstrust.org.
7. See http://www.nature.org/wherewework/northamerica/states/kentucky.
8. See http://www.townbranch.org.
9. Town Branch Trail's publications are available through its Web site (see n. 8 above).
10. See http://www.kyrailtrail.org.
11. KRTC, "Amended and Restated Bylaws," art. 7, sec. 2, available at http://www.kyrailtrail.org/faqs.html.
12. See http://www.kyrailtrail.org/groups.
13. See http://www.kyagr.com/marketing/farmland/index.htm.
14. See http://www.nrcs.usda.gov/programs/frpp.

References

American Farmland Trust (AFT). 1999. "American Farmland Trust Cost of Community Services Study for Fayette County." Commissioned study. Lexington, KY: Bluegrass Conservancy.

Lexington-Fayette Urban County Government. 2005. *Historical Aerial Photos of the Lexington–Fayette County Area.* Lexington, KY: Division of Planning.

U.S. Department of Agriculture. 2006. "Historic County Agriculture Statistics: 1909–2004." Kentucky Agricultural Statistics Service, Lee Brown, www.nass.usda.gov/ky.

World Monuments Fund. 2005. "Bluegrass Cultural Landscape of Kentucky: Central Kentucky, United States." *World Monuments Watch, 100 Most Endangered Sites in 2006.* http://wmf.org/USA_Ken_bluegrass_2007.html.

Dan Carey and Pradyumna P. Karan

Chapter 10

Farmers' Efforts toward an Environmentally Friendly Society in Ogata, Japan

Shinji Kawai, Satoru Sato, and Yoshimitsu Taniguchi

Organic or semiorganic rice[1] made up barely 20 percent of Japan's market in 2001 (MAFF 2002), and, presently, only a handful of local governments plan to introduce cost-sharing incentives for environmentally friendly agriculture. However, in Ogata Village in the northern prefecture of Akita (see figures 10.1–10.2), 80 percent of the rice is organic or semiorganic, according to a 1998 Akita Prefectural Agricultural College [Akita kenritsu nogyo tanki daigaku] survey (see table 10.1). This outstanding figure[2] was realized through Ogata farmers' unremitting struggles to overcome the political, economic, and social pressures that mounted over thirty years of settlement on Ogata's newly reclaimed land.

This chapter will look at how various independent grassroots movements in Ogata played a role in creating an environmentally friendly farming society and how they together formed Ogata Environment Creation 21 (OEC21), which promises a new era of environment-centered thinking supported by agricultural industries.

Political and Economic Pressures

Local and global agricultural policies have greatly affected Japan's rice farmers, both in recent years and at Ogata's founding in the 1960s. In 1964, a seventeen-thousand-hectare, ¥85.2 billion reclamation project turned Japan's second largest lake, Hachiro, into Ogata Village (Akita Daigaku Hachiro-gata Kenkyu Iinkai 1968) (see figure 10.3). Lake Hachiro used to support around twenty thousand residents in fishing-related industries[3] and provided half of Akita Prefecture's marine products (Hachiro-gata Kantaku Jimusho 1969). It is now a distinctive geographic feature, a "closed-water system" that includes the remnants of Lake Hachiro, irrigation channels, and the river systems of the surrounding area.

Ogata's reclaimed land was intended to be Japan's food basket. When the project started, Japanese rice production remained below national needs. National agricultural policies of the time sought to make Ogata a model for

Table 10.1. Environmentally Friendly Agriculture Practices in Ogata Village (1998)		
Farming Method	**Area (ha)**	**% in Ogata**
Reduced spray and chemical fertilizers	5,110.0	74.1
No spray and no chemical fertilizers	520.8	7.5
SBNA (single basal nursery fertilizer application)	1,214.7	17.6
Side-dressing	1,052.3	15.3
No-puddling	132.7	1.9
No-till	36.8	.5
Direct seeding	7.2	.1
Use of ducks for weed and pest control	71.5	1.0

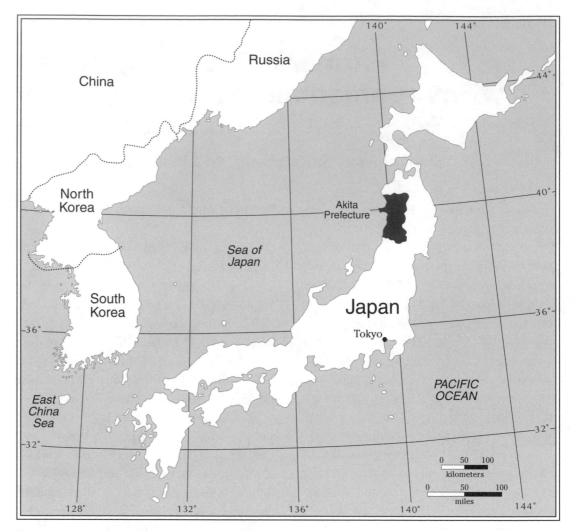

Figure 10.1. Akita Prefecture, Japan.

future Japanese agriculture (Tashiro and Suzuki 1982). Large rice fields, mechanized operations, and cooperative farm management were the model's key characteristics.[4] Starting in 1967, and continuing for twelve years, 589 selected farm families migrated from thirty-six prefectures, including the Tokyo metropolitan area (JA Ogata-mura 2002).

As Japanese rice production steadily increased toward the end of the 1960s (the national rice surplus peaked in 1970 at 7.2 million tons [Rice Databank Co. 2003]), the government quickly reversed its policy and mandated rice-field acreage reductions.[5] In 1973, the Ogata Reclamation Office [Hachi-ro-gata kantaku jimusho] decided to grant

farmers fifteen hectares on the condition that half would be planted with rice and the other half with other crops (Tozawa 1993). Confused Ogata farmers protested, motivated not only by their political beliefs but also by their farm-related debts. The first four Ogata migrant groups typically owed initial twenty-five-year loans of ¥60 million (Shimizu 1972). Ogata farmers' protests against government intervention drew attention from other Japanese farmers as well as ordinary citizens. In the late 1970s, the national government began to take strict measures, including restrictive rice-production policies and penalties against uncooperative farmers. Ogata farmers split on how to deal with the problem. The

Shinji Kawai, Satoru Sato, and Yoshimitsu Taniguchi

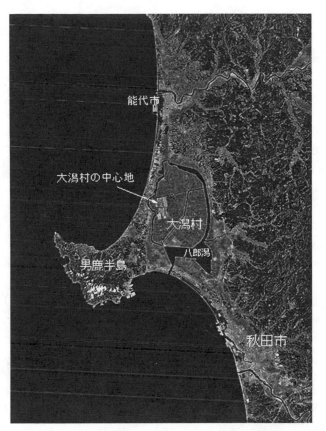

Figure 10.2. Ogata Village, Akita Prefecture (satellite photograph).

Figure 10.3. Birth of Ogata Village (satellite photographs, 1956 and 1976).

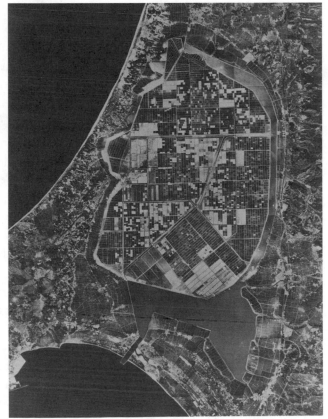

pro-government group sold its rice through the Ogata Country Elevator Public Corporation [Ogata-mura kantori erebeta kosha] at the government-set price, while the anti-government group marketed its rice through private (and, at the time, illegal) channels. The antigovernment group developed its own marketing channels geared toward rice consumers' demands, but the pro-government group struggled with obligatory non-rice-crop production yet was still not able to receive the subsidies granted to other Japanese farming communities. Strong tensions existed between the two groups well into the late 1990s.

The Staple Food Control Act,[6] which had supported rice as a national staple, was modified through globalization processes and rice overproduction. Starting in 1969, the government reduced its commitment to buy rice[7] and began to change rice-marketing systems. The new systems diversified rice values and allowed designated dealers, the agricultural co-ops, to respond to consumer demands for higher-quality rice. Rice prices rose sharply from the early 1970s to the mid-1980s owing to the 1974 oil shock and the associated hyperinflation.[8] The gap between the international and the domestic rice prices grew further with the appreciation of the Japanese yen against the American dollar.

Rice prices stayed relatively high into the mid-1990s but then began to drop with the abolition of the Staple Food Control Act and the 1995 introduction of the New Food Act. The New Food Act basically allowed rice to be marketed just like any other commodity. Japan's 1995 entrance into the World Trade Organization also put direct and indirect pressures on Japan's rice-marketing policies. Falling rice prices particularly affected large-scale farmers like those in Ogata. Since 1995, rice farmers' estimated yearly incomes have begun to drop, in some cases by as much as ¥1 million.[9]

Environmental and Social Pressures

Environmental pollution intensified and began to be widely publicized from the 1960s as Japan grew increasingly industrialized.[10] In Ogata, where the closed-water system encouraged water pollution, environmental hazards were first noticed in the 1970s by a few individual farmers and housewives. In 1974, a grocery store run by JA Ogata Village [JA Ogata-mura], the village agricultural co-op, removed all the synthesized laundry detergents from their shelves. In 1983, women from the co-op also had the herbicide CNP (chlornitrofen), which contains dioxin, banned.

The story of one Ogata farmer illustrates how organic principles first took hold in the area. Mr. Gotsu encountered a shocking scene in June 1974. Numerous young black-browed reed warbler birds (*Acrocephalus bistrigiceps*) lay dying in agony along the rice-field ridges. To control rice leaf beetle (*Oulema oryzae*) outbreaks, Mr. Gotsu had had an organic phosphate pesticide sprayed by helicopter over the entire forty-six hundred hectares of Ogata Village rice fields. After witnessing the birds' deaths, he refused to allow aerial cover-spray practices on his land and began to seek ways to reduce chemical sprays as well as chemical fertilizers. In 1982, Mr. and Mrs. Gotsu started to grow organic rice, hand-weeding their fields.

Cooperation between farmers, agricultural researchers, and industry became more relevant around 1990, with a number of technological breakthroughs. Although divisions between farmers persisted well into the late 1990s, the sharp drop in rice prices in the late 1990s helped farmers come together to seek solutions to the crisis. In December 1997 and January 1998, a significant panel discussion entitled "Inasaku nogyoshano iki-nokoru michi" [The road toward agricultural revival] was held. It was initiated by second-generation Ogata farmers at the suggestion

Shinji Kawai, Satoru Sato, and Yoshimitsu Taniguchi

of a young social scientist from the Akita Prefectural Agricultural College in Ogata. At two packed sessions, pro- and antigovernment groups held active discussions ("Tsuchi to ikiru" 2001). One of the guest speakers, Dr. Nakajima, recognized Ogata farmers' advanced environmental awareness. Forum organizers decided to investigate the current status of Ogata's environmental agriculture by sending a questionnaire to all Ogata farmers. This would be a turning point for the Ogata environmental movement, which until that point consisted solely of small and uncoordinated groups.

The Dynamics of Environmentally Conscious Farmers in Ogata

As we have seen, multiple pressures tugged Ogata farmers in many directions as they mobilized politically. Mostly they formed many small groups rather than one organized, large-scale movement. However, as the following sections show, the pathway to OEC21 began to draw them together. Numerous personal interviews, as well as archival research, tell the story.

Pro-government farmers had long struggled to find suitable crops for Ogata's environment and to develop niche market crops and innovative farming methods. Despite significant economic pressures, their efforts started to bear fruit in the 1990s. Mr. Yamazaki, in particular, struggled to reverse the pressures on the Ogata community. Mr. Yamazaki came to Ogata as a fourth-wave migrant from Hokkaido in 1970. His rice field happened to be adjacent to the experimental plot of the Akita Prefectural Agricultural Experiment Station [Akita kenritsu nogyo shikenjo]. One spring day in 1989, he saw a group of researchers planting soybeans with a newly developed no-till planter. His interest was sparked, and soon he applied the no-till principle to help eliminate puddling problems in his rice field.

Rice fields must be made level through puddling prior to planting, and Mr. Yamazaki disliked the strenuous, cold, and dirty spring work. He thought that no-till[11] was the answer and quickly developed a prototype no-till rice-planting machine (Yamazaki 2001). In 1991, ten Ogata farmers, including a soil scientist, formed the Ogata Low-Input Sustainable Agriculture [Ogata-mura teitonyu jizokagata-nogyo kenkyukai] study group, or O-LISA, whose aim was to make no-till rice farming feasible.

Fertilization demands initially hindered the time- and laborsaving benefits of no-till technology. Quick-release nitrogen fertilizers must be broadcasted after transplanting. Applying postplanting fertilizer is demanding work, yet it is mandatory to assure optimal vegetative growth before tillering. Mr. Yamazaki helped complete the invention of another important technology: a single basal nursery fertilizer application (SBNA). With the help of a coated fertilizer developed by a fertilizer manufacturer (Yamazaki 2001), he created an assembly line to mass-produce nursery beds in his shed. SBNA technology provides very efficient fertilizer application and also prevents nitrogen from leaching into the soil. This innovation quickly spread beyond Ogata and is now becoming an important technology for rice farmers in Akita and surrounding prefectures.[12]

Another challenge was to make no-till machines available commercially. In 1994, the Ogata group went to two different agricultural-machinery companies to ask them to produce a no-till rice planter. One company modified the idea and came up with a machine that can be used for no-till, no-puddling, and conventional rice fields. In 1997, O-LISA formed into a machinery cooperative and purchased two no-till machines to share. The soil scientist and others studied the technology more closely and found that, in terms of total nitrogen and phosphate, no-till can actually create cleaner outflow drainage water than the inflow irrigation water (Kaneta 2002). No-till in Ogata also promotes higher yields, better drainage, fewer suspended solids, and less methane formation and provides better soil structure for rotating crops with rice. Besides these more practical farming benefits, the population of the culturally beloved red dragonfly (*Sympetrum frequens*) is eight times larger in no-till fields than in conventional fields (Kaneta 2002).

No-till technology has not, however, spread beyond this group (Sato and Taniguchi 2002) as the initial investment is high and there are no immediate economic rewards, the rice produced selling for the same price as conventional rice. One of the disadvantages of no-till rice is that the expensive no-till machine can be used only in large fields. The O-LISA group hopes to further cut costs by eliminating the use of the now-mandatory herbicides. It is investigating the allelopathic use of the legume hairy vetch (*Vicia villosa*). Once the cropping patterns of hairy vetch and rice are sorted out, this will be another much-needed breakthrough for environmentally friendly agriculture.

In addition to the persistent farmers, housewives also contributed significantly to the environmentally friendly practices movements. The housewives' grassroots movement was critical as it helped educate the community, not only about farming, but also about how to live in a closed-water-system environment. The JA Ogata Village women's group started a grassroots movement in the early 1970s. Soon after the group learned how synthetic detergents could affect aquatic ecologies,[13] it began to encourage the community-wide use of environmentally safe soap and in 1974 persuaded the co-op supermarket to ban synthetic detergents. The co-op supermarket was the single largest outlet for groceries in the village. Furthermore, in 1982, another women's grassroots movement—an offshoot of the Organic Farming Study Group [Yuki-noho kenkyukai]—requested that Ogata Vil-

lage's council investigate pesticide residues in drinking water and ban herbicides containing dioxin (Gotsu 1991). A CNP ban was introduced in 1983. Ogata farmers are now well aware that Lake Hachiro drinking water is part of the same closed-water system used for agricultural purposes.

In 1990, another epoch-making movement began. Some thirty households of antigovernment growers acted together to raise public concerns about a golf course planned by the Ogata Village administration. The golf course was to be shared with eleven surrounding villages and towns. The core group against the project consisted of more than a dozen women, one of whom became the head organizer. Ogata farmers and consumer groups banded together and in 1990 hosted a national meeting protesting golf course developments (Hanatsuka 2003). Fifteen of the seventeen districts in the Ogata farming community spoke against a golf course near their residential area. After a year-long struggle, the village administration finally gave up on the idea. The opposition, which thought it would lose, was thrilled to find that ordinary people could reverse village government plans. From this victory would emerge numerous changes in civil movements.[14]

A group of activist housewives emerged from the anti–golf course movement. Nine women, empowered by their success at stopping the "unwanted" golf course, sought to do something useful for their community. One day, one member received a packet of handmade soap from a soap-making group. This triggered the women to start making environmentally safe soap with used cooking oils from the village schools and an agricultural co-op restaurant. They started to sell their soap, which they called Taro no yume [A dream of Taro],[15] in 1992 and the same year developed a plant to produce powdered laundry detergent in a co-op storage building. The powdered detergent was approved by the Japanese Industrial Standards Commit-

Shinji Kawai, Satoru Sato, and Yoshimitsu Taniguchi

tee [Nihon kogyo kijun-chosaki] four months later. The women also obtained for their products the *eko-maku* [eco mark] granted by the Japan Environment Association [Nihon kankyo-kyokai], which is affiliated with the Ministry of Environment. They continue to receive special attention from surrounding communities as well as communities in other prefectures.

The Formation of OEC21

Soon after a rural sociologist came to the Akita Prefectural Agricultural College from Tokyo in 1992, he encountered three second-generation, pro-government Ogata farmers at a local restaurant. They were loudly discussing the grim future of the government-controlled rice-marketing system. The rural sociologist decided to help the young farmers learn about direct rice marketing. They began to run regular group study meetings. The sharp drop in the price of rice in 1995 had caused increased anxiety about the future of Ogata rice farming. Ogata's environmentally friendly agriculture would be in jeopardy if profits from organic and semiorganic rice continued to fall. Further, other environmentally friendly practices had won little public recognition and few economic returns. The environmental benefits of Ogata's farming techniques could no longer be confined just to individuals or farmers' circles if Ogata farmers hoped to make the profits necessary for their survival. Primarily young, second-generation Ogata farmers and researchers started to seek ways to gain public support for environmentally friendly agriculture.

In Ogata, it is not difficult to find farmers involved in a dozen different study circles and other business and recreational activities.[16] However, the entire village had yet to join together for a common purpose. The rural sociologist, who had led various grassroots movements, including organic farming

groups, sensed that it might be difficult to bring Ogata's residents together through the typical Japanese structured-group style. He therefore decided to try to link existing environmentally minded groups under one umbrella group. The idea was well received by many of the groups he contacted, and leaders from dozens of already established groups formed the OEC21 declaration committee.

Concurrently, a group of researchers had started to consolidate research on Ogata's environment. Soil scientists, biologists, hydrologists, and social scientists joined together as part of an Akita Prefectural Agricultural College project to analyze the effects of environmentally friendly agriculture in Ogata. Later, the Japan Society for the Promotion of Science [Gakujutsu shinkokai] granted the group research funds for a three-year comprehensive environmental study. The accumulated data confirmed the benefits of Ogata's environmentally friendly agriculture and raised important issues that needed to be dealt with.

In June 2001, OEC21 officially commenced. A declaration and data book were prepared to show the current status of environmentally friendly agriculture in Ogata. The publication was put together through the efforts of farmers, housewives, the village government, and researchers. Currently, OEC21 consists of thirteen groups[17] with more than five hundred associated members, including all the farmers and housewives described in this chapter.

Since OEC21's initiation, several main events have strengthened ties between insiders and outsiders. In November 2001, a devoted local soil scientist discussed Ogata innovations at an international symposium (held in Akita City and sponsored by the Akita Prefectural University) titled "Jizokukano na nogyo eno michi" [Agricultural innovation for sustainability]. In June 2002, guests from four Akita and Tokyo consumer groups participated in a field tour of Ogata. In November 2002, two major Tokyo consumer groups discussed food-related environmental issues with Ogata farmers. In March 2003, Dr. Koike from Shiga Prefectural University [Shiga kenritsu daigaku] was invited to present one of the first cases of a local government's cost-sharing scheme for environmentally friendly rice farming around Lake Biwa in Shiga Prefecture.

Reversing Pressures: OEC21's Mode for Action and Future Implications

An important goal of the movement was to focus multiple stakeholders on one shared purpose: the environment. In particular, as indicated in articles 5 and 6 of the group's constitution, organizers hoped to show how better environmental practices could lead to both improved farming and improved living standards, thus creating a real model for sustainable farming communities (Ogata-mura Kankyo Sozo 21 2001). OEC21's grassroots movement indeed involves a wide range of stakeholders: farmers, housewives, researchers, extension workers, village council members, agricultural co-op members, and consumer groups. It hopes to affect the surrounding communities by creating a working, environmentally friendly farming society.

Ogata's OEC21 grassroots movement developed rapidly, but it should be noted that individualism and social divisions among the Ogata community still persist. OEC21 consists of a large variety of genuinely interested people who care about their farming and about life in Ogata. The understanding of environmental problems and the degree of commitment vary widely from one member to another. However, this loosely bonded grassroots movement functions well in a society like Ogata's where strong-willed individuals came from different parts of Japan to form a new community.

Currently, profit margins for organic and semiorganic farming have become smaller (Ogata-mura Kankyo Sozo 21 2001), and the costs of other proved environmentally friendly agricultural exercises such as no-till and no-puddling are not reflected in retail prices. Some Ogata farmers are losing the incentive to make further progress toward environmentally friendly agriculture. Social scientists involved in OEC21 have suggested that environmentally friendly farmers get social recognition and monetary compensation, in the form of either subsidies or increased retail prices for their products (Sato and Taniguchi 2002).

Therefore, OEC21's immediate goal is to create a consensus on environmental cost sharing. Through interactive meetings, OEC21 is trying to appeal to consumer groups and policymakers. Elsewhere in Japan, Shiga Prefecture has already introduced direct payments to farmers who institute environmentally friendly farming techniques, but that initiative was spurred by Shiga Prefectural University and the Shiga prefectural administration (Koike 2003). However, in Ogata, farmers have done most of the organizing work toward environmental cost sharing.

OEC21 is still in the early phase of the process of creating economically feasible, environmentally friendly agriculture in Ogata, and it may take some time to win support for the movement at the prefectural and national levels. However, the group's position has already been confirmed by scientific evidence, and its goals have been publicized via the Internet, a data book, and various events.

Conclusion

Ogata Village was formed in 1964 as a model Japanese farming community. However, the five hundred farm households who migrated there from around the country soon faced various political, economic, and social pressures that affected their farming and their lives. The Ogata farming community, long divided over government policies, came together to

Shinji Kawai, Satoru Sato, and Yoshimitsu Taniguchi

confront the numerous pressures it was facing. Living in a pollution-prone, closed-water system made Ogata farm households aware of environmental risks to their farming and health. The village co-op's women's group played a critical role in educating the community. As early as the mid-1970s, it started to encourage the use of environmentally safe soap over synthetic detergents and push for chemical-spray restrictions.

Other major negative pressures, at least initially, were the rice-field acreage-reduction policies and government restrictions on rice marketing as well as individual pressures. These pressures, however, worked in many positive ways. Farmers outside the government system developed higher-quality niche rice and earned higher profits than farmers dependent on the government rice prices and subsidies. Consequently, the success of the independently marketed rice led other Ogata farmers to seek a better understanding of consumer demand. Pro-government farmers, along with scientists and farming industry officials, developed more efficient ways to plant rice, which eventually provided a cleaner environment in rice fields and led to a breakthrough in the reduction of chemical fertilizers used.

Finally, sharp drops in the price of rice since 1995 became the primary pressure to form OEC21. Farmers knew that they had to do something, but they could not find the answer by themselves. An experienced social scientist unified the community. Together, the Ogata community began working toward a common goal.

Initially, environmentally friendly measures were tied to farmers' responses to negative pressures. The true benefits were not fully realized until researchers revealed Ogata to be one of the most progressive and environmentally friendly farming areas in all of Japan. Significantly, OEC21 is a grassroots initiative begun by farmers and housewives. Natural- and social-science researchers were simply facilitators who helped provide the hard facts and networks that farmers could use to improve their farming practices. Now, when farmers and housewives want to appeal politically or economically to the public, they are equipped with concrete evidence to promote their movement. OEC21 provided a theater in which all the stakeholders held a common goal: an environmentally friendly society. With OEC21 and their own significant environmental activism and environmentally friendly farming practices, Ogata's farmers can, indeed, claim once again that they are a model Japanese farming community.

Notes

1. In 1999, the Ministry of Agriculture set the standard for organic foods. Semiorganic foods are loosely defined as those using less than half the recommended farm-chemical sprays allowed in each prefecture. Consequently, the doses and types of chemicals used on semiorganic crops vary regionally. However, farmers in the northern prefectures use considerably fewer chemicals than do farmers in the southern prefectures.

2. Survey data reported in JA Ogata-mura (2003) show no significant change in the figures except the rapid increase in the use of SBNA (single basal nursery fertilizer application) and the acreage of certified organic vegetable fields.

3. Prior to the reclamation project, about 200,000 pieces of fishing gear were gathered as evidence to obtain compensation from the Japanese government for diminished fishing rights (Chiba 1972).

4. The average Japanese rice farm at that time was 0.9 hectare, whereas Ogata farmers started out with a 10-hectare rice field. Many tractors were introduced, but they often failed to function in the muddy paddy fields. Machinery cooperatives introduced by the Ogata Reclamation Project soon disappeared.

5. National rice consumption was about 12 million tons around 1970 (Kitade 2001).

6. The Staple Food Control Act, introduced in 1942, allowed the government to set rice prices and control rice marketing and storage while rice production still stagnated and rice prices fluctuated erratically prior to World War II. The act encouraged farmers to produce rice and was of-

ficially in effect until the 1995 New Food Law was introduced. After World War II, rice production boomed, not only because of government protection, but also because of high-yielding varieties, enhanced mechanization, and increases in input materials such as chemical fertilizers and sprays. In addition, the Agricultural Basic Law, introduced in 1961, encouraged the development of agricultural infrastructure (the improvement of roads and irrigation and drainage systems, the enlarging of fields, and restructuring for mechanization). Better infrastructure brought better labor efficiency.

7. The quantity of government-purchased rice was reduced from 80.0 percent in 1970 to 27.3 percent in 1995 (Kitade 2001), and it further declined to less than 4 percent in 2002 (Komeno Jukyu Kakaku-joho ni kansuru Iinkai 2003).

8. Thereafter, consumers had to bear the increasing rice price as the gap between the purchase and the retail prices of rice widened from 10 percent in 1980 to 40 percent in 1990 (Kitade 2001).

9. The average market price for Akita Komachi (a variety of rice from Akita) was above ¥22,000 per sixty kilograms prior to 1995, dropping to around ¥16,600 per sixty kilograms in 2002 (Kome Kakaku Keisei Senta 2003).

10. In 1974, Ariyoshi's famous *Fukugo osen* [Multiple contamination] was published (initially as a serial novel in the *Asahi shimbun* that appeared between October 14, 1974, and June 30, 1975), and this helped raise public awareness of environmental problems in Japan.

11. Here, the technology can be called "no-till" or "minimal tillage." Staff at the Akita Prefectural Agricultural College and the Akita Prefectural Agricultural Experiment Station and members of the O-LISA study group (see below) together have decided that no-till for rice farming in Ogata will be defined as a maximum tillage of five centimeters in depth and five centimeters in width.

12. In Akita Prefecture, SBNA technology has been applied to around six thousand hectares as of 2000 (Yamazaki 2001).

13. In 1972, some members of the group visited polluted Lake Biwa and realized that the same thing could happen to Lake Hachiro.

14. For example, of the members of a soap-making group formed after the movement (see below), one became the first female mayor in 2001, another became the first female village council member the same year, and another was the head of the co-op's women's group until 2003.

15. Taro is a legendary dragon living in Lake Hachiro.

16. For example, some members of O-LISA are also members of the Organic Farming Group [Yuki-noho kenkyukai], the Ogata Village Natural Farming Study Group [Ogata-mura shizen-noho kenkyukai], the Flower Producers' Group [Khaki bukai], the Sea Salt Farming Group [Shio no kai], the Vegetable Growers' Club [Yasaikurabu], and the Brussels Sprouts Group [Puchiberu yasai no kai] besides groups devoted to such recreational activities as fishing, skiing, running, and theater.

17. The thirteen groups are ALM (Akita Land Marker), the Paddy-Upland Cropping System Group [Denpata rinkan gurupu], the Yuuki Organic Fertilizer Company [Kabushikigaisha yuki], the Hachiro-ko Cooking-Oil Recycling Group [Haiyu risaikuru no kai hachiro-ko], the Ogata Village Natural Farming Study Group [Ogata-mura shizen-noho kenkyukai] of JA Ogata Village, O-LISA, the Ogata Village Rice Seeds Co-op [Ogata-mura suito-shushi kumiai], Iki-iki Farm [Iki-iki nojo], the Soil Environment Database Working Group [Dojo-kankyo detabesu], the Environment Assessment Committee of the Ogata village administration [Ogata-mura chonaini tsuiteno kankyo-kento Iinkai], the "Eco-Office Plan" Section of JA Ogata Village [JA Ogata-mura eko-ofisu puran], the Hachiro-ko Water Quality Enhancement Group [Hachiro-ko suishitsu-kaizen], and the Rice Fields' Living-creature Monitoring Group [Tanbo no iki-mono-chosa].

References

Akita Daigaku Hachiro-gata Kenkyu Iinkai. 1968. *Hachiro-gata: Kantaku to shakai hendo* [Lake Hachiro: Reclamation and social changes]. Tokyo: Sobunsha.

Ariyoshi, Sawako. 1974. *Fukugo osen* [Multiple contamination]. Tokyo: Shinchosha.

Chiba, J. 1972. *Hachiro-gata: Aru daikantaku no kiroku* [Lake Hachiro: A report on a large reclamation]. Tokyo: Kodansha.

Gotsu, K. 1991. *No no michi massugu, Gotsu Tsuneo tsuito-shu* [The straight road to farming: Gotsu Tsuneo eulogy]. Akita: Mumyosha.

Hachiro-gata Kantaku Jimusho. 1969. *Hachiro-gata kantaku jigyo-shi* [Lake Hachiro Reclamation Project records]. Akita: Hachiro-gata Kantaku Jimusho.

Hanatsuka, Toshiko. 2003. Telephone interview with the author. March 27.

JA Ogata-mura. 2002. *Ogata-mura ni tsuite* [In-

Shinji Kawai, Satoru Sato, and Yoshimitsu Taniguchi

formation on Ogata Village]. Available at
http://www.ogata.or.jp/ja.
———. 2003. *Heisei 15-nendo JA Ogata-mura
einoshiryo* [Heisei 15th fiscal year JA Ogata
Village farm management data]. Akita: JA
Ogata-mura.
Kaneta, Y. 2002. "Development and Extension of
New Agro-Technologies for Sustainable Agriculture." In *Agricultural Innovation for Sustainability: Characteristics, Constraints and
Potential Contribution of Participatory Approaches,* ed. S. Sato et al. Tokyo: Norin Tokei
Kyokai.
Kitade, T. 2001. *Nihon nosei no 50-nen: Shokuryo
seisaku no kensho* [Fifty years of Japanese ag-
ricultural policies: Examination of the food
policy]. Tokyo: Nihon Keizai Hyoronsha.
Koike, Tsuneo. 2003. "Shintenkai suru kankyo-
hozen-gata inasakuchi" [The new development
of an environmentally friendly rice farming
region]. Lecture presented at the Agricultural
Training Center [Nogyo kenshu senta], Ogata
Village, March 2.
Kome Kakaku Keisei Senta. 2003. "Nensan goto no
Nyusatsu Deta" [Bidding records by fiscal year].
Available at http://www.komekakakucenter.jp/
dataold/index.html.
Komeno Jukyu Kakaku-joho ni kansuru Iinkai.
2003. "Kome no jukyu, kakaku-doko ni
kansuru joho, shiryo-hen (dai 62-kai), Heisei
15-nen 1-gatsu" [Information on supply-
demand and price of rice: Databook (no. 62),
January, Heisei 15th year]. Available at http://
www.zenchu-ja.or.jp/food/ja-zenchu/suiden/
shiryohen/komeinfo/h1501.pdf.
Ministry of Agriculture, Forestry and Fisheries
(MAFF). 2002. "Tokei joho: tokei joho kankyo
hozen-gata nogyo ni yoru nosanbutsu no seisan,
shukka-jokyo chosa kekka no gaiyo" [Statistical
data: Summary of the survey results on agricultural production and marketing by environmentally friendly farming]. Available at http://www
.maff.go.jp/toukei/sokuhou/data/hozen2001/
hozen2001.pdf.
Ogata-mura Kankyo Sozo 21. 2001. "Dai 1-bu:
Nogyo: Hanbai, keiei" [Pt. 1: Agriculture: Sales
and management]. In *Akita-ken Ogata-mura
Kankyo-sozo 21, Deta-bukku 2001-nendo ban*
[Akita Prefecture Ogata Village Environment
Creation 21, databook fiscal year 2001 edition].
Available at http://www.ogata.or.jp/kankyo21/
databook/chapter1/chapter1_7.html.
Rice Databank Co. 2003. "Kome seisanryo to
mochikoshi zaikoryo no suii" [Trends of rice

production and rice reserves]. Available at
http://www.japan-rice.com/data/seisan-zaiko
.htm.
Sato, S., and Y. Taniguchi. 2002. "Local Problem-
Solving for Building Sustainable Agriculture
in the Case of Ogata, Japan: Goals, Process,
Organization and Cost Payment." In *An In-
terdisciplinary Dialogue: Agricultural Production and Integrated Ecosystem Management of
Soil and Water.* Proceedings of an OECD Co-
Operative Research Program Workshop, November 12–16, 2002, Ballina, NSW, Australia,
printed and distributed by the workshop committee.
Shimizu, T. 1972. *Ogata-mura: Janarisuto no mita
moderu noson* [Ogata Village: A model farming
community]. Akita: Ndanda Bunko.
Tashiro, Y., and T. Suzuki. 1982. *Suiden riyou sai-
hen-ka no hachiro-gata nogyo* [Lake Hachiro
agriculture under the rice-field acreage reduction policy]. Nihon no nogyo [Japanese agricul-
ture] 140, 141. Tokyo: Nosei Chosa Iinkai.
Tozawa, T. 1993. *Mohitotsu no Ogata-mura* [An-
other view on Ogata Village]. Akita: Akita Bun-
ka Shuppansha.
"Tsuchi to ikiru: Ogata-mura no chosen, no. 2"
[Living with the dirt: Ogata Village's challenge].
2001. *Mainichi shimbun,* June 3.
Yamazaki, M. 2001. "1.3. Nogyo-kikai: 1.3.1.
Watashi to nogyo-kikai [1.3. Agricultural machinery: 1.3.1. Agricultural machinery and I].
In *Atarashii suiden nogyo* [New rice farming],
ed. S. Shoji. Akita: Imano Insatsu Kabushiki-
gaisha.

Chapter 11

The Administrative Process of Environmental Conservation and Limits to Grassroots Activities

The Case of Kyoto

Masao Tao

In a democracy, interested parties, or stakeholders, attempt to affect the policymaking process, to their own benefit. The influence of some stakeholders, however, is so great as to distort the policymaking process, turning it into one promoting private, not public, interests. Hughes (2003) takes the position that, when this happens, the public sector should mirror the private, that is, that, just as bad management should be replaced, so should easily influenced administrators. This position—which assumes that, in government as in business, there is one best course of action—cannot be supported. In the real-life business of public policy–making, any number of stakeholders can have perfectly valid positions that must be taken account of. This makes the public business of government much more difficult to manage than the business of private enterprise (Denhardt and Grubbs 2003). This situation holds at the level of local governments as well as national.

The policymaking process can be analyzed in terms of decisionmaking theory. According to Allison (1971), policymaking should be regarded neither as a decision by one entity acting rationally on behalf of the whole (a nation or a government) nor as an organizational process that takes into account all the factors needed to reach an optimal solution. In recent years, this view—that policymaking at whatever level is not a rational process—is beginning to prevail (Hilsman 1967). According to the garbage-can model (Cohen, March, and Olsen 1972), decisionmaking generally is not rational. Rather, it is distorted by various parties external to the process pursuing their own best, and often conflicting, interests. This model was further developed by Kingdon (1984), who emphasized that irrational processes are natural, that the clash of interests is inevitable, and that cooperation among interested parties is highly unlikely. Anderson (1983) pointed out the difficulty involved when organizations attempt to winnow the decisionmaking field to a single issue, a process that is even more difficult for local governments, which must balance the interests of many more stakeholders. Agreement is rarely reached in such cases (Neustadt 1960).

Until recently, influence on local government policymaking has been restricted to such traditional interest groups as trade unions, professional organizations (e.g., the Japan Dentists' Association and the Japan Doctors' Association), and business groups. Grassroots movements have, however, steadily been gaining a higher profile (Smith 1997). And it is precisely those cases where the interests of the traditional interest groups cannot be reconciled that grassroots movements have taken center stage in the policymaking process, offering ordinary citizens an unprecedented opportunity to exert political influence. A typical example is a policymaking situation that must take account of the

Photograph 11.1. Old Kyoto. (Courtesy Masao Tao.)

Photograph 11.2. Traditional houses, Kyoto. (Courtesy Masao Tao.)

interests of both proponents of industrial development and proponents of landscape preservation. Both positions make important arguments, but both also allow little room for compromise. It is in such a situation—where the view of the general public becomes decisive—that grassroots movements have had the most influence.

The conflict between the interests of development and the interests of preservation are most clearly seen in Japan in the city of Kyoto. Kyoto is the second-oldest city in Japan, after Nara. Its foundation was laid about twelve hundred years ago, and it still contains many historic buildings, clustered largely at the bases of the mountains that surround it on three sides (see photograph 11.1). It also contains numerous world heritage sites. For example, the old rows of houses in the central area of the city have survived to this day, and they symbolize Kyoto's long history. But Kyoto is also a modern city. It has a population of close to 1.5 million (more than 2 million if the greater metropolitan area is taken into consideration), it is home to many major Japanese corporations, including Rohm, Nintendo, Murata Manufacturing, and Kyocera, and modern high-rises dot the skyline. It is also what is called a "cabinet-order-designated city," meaning that the city government has authority equal to that of a prefectural government.

The city administration is fully cogni-

zant of the need for both preservation and development and of the need to achieve both simultaneously. It is, however, faced with a difficult balancing act, one that it has not so far mastered. While laws and ordinances protect temples and shrines and control new construction (e.g., in the central part of the city, north of the Kyoto Station Building, building height is restricted to forty-five meters), modern construction still dots historic areas, the result of policy decisions that are reactive rather than proactive. Few old-style charming cityscapes remain intact (see photograph 11.2).[1]

Other cities in Japan face the same dilemma. But it is particularly acute in Kyoto, whose tax revenue is among the lowest of Japanese cities (Kyoto City Information 2002). If tax revenue is to increase, development must occur, but that development can be only at the expense of preservation. In what follows, I explore both the factors contributing to the inefficiency that to date has characterized policymaking in Kyoto meant to address this dilemma and the role that grassroots movements have played in the policymaking process.

Factors Specific to Kyoto

Five factors, outlined below, contribute to the particularly acute conflict between preservation and development in Kyoto, a conflict that preservation is losing.

Factor 1: Geographic Restrictions Because Kyoto is surrounded on three sides by mountains (in ancient times a configuration that was considered propitious), it can expand only into the flatlands to the south (see figure 11.1). That is, while new residential construction is possible in the mountains, business and industrial development can occur only in the flatlands. (Even if the mountains were leveled—a step that Kobe City, e.g., took—the new flatland would not be readily accessible and, thus, could not be used for industrial construction.) Another geographic restriction is that, alone among metropolitan areas in Japan with a population exceeding 1 million, it is not on the sea and, thus, has no coastal areas suitable for reclamation. In short, Kyoto has no developable periphery because of the necessity of preserving historic sites. To enable development, old urban areas of the city must be continuously demolished and rebuilt, making the preservation of the old cityscape impossible. A previous urban-development plan divided the city into northern and southern regions and tried to foster development in the southern region, which has more open space, but offices are still clustered in the northern region, which has more convenient transport links. There are also plans to move the city administration from the old city hall in the northern region to a new high-rise in the southern region. However, financial constraints may prevent this.

Factor 2: The Administrative Culture Japanese local governments are often described as "department stores" or "supermarkets" since, with only the occasional exception, all administrative departments (residential services, transportation, education, etc.) are located in the same building (Tao 1981). (In fact, Tatsuo Miyazaki, a former mayor of Kobe, once remarked that the city government functioned like a large conglomerate, an observation that is applicable to other cities as well.) The result is what Mineno (1995), describing the administrative culture of Kyoto, characterizes as decentralized decisionmaking. That is, coordination among departments is generally avoided. Each department pursues its own decisionmaking process, the mayor ratifying policies only after the fact. This is called *genkyoku-shugi* [the department-oriented principle]. The potential for the various departments to institute conflicting policies is obvious. And the administrative environment ensures that conflict resolution is difficult and time-consuming. Those mayors who have attempted to strengthen their position (see factor 3 below) in order to pursue effective conflict resolution have had little success.

Factor 3: Lack of Leadership Kyoto is run by an elected mayor and city council. If the mayor's party is in the majority in the city council, his policy agenda can, in most cases, be implemented. Since World War II, however, few mayors have stayed in power long enough to push their initiatives through (Muramatsu 1981). The increasing autonomy of departments since the last strong mayor, Funahashi Motoki, left office in 1981 has also undermined the mayor's position. Further weakening it is the political situation in the city, where the forces for preservation and those for development are equally balanced. Because few recent mayors have been backed by a stable council majority, some have attempted to gain a consensus on policy by supporting moderate positions acceptable to both conservative and progressive parties, but they have seen little success. Most often, decisions are based on political considerations, some mayors, for example, supporting development in order to gain the support of business. The lack of strong leadership in Kyoto is clear.

Factor 4: Intricate Interests As in any large city, a variety of interest groups attempt to influence public policy. Like other city administrations, Kyoto's must juggle the competing interests of business groups, labor unions, and professional associations.

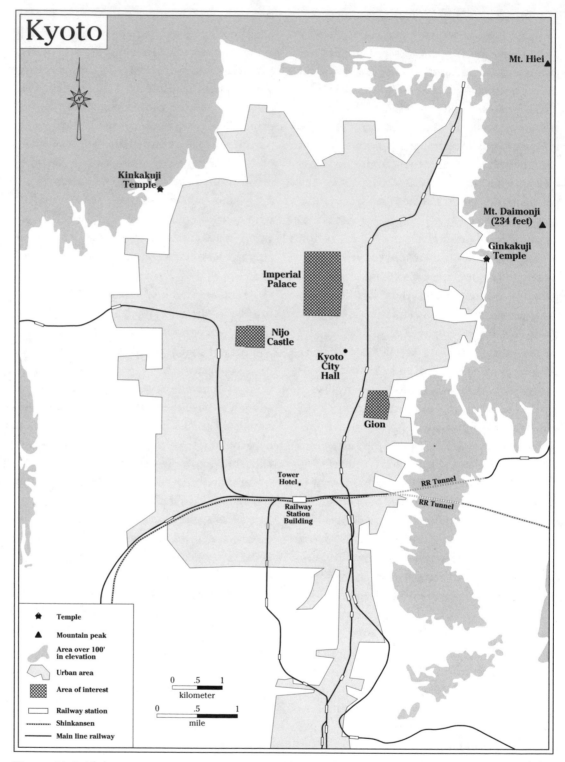

Figure 11.1. Kyoto.

But, thanks to the city's historic legacy, it also has those of religious groups and the tourism industry to contend with, the influence of the former being particularly strong. (For example, a proposal to impose a new tax on shrines/temples—a political misstep that contributed to the mayor's decision not to run for reelection [Mineno 1997]—was withdrawn owing to opposition by religious groups.) Thus, in Kyoto, more than in other cities, a sensitive balance must be maintained between competing interests, including, increasingly, public opinion. Accordingly, the mayor tends to take up a neutral position with regard to the issues of preservation and development.

Factor 5: The Indifference of Ordinary Citizens The 1.5 million residents of Kyoto perceive life there as inconvenient compared with that in other large cities (Noda 2000) and, thus, tend to prefer development over preservation. Also working in the favor of the development interests is the fact that only the smallest proportion of the general public is politicized. Development is perceived by the vast majority as a problem affecting only a very few adversely, and grassroots opposition movements are seen as pawns of the preservation interests, distasteful because self-serving. People also wish to avoid being drawn into a political struggle. Not surprisingly, therefore, preservation has not become part of the mainstream political agenda. Those seeking elected office court, not the preservation interests, but, as we have seen, those groups whose support is more capable of generating votes. When opposition movements do arise, they are rarely successful. A case in point—discussed at length below—is the opposition to the reconstruction of the Kyoto Station Building.

Thus, a consideration of the policymaking process in Kyoto must assume the existence of a number of political actors, be they individuals, groups, or institutions. The first actor is the mayor, whose position gives his opinion decisive importance—or at least the appear-

ance thereof (as we have seen, mayors' decisions are often made in service of the political considerations of others). The second actor is City Hall, which follows, but thinks and acts independently of, the mayor. The third actor is the interest groups that directly or indirectly influence municipal policymaking. The fourth actor is the city council. The fifth actor is the general public, which tends not to take organized political action, but which does play an important role during elections, unaffiliated voters often accounting for more than 30 percent of those voting. The way in which these five actors influence the decisionmaking process in Kyoto is similar to the way in which a complicated power game works. That is because four of the five are groups, and groups' decisionmaking processes are not always logical, making their actions difficult to predict. Complicating matters is the fact that rarely does one particular group dominate in the competition for influence, with the result that policymakers are faced with the almost impossible task of making everyone happy. Overall, the system has so much inertia that movement in any direction becomes almost impossible.

The Dynamics of Decisionmaking in Kyoto

As a cabinet-order-designated city, Kyoto functions largely independently of Kyoto Prefecture and even the central government. Consequently, the decisionmaking process that has developed there is unique. As we have seen, the various City Hall departments operate largely independently of each other and of the mayor and the city council. A network of influence thus connects each department to the various actors outlined above—other departments, interest groups, the city council, and representatives of the general public. It is through consultation with these actors that policy is determined. Once the details of a policy are worked out, a proposal [*ringi*] is circulated among all department managers. If they all approve it, the proposal is accepted. The *ringi* system is a decisionmaking method unique to Japan.

Because they are reluctant to interfere with one another, departments may decide on and implement conflicting policies (e.g., one department's plan to preserve an old town may be undermined by another's plan to build a highway near that town). Even though *chogis* (meetings between the mayor, the vice mayor, and the department managers at which major management decisions are made) are held periodically, affording participants the opportunity for comment, decisions are rarely overturned—not even by the mayor, who, as we have seen, functions these days largely as a rubber stamp. (According to Murakami, Mayama, and Tao [1993], among mayors of cabinet-order-designated cities, the mayor of Kyoto has the least practical power.) The most likely policy roadblock is the Financial Department, which will reject financially impractical plans.

Attempts to coordinate policy can be made by the mayor and his staff or by the Planning Coordination Department, but they are rare. For one thing, only the strongest mayors are capable of brokering such deals, and few recent mayors have been able to. For another, the Planning Coordination Department (a fixture in other city administrations) is usually eliminated during the tenure of a weak mayor by the mayor himself at the request of local interest groups (whose support he requires). (Not surprisingly, strong mayors tend to have active Planning Coordination Departments [Kimimura 1981].) Trade-offs among departments are very occasionally made to square conflicting policies. Generally, however, departments are left to follow their own lead, to the detriment of the planning process.

Very occasionally the citizens of Kyoto are roused to take action, but mostly only when spurred by an issue that galvanizes the entire city. Anything less than a crisis affecting the identity of Kyoto, awakening their pride in living in an ancient capital, tends to pass unnoticed. Because it takes time for issues to attract citywide attention, citizen protest figures only late in the policymaking process. It can, nevertheless, result in the mayor and the city council members—always sensitive to their position as elected officials—changing their minds. For example, one recent mayor, up for reelection, abandoned plans to build a new, modern bridge across the Kamogawa River, which runs through the historic city center, in the face of opposition from citizens who felt that it would be out of character architecturally.

Historically, grassroots citizens' groups have tended to be led—and dominated—by more prominent members of the community (e.g., chairmen of neighborhood associations, representatives of interest groups). Such ties, however, can lead to the groups' motives being perceived as self-interested. One way around this problem is for a grassroots movement to align itself with more than one interest group. In recent years, however, groups of citizens acting on their own have sprung up, gaining ever-increasing influence on the municipal administration. Nevertheless, because their involvement still comes late in the policymaking process, the situation that has aroused them is usually quite serious. This is particularly true in the case of preservation. Few ordinary citizens think seriously or regularly about it, and policymakers accord it low priority since no political party champions it. Also, while leftist political parties do tend to be interested in preservation, they rarely take a stand on the issue because doing so brings them no direct political benefit. Also exacerbating the problem is that existing regulations are largely ignored (e.g., permission is not sought before clearing land to make way for new construction). So, even though preservation issues quickly become political issues, a great deal of environmental damage has usually already been done by the time notice is taken of them.

The coordination of policies promoting development and policies promoting preservation will be achieved only when the two most important political actors—the mayor and the citizens of Kyoto—effectively exercise their power (e.g., through petition campaigns). That they have yet to do so is at the root of most recent preservation problems, including the construction of the Kyoto Tower Hotel, the construction of the illegal housing development known as Mohican Mountain, the construction of a golf course in Higashiyama, the reconstruction of the Kyoto Hotel as a high-rise, and the reconstruction of the Kyoto Station Building. At the heart of these preservation failures is the same pattern of preemptive development by contractors and late action by opposition forces within the administrative organs. There is even some suggestion that, during his tenure as mayor, Takayama Syoji supported development in all these cases. The result was disorder in City Hall and defeat for the forces of preservation. Typical of the confrontation between development and preservation, and, thus, instructive, is the reconstruction of the Kyoto Station Building, to which I turn in the next section.

The Reconstruction of Kyoto Station Building

The Kyoto Station Building—often described as a huge wall separating the city's northern and southern regions, on whose boundary it is located—is the largest building in the city (see photographs 11.3–11.5). Its design is multipurpose, the structure incorporating a railway terminal, a parking garage, a theater, a department store, and an exhibition hall. Opposition to its construction centered on its architectural inappropriateness, the originally proposed great height of the building destroying the traditional urban skyline.[2]

Photograph 11.3. (above) Kyoto Tower and Station Building. (Courtesy Masao Tao.) Photograph 11.4. (below) Kyoto Station Building. (Courtesy Masao Tao.)

Photograph 11.5. A view of the huge Isetan Department Store above the Kyoto Railway Station. (Courtesy of Todd Stradford.)

Renovation of the existing structure (built in 1952) was initially considered. But it was so run-down that the cost would have been prohibitive, and the sentiment grew, both in City Hall and among the members of the Kyoto Business Association [Kyoto syoko kaigisyo], that a new building should take its place. In 1989, a company (half owned by the government [the third sector]) for reconstruction was provisionally established. At the same time, the period's economic growth (Japan was then in the midst of a boom later referred to as the "bubble economy") fueled a similar growth in construction plans, in the sense that what was originally conceptualized as a relatively modest structure (roughly 15 meters tall) became a 130-meter-tall high-rise. (It was at this point that the design became multipurpose.) Business interests and conservative political parties were enthusiastic about the plan.

Grassroots opposition to the project was initially based on the perceived need to protect small businesses in the immediate vicinity—and, thus, perceived as self-interested. Soon, however, concern grew among people interested in preserving the cityscape, but the citizens' groups that became involved in the movement remained fragmented. Because it did not present a united front, the opposition continued to be perceived as self-interested. In an attempt at coordination, the various grassroots opposition movements tried to form an alliance but were unable to do so, and their influence on the policymaking process was, therefore, limited.

At this point in time (1987–1990), the city administration (in particular, departments concerned with economic policy that favored development) joined business interests in favoring the construction of a high-rise building, as did the city council, where, at first, only the Communist Party opposed the plan. But, as grassroots opposition movements grew in certain districts, even conservative council members began to change their minds. And the opposition was given a boost when business interests succeeded in doing away with the city's restriction on the construction of buildings taller than forty-five meters. By 1991, except for the business interests, the mayor stood largely alone in his support of the plan. In fact, in that year's election one of the Communist candidates collected only three hundred fewer votes than he did. Finally, the politically sensitive issue of preservation was being discussed publicly and the view of the general public attracting attention.

In spring 1991, a building design competition produced two finalists, one plan calling for a structure 59.8 meters tall, the other for one 120 meters tall. The screening process was held by a third party under the observation of the city, but, because the mayor made clear his position that the design chosen should be "suitable for Kyoto" (thus, distancing himself from the business interests), the shorter of the finalists was selected. Murakami described the winning design as "relatively gentle," "intermediate between the pro-development and pro-preservation groups" (1993, 89).

Construction was completed in 1997, the building opening in September of that year. Although it is somewhat different architecturally than all other buildings in the city, it does not, according to Murakami, undermine the image of Kyoto as "an ancient capital without incongruous skyscrapers" (2003, 109). The fact that a design competition was held and that the originally planned height of the building was reduced by half can be considered an achievement of the citizens' opposition movement. Still, the historic 1952 structure was torn down and replaced, its replacement a reminder of the essential powerlessness of grassroots efforts in the city.

Conclusions

If the forces of preservation are to prevail in Kyoto, what is needed is a strengthening of grassroots opposition movements and a revamping of the policymaking process.[3] How is this to be achieved? First, the generally pro-development attitude of the general public must be changed.[4] Second, the activities of the various grassroots groups must be coordinated and the focus of those groups expanded to citywide issues. Third, direct links to council members and especially the mayor must be established so that grassroots groups can function as proper interest groups. Fourth, a mayor who is a strong leader and committed to preservation must be put, and kept, in power so that a consistent, and preservation-friendly, approach to policy can be developed.

It must also be kept in mind, however, that there are limitations to grassroots activities, particularly those favoring preservation over development. It is perfectly possible that, despite opposition movements' best efforts, the general public will side with the interests of development. Grassroots activities are not, after all, confined to preservation.

Notes

1. According to Okada (1997), the majority of Kyoto residents feel that the city is becoming increasingly less beautiful. A similar sentiment exists among tourists: "Overseas visitors to Kyoto are impressed by the temples and shrines, but are distressed by the chaotic townscape not worthy of an ancient capital, where electric cables are exposed and historic buildings are demolished to make room for office blocks" ("Kanko rikkoku no tame" 2003).

2. For histories of this incident, see Murakami (1993, 2003), Okada (1998), and Kerr (2001).

3. Mie Prefecture affords a typical example of administrative reform spearheaded by a strong leader. During his tenure in office, Kitagawa Masayasu, the prefectural governor, introduced a new form of public management whereby he bypassed the traditional interest groups and appealed directly to the general public.

Kyoto itself in 1997 established a committee (on which I serve) to develop a plan to reform the city's policymaking process. The basic idea is that the first step in the policymaking process must be to develop a coordinated, citywide plan to guide

subsequent proposals and aid in adjudicating between them when they conflict.

4. There has, in fact, been progress made in this direction. Interest in preservation among the general public has been growing over the last decade, most likely tied to the growth in the number of nonprofit organizations supporting preservation after Kyoto hosted the third session of the conference of the parties to the UN Framework Convention on Climate Change in 1997.

References

Allison, G. T. 1971. *Essence of Decision: Explaining the Cuban Missile Crisis.* Boston: Little, Brown.

Anderson, P. A. 1983. "Decision Making by Objection and the Cuban Missile Crisis." *Administrative Science Quarterly* 28:201–22.

Cohen, M. D., J. G. March, and J. P. Olsen. 1972. "A Garbage Can Model of Organizational Choice." *Administrative Science Quarterly* 17:1–25.

Denhardt, R. B., and J. W. Grubbs. 2003. *Public Administration: An Action Orientation.* 4th ed. Belmont, CA : Wadsworth/Thomson Learning.

Hilsman, R. 1967. *To Move a Nation.* New York: Doubleday.

Hughes, O. E. 2003. *Public Management and Administration: An Introduction.* 3rd ed. New York: Palgrave Macmillan.

"Kanko rikkoku no tame no moo hitotsu no shiten, hokoreru kyoodo, kihon ni keikan hozen nado miryoku wo tsukuru genten to ha" [Another viewpoint on tourism: Pride in one's home country, the basis for making attractive points such as landscape conservation]. 2003. *Nihon keizai shimbun,* February 22.

Kerr, A. 2001. *Dogs and Demons: Tales from the Dark Side of Japan.* New York: Hill & Wang.

Kimimura, M. 1981. "Sogo keikaku to kikaku chosei bumon" [General plan and the planning coordination department]. In *Kyoto-shi seiji no dotai* [Dynamics of Kyoto City administration], ed. I. Miyake and M. Muramatsu. Tokyo: Yuhikaku.

Kingdon, J. W. 1984. *Agenda, Alternatives, and Public Policies.* Boston: Little, Brown.

Kyoto City Information. 2002. *Kyoto-shi zaisei no aramashi* [Summary of financial administration]. Kyoto: Kyoto City Information.

Mineno, Y. 1995. "Chiho-jichitai ni okeru yosan

hensei" [Budget planning in local government]. *Rebaiasan* [Leviathan] 16:145–68.

———. 1997. "Chuoo seifu tosei ka ni okeru chiho seifu no ko do" [Actions of local government under restrictions of the central government]. *Seisaku kagaku* [Policy science] (Ritsumeikan University) 14, no. 2:37–47.

Murakami, H. 1993. "Kyoto no keikan gyosei to seisaku katei" [Landscape administration and the policymaking process in Kyoto]. *Toshi mondai* [Urban problems] 84, no. 4:71–97.

———. 2003. "Kyoto no keikan-seisaku to Shin-Kyoto-eki-biru" [Policies of preservation and Kyoto Station Building]. In *Nihon no chiho-jichi to toshi seisaku* [Local government and urban policy in Japan]. Kyoto: Horitubunnka-sya.

Murakami, H., T. Mayama, and M. Tao. 1993. "Daitoshi no toshiseisaku to shokuin ishiki hikaku gyosei chosei wo moto ni shite" [Policies in large cities and awareness of municipal staff, based on a comparative administration survey]. Working paper. Kyoto: City Hall.

Muramatsu, M. 1981. "Shico to gyouseisoshiki" [Mayor and administrative organ]. In *Kyoto-shi seiji no dotai* [Dynamics of Kyoto City administration], ed. I. Miyake and M. Muramatsu. Tokyo: Yuhikaku.

Neustadt, R. E. 1960. *President Power*. New York: Wiley.

Noda, H. 2000. "Rekishi-toshi to keikan-mondai: 'Kyoto rashisa heno manazashi'" [Historical city and landscape issues: "Eyes for Kyoto identity"]. In *Rekishi-teki kankyo no syakai-gaku* [Sociology in historical environment], ed. Arata Katagiri. Tokyo: Shinyo-sha.

Okada, T. 1997. "Seikatsu kukan to shiteno kanko toshi wo mezashite: Kyoto-shi kanko choosa hookokusyo" [Tourist city as living space—Kyoto City tourism report]. Kyoto: City Hall.

———. 1998. "21 seiki ni habatakeruka Kyoto-eki-biru" [Kyoto Station Building leaping into the twenty-first century]. Working paper. Kyoto University, Faculty of Economics.

Smith, D. H. 1997. "Grassroots Associations Are Important: Some Theory and a Review of the Impact Literature." *Nonprofit and Voluntary Sector Quarterly* 26:269–306.

Tao, M. 1981. "Chiho jichitai ni okeru kanri-katei no tokushitsu ni kansuru ichi kosatsu" [Prospect of a characteristic for a management process in local government]. *Kyoto Furitsu Daigaku kiyoo* [Kyoto Prefectural University bulletin] 33:124–39.

Chapter 12

The Grassroots Movement to Save the Sanbanze Tidelands, Tokyo Bay

Kenji Yamazaki and Tomoko Yamazaki

During the period of rapid economic growth following World War II, roughly 250 square kilometers of land were reclaimed from Tokyo Bay. This reclaimed land was used for both residential and commercial development, and both the factories and the homes built there dumped their wastewater directly into the bay. The result was unprecedented levels of pollution.

It was local inhabitants, mostly fishermen, who began the movement to revive Tokyo Bay. This should come as no surprise since, historically, fishermen have often been the bellwethers of the environmental movement in Japan. They were the first to be affected by pollution from the Ashio copper mine (Ooga 1972), by the two outbreaks of Minamata disease (Ishimure 1965; Ui 1971), and by the outbreak of Yokkaichi asthma (Tajiri 1972).

This chapter focuses specifically on the Sanbanze Tidelands (photograph 12.1). Sanbanze is located in the innermost part of Tokyo Bay in Chiba Prefecture, facing the cities of Funabashi, Ichikawa, and Urayasu. It has an area of sixteen hundred hectares. About a hundred species of marine creatures, including flatfish, goby, short-necked clams, and other shellfish, reside in the tidal area. Sanbanze has, since the Edo period (1603–1868), been one of the best spots in the bay to net fish and shellfish. It is also a wintering site for ducks and other migrating birds. Eighty-nine species of birds have been confirmed to winter in the area. Other such areas once existed in Tokyo Bay, but today only Sanbanze remains

untouched by the prefectural government's reclamation projects.

Tokyo Bay generally has been badly polluted as its coastline areas have been developed. The essays collected in Tajiri (1988) report on how it has been affected by industrial development and call for the abandonment of reclamation and, in its place, the conservation of the ecosystem and the reinforcement of disaster-prevention measures. Sanbanze is a good example of what Tokyo Bay would be like after such a program. It has not been reclaimed; thus, it has been sufficiently fertile to filter out the effects of pollution.

The Reclamation and Transformation of Tokyo Bay's Environment

Tidelands along the shores of Tokyo Bay are easy to reclaim and have long been favored for urban development. The reclamation of Tokyo Bay began in the early seventeenth century when Shogun Ieyasu settled in Edo Castle and constructed the outer moat (known as Sotobori); the soil that was removed during the process was used as landfill on the shores of Hibiya Inlet. The scale of this reclamation project, however, was not so great as to completely destroy the ecosystem of Tokyo Bay. Edomae-no-Umi [the sea in front of Edo] was the source of protein for the people of Edo (as Tokyo was then called). In the early years of the Edo period (1603–1868), Shogun Ieyasu summoned some fishermen from Settsu (a region near Osaka) to Edo and gave them land

Photograph 12.1. An aerial view of the Sanbanze Tidelands. (Courtesy of Kenji Yamazaki.)

to live on at the delta of the Sumida River, a district currently known as Tsukuda. These fishermen had the privilege of presenting fresh whitebait to the shogun every year, and they were given the exclusive right to fish the fronting sea (see figure 12.1). Tokyo Bay was a bountiful food source, supplying animal protein to the 1 million people of Edo. However, at the same time, it had become a dumping ground for waste and garbage, and large-scale reclamation projects that destroyed its most productive tidelands—

breeding grounds for fish where, because of photosynthesis, the dissolution of organic material is greatly enhanced—rendered it no longer self-cleansing.

Reclamation in order to create industrial sites started in earnest in 1912, the very end of the Meiji era (1868–1912), and 570 hectares just offshore of Kawasaki City were filled in by 1931 (Hanayama 1983). The development of modern technologies expedited the reclamation of shallow inshore areas. The dredged soil from the construction of port

Kenji Yamazaki and Tomoko Yamazaki

Figure 12.1. A woodblock print of what is now the Tsukuda district as it appeared during the Edo period. The image depicts a festival. The text in the banner reads: "Fishermen from Sumiyoshi [Settsu] are working at Tsukuda [Edo]."

facilities was used to reclaim new land for industrial use. In this way, the simultaneous construction of port and industrial facilities became a reality.

The pollution caused by the Asano Cement Corporation was one of the primary factors triggering the Kawasaki Reclamation Project. The fine dust and smoke coming from the company's Fukagawa factory prompted local protests. This issue was taken up by the imperial parliament, and, in 1911, Asano Cement promised in writing to halt production in Fukagawa within five years. At the same time, the company decided that it would be much more profitable to construct a larger-scale, modern factory. Asano Soichiro, Shibusawa Eiichi, and Yasuda Zenjiro formed the Tsurumi Reclamation Association [Tsurumi umetate kumiai] in 1912. A dredge was purchased from Britain, and 586 hectares of re-

claimed land were constructed at the mouth of the Tsurumi River, the seaward-extended region of Kawasaki. A pier was constructed at the site that could accommodate ten thousand ton-class ships. On the site itself, Asano Cement, Nippon Kokan, Asahi Glass, and other heavy chemical industries established plants: the nucleus of what would become the Keihin Industrial Zone.

The large-scale reclamation of Tokyo Bay commenced after World War II, the projects initially being supported by public funds (a method of financing known as "the Tokyo scheme" [see figure 12.2]). Land reclaimed during the period 1946–1965 was used to build wharves capable of accommodating large ships, for example, a coal-offloading wharf at Toyosu and a grain-offloading wharf at Harumi. Reclaimed land near the center of Tokyo was sold to gas and electric companies, which built gas-distribution facilities and power stations on it. The money raised through such sales was used to fund additional projects.

In Chiba Prefecture, by contrast, reclamation projects during the period 1966–1975 were supported by the advance sale of the land to be reclaimed (a method of financing known as "the Chiba scheme" [see figure 12.3]), its bankruptcy preventing the prefecture from obtaining any loans. The money thus collected was used to fund reclamation and also to buy back fishing rights (which went for a much lower price than did the reclaimed land), making the development process easier for business and profitable for the prefecture. Thus, reclamation proceeded rapidly. It was the Mitsui Real Estate Company—now the largest real estate company in Japan but just a small subsidiary of the Mitsui combine before the reclamation projects were carried out—that took the initiative in reclaiming the Chiba side of Tokyo Bay (see figure 12.4).

The reclaimed area of Tokyo Bay, based on the fishery census, is shown in table 12.1.

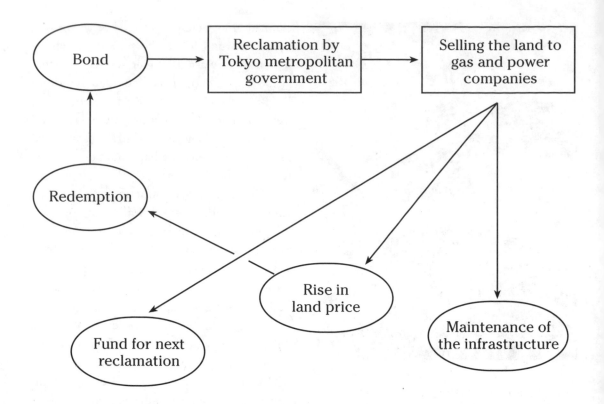

Figure 12.2. Land reclamation: the Tokyo scheme.

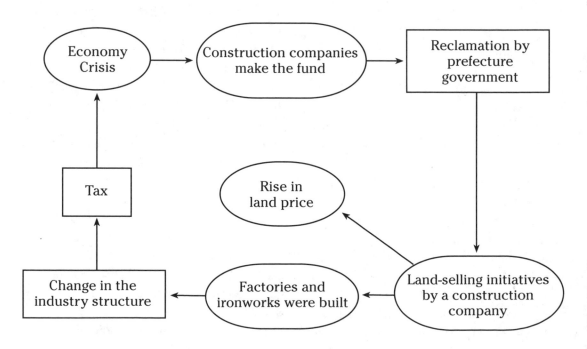

Figure 12.3. Land reclamation: the Chiba scheme.

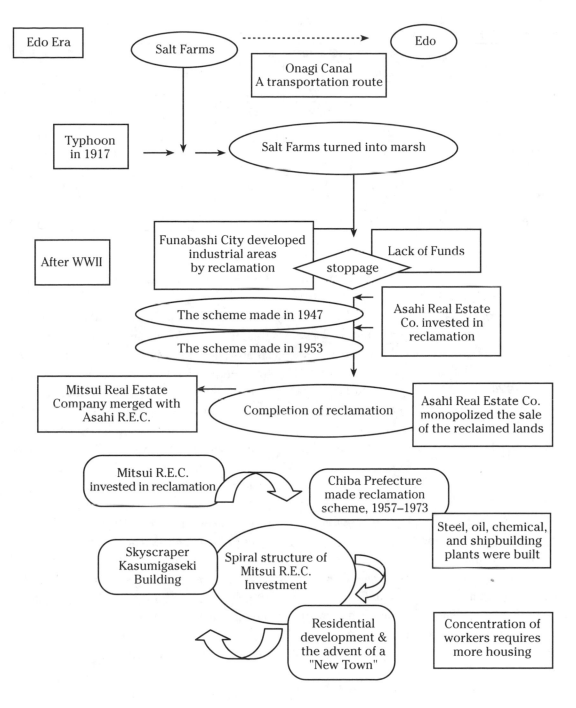

Figure 12.4. Mitsui Real Estate Company's investment in the reclamation scheme of Funabashi City.

Table 12.1. Reclaimed Land (in hectares)

	1963–1967	1968–1972	1973–1977	1978–1982	1983–1987	1988–1992	1993–1997
Japan	21,592	12,307	13,051	8,085	7,034	6,227	3,514
Chiba	1,365	3,533	267	1,3270	485	5	43
Tokyo	787	0	522	572	466	12	71
Kanagawa	574	311	343	236	171	20	163
Tokyo Bay Area	2,727 (12.6)	3,844 (31.2)	1,132 (8.6)	2,178 (26.9)	1,122 (16)	37 (.6)	277 (7.9)
Setonaikai	11,493 (53.2)	4,876 (39.6)	4,423 (33.9)	1,792 (22.2)	2,207 (31.4)	2,579 (41.4)	1,043 (29.7)

Note: Figures given in parentheses are percentages.

From 1968 to 1972, reclamation projects were concentrated in the Inland Sea (Setonaikai) and Tokyo Bay, reclaimed land in those two areas accounting for 39.6 and 31.2 percent, respectively—or roughly 70 percent—of all reclaimed land in Japan. Furthermore, from 1978 to 1982, the reclaimed land area of Tokyo Bay was larger than that of the Inland Sea even though the Inland Sea is approximately twenty times larger than Tokyo Bay.

The impact of reclamation—and particularly postwar reclamation—on the Tokyo Bay fishing industry has been profound. The fishermen working the bay—most of whom historically made a living off the shellfish and seaweed in the fronting sea—have been deprived of their fishing grounds and, thus, their livelihood. The result has been a drastic reduction in the fishing population (see figure 12.6). For example, in 1954, when the first fishing industry census was conducted, there were 14,673 fishing households along the coast of Tokyo Bay. By 1998, that number had dropped to 2,306. Still, four fishing communities—Shiba in Kanagawa Prefecture and Funabashi, Kaneda, and Futtsu in Chiba Prefecture—have retained their fishing rights, and all of them continue to fish, as does Oota in Tokyo despite having abandoned its rights. However, owing to reclamation, Shiba and Futtsu have had to develop new fishing methods (Yamazaki 1988).

The Sanbanze Tidelands

The Sanbanze Tidelands, a shallow inshore area fronting Funabashi, formed a pristine landscape in the Meiji era (see figure 12.7). In 1966, Chiba Prefecture sent notification to the Funabashi Fishermen's Association of plans to reclaim Sanbanze. Area fishermen initially protested but, in 1969, decided to waive their inshore fishing rights. Reclamation did not, however, begin immediately, the prefecture being occupied with other reclamation projects (e.g., Makuhari). Plans were further delayed by the 1973 worldwide oil crisis, which changed the industrial structure of Japan, discouraging the construction of large-scale material-supply plants. The delay afforded citizens' groups the opportunity to mount—in partnership with area fishermen—a grassroots movement advocating the preservation of Sanbanze, a cooperative effort that served as a model for the environmental conservation movement in Japan.

The main fishing grounds fronting the tidelands—all currently endangered—are Sanmaizu at the Arakawa River mouth, Sanbanze, Banzu at Kisarazu, and Futtsu Point. The main catches there are sea bass, black porgy, and mullet, which are shipped live, not only to the Tokyo market, but also to the markets in the Hanshin and Nagoya areas. In addition, sardines caught in the morning are delivered to the shelves of retailers for sale in

Kenji Yamazaki and Tomoko Yamazaki

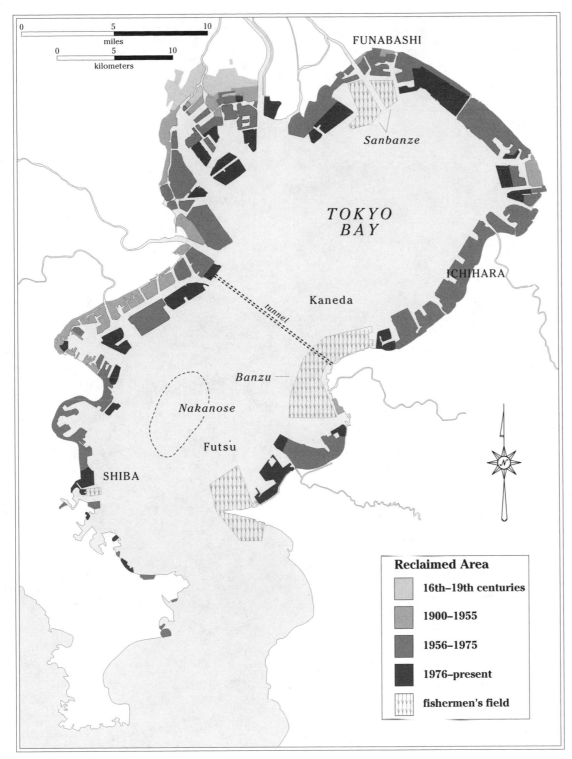

Figure 12.5. Stages in the reclamation of Tokyo Bay since the Edo period.

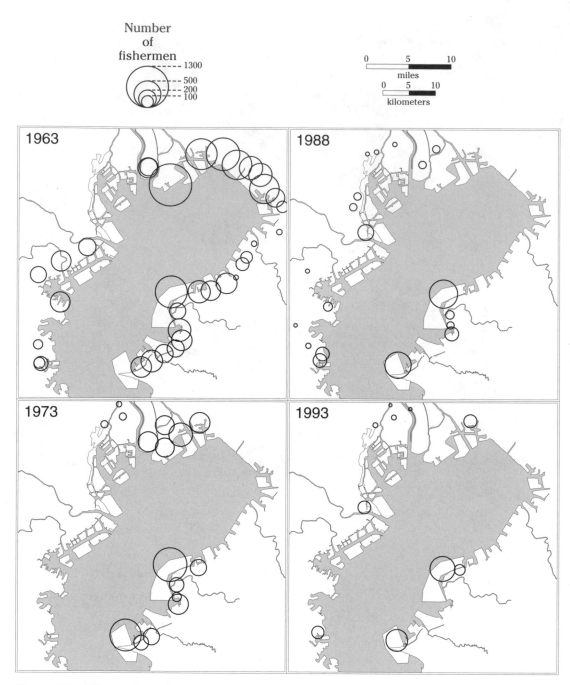

Figure 12.6. The reclamation of Tokyo Bay: number of fishermen.

the early afternoon. Thus, the close proximity of the tidelands to the markets is fully capitalized on—even to the extent that market trends are radioed to working fishing boats, enabling them to sell their catch directly to markets offering the best prices, instead of going through mediators.

The fishermen working Sanbanze harvest

shellfish in the summer and seaweed [*nori*] in the winter. Of the shellfish, the main catch is asari clams. The area is so productive that even the widespread blue tide[1] that devastated the tidal flats in September 1985 shut down fishing operations for only four months. Because the area that can be fished is so small, the cooperative that coordinates activity there

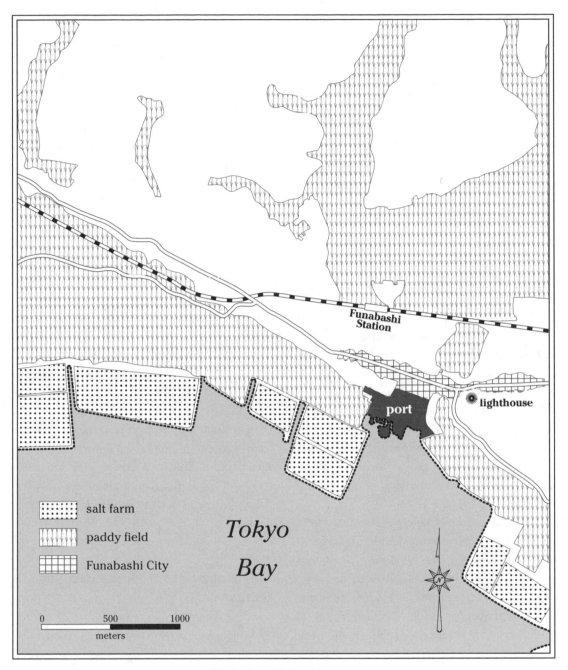

Figure 12.7. Meiji era map of the Sanbanze Tidelands.

strictly controls the size of operators' catch and the sectors they are allowed to fish. Harvesting shellfish requires little capital and can be pursued by a lone individual, making it popular among the older fishermen. In the past, fishermen waded the flats dragging a shovel. Today, the shovel is operated from a boat, and the shellfish that are reeled in pass through a sieve that retains only those large enough to harvest.

Seaweed is harvested at Sanbanze from November to April. The production process is, in contrast to shellfish harvesting, an expensive one (e.g., the cost of the machines used to wash and dry seaweed ranges between ¥20 and ¥60 million [roughly $200,000–

$600,000], the more expensive machines producing better-quality seaweed). And, because only short-term fishing rights can be obtained, fewer and fewer fishermen are willing to make the investment. Ironically, the few fishermen who continue to pursue the production of seaweed have seen per capita production increase because there is so little competition for the most highly productive seaweed beds. A cooperative system for the sale of seaweed was organized in Funabashi. Because it promotes the upgrading of harvesting and processing techniques and, thus, the production of a high-quality product, it has had a decisive influence in establishing the market price.

Because the environment of Sanbanze has deteriorated steadily since the 1990s—the result of past reclamation, the dredging through the bay of a sea route, contamination of the tidal flat by polluted river water, and recurring blue tides—the catch of shellfish there has declined, as has the quality of seaweed produced. The preservation of Sanbanze is, therefore, vital, from the standpoint of both the fishing industry and the environmental movement (see figure 12.8).

Consider, for example, the decrease in the shellfish catch (see figure 12.9). The overall fish catch in 2001 was 43.5 percent of that in 1984. When the catch is broken down by type, however, it becomes evident that the decrease is even worse for shellfish: the total catch was only 9 percent of that of the peak year, that of asari clams 8 percent, and that of trough shells 22.6 percent. Fishermen used to be able to catch a large amount of shellfish in a short time, but now it takes a whole day to collect at most a few kilograms. This has discouraged fishermen from digging for shellfish, which, in turn, has led to the shrinking of the shellfishing industry. Where there were 148 operators in 1943, there were 119 in 1993 and 79 in 1998.

A comparable case is the seaweed industry. While there was no decrease in the pro-

duction of seaweed sheets between 1984 and 2001, profits dropped 25 percent over that period. This suggests that the deterioration of the fishing grounds has affected the quality of the seaweed harvested, making the production of high-quality seaweed sheets more difficult and, thus, an expensive industry even more expensive. Not only does the situation discourage new investment/reinvestment, but, as in the shellfishing industry, the number of operators has been steadily decreasing in recent years, from 68 in 1983, to 38 in 1988, 26 in 1993, and 20 in 1998. Consequently, the Funabashi cooperative no longer has its former influence in setting market prices.

An instructive comparison is the trawl-fishing industry. The number of operators has, largely, increased: from 16 in 1983, to 19 in 1988, 25 in 1993, and 21 in 1998. And the catch in 2001 increased to 122 percent of that in 1984. One reason, of course, for the increased catch is the increased number of operators. But another is, possibly, a subtle improvement in the environment of Tokyo Bay overall. Still, the evidence is inconclusive—for example, the catch of sea bass has increased but that of flounder decreased precipitously.

The Grassroots Movement to Save Sanbanze

Although reclamation work at Sanbanze was scheduled, it has not, in fact, been carried out. The local grassroots movement played an important role in stopping it. And there is optimism that the movement's influence will continue in the sense that its work will provide a model for the conservation of Sanbanze and the restoration of Tokyo Bay.

It was the Chiba prefectural governor, Domoto Akiko, who revived the grassroots movement to save Sanbanze. In 2001, hoping to get citizens involved in the policymaking process, she organized what is known as the

Kenji Yamazaki and Tomoko Yamazaki

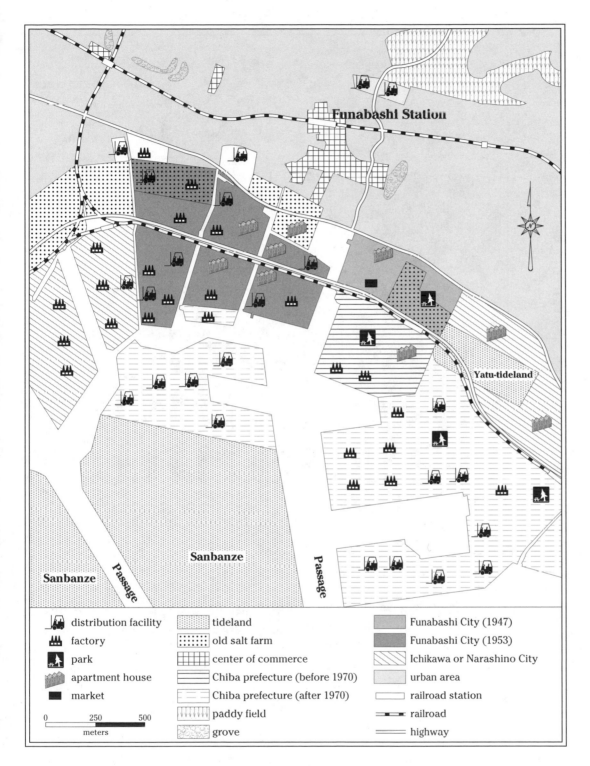

Legend

distribution facility	tideland	Funabashi City (1947)
factory	old salt farm	Funabashi City (1953)
park	center of commerce	Ichikawa or Narashino City
apartment house	Chiba prefecture (before 1970)	urban area
market	Chiba prefecture (after 1970)	railroad station
	paddy field	railroad
	grove	highway

0   250   500
meters

Figure 12.8. Reclamation area and present coast of Funabashi.

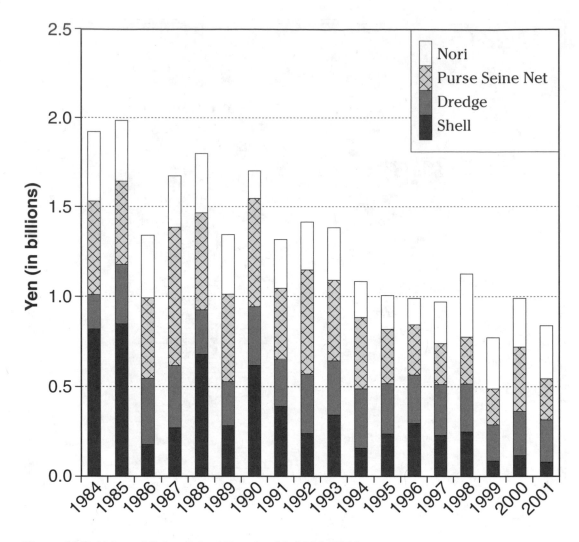

Figure 12.9. Value of fish catch at Funabashi, 1984–2000.

Sanbanze Roundtable Conference, a public roundtable discussion of ways to preserve the tidal flats. Among the participants in the thirty-odd meetings subsequently held were local activists, fishermen, academics, and government officials. At the conclusion, it was decided to establish the Council for Creating Plans to Restore Sanbanze [Sanbanze saisei keikaku kento kaigi] to develop conservation programs based on the ideas discussed at the roundtable sessions.

What accounts for the greater success of the Sanbanze grassroots movement than earlier environmental movements? Two factors, we think. First, earlier environmental movements tended to be narrowly focused on one solution to the problem with which they were confronted, alienating even sympathizers who happened to view things differently. The Sanbanze movement, by contrast, invited the participation of as many people, with as many different ideas, as possible. Second, whereas earlier movements became politicized and radicalized as they grew more powerful, the Sanbanze movement did not, seeing its mission instead as providing information (i.e., a variety of perspectives) and encouraging citizen participation in the conservation movement.

The process by which the Sanbanze environmental movement grew can be characterized as the repetition of the following series of steps: (a) the investigation of a problem; (b) the reporting of results; (c) the presentation of relevant issues and a plan of attack; (d) the holding of a town meeting; (e) organizational expansion and solidification; and (f) the identification of a new problem and its ensuing investigation. This series of steps can be shown to have been repeated four times to date.

The First Stage (1983–1987)

In March 1983, as we have seen, a group of fishermen organized the grassroots group the Tokyo Bay Conference in response to a plan to reclaim a portion of Tokyo Bay bordering on Ichikawa. In June 1985, intent on reviving the plan, the Chiba prefectural parliament approved a new reclamation proposal, and, in September of that year, the Chiba municipal government drew up the so-called Blue Plan for the reclamation of Sanbanze (the newly created land to be used for a sewage treatment plant and highway construction), with construction scheduled to begin in 1988. These developments prompted the organization of several citizens' movements.

For example, in October 1988, after conducting a study of Sanbanze concluding that the tidal flats were indispensable to the fishing industry and equally important for environmental education and recreation, the Ichikawa Sea Study Circle [Ichikawa-no-umi kenkyukai] held a symposium, titled "Do Not Fill Up Tokyo Bay Any More" [Koreijo umeru-na Tokyo-wan!], critical of the Blue Plan. Also, a Mr. Oono, the leader of the Tokyo Bay Conference, translated "The San Francisco Bay Plan" and, at his own expense, published it (see Oono 1984);[2] two years later, he published an essay of his own about how a grassroots movement like the one that fought to conserve San Francisco Bay was needed to save Sanbanze (see Oono and Oono 1986). Other efforts to involve as many local resi-

dents in the movement as possible included the Funabashi Sardine Festival [Funabashi iwashi matsuri], held in October 1984, and the Funabashi Sanbanze Festival [Funabashi Sanbanze matsuri], held annually from August 1986 to August 1990—both organized by the Tokyo Bay Conference—whose focus was educating people about the reality of Tokyo Bay, especially Sanbanze. As an important part of the Sanbanze Festival—held on the last Sunday of summer vacation so as to attract schoolchildren—fishermen associated with the Tokyo Bay Conference let the children board their boats to be taken to Sanbanze so that they could see how clear the water was and experience firsthand the marine animals that live there. The parents, meanwhile, were entertained onshore with jazz, beer, and tempura made from fish caught in Tokyo Bay.

As a result of all this grassroots activity, public pressure grew so great that the prefectural government had no choice but to cancel its Sanbanze reclamation plan.

The Second Stage (1988–1996)

The Sanbanze Study Circle [Sanbanze kenkyukai] was formed in January 1988, an offshoot of the Ichikawa Sea Study Circle. True to its motto, "Learn from the field" [Genba kara manabu], it applied for and received a two-year Toyota research grant supporting an in-depth study of Sanbanze. The results of the study were instructive. First, it taught us that our attention should be focused not just on the fish that Sanbanze supports but also on the entire, interrelated food chain. Furthermore, it revealed that oxygen demand generally is more important than biological oxygen demand specifically in understanding the effects of pollution, and it introduced the notion of "intimate aqua space" [sinsui-kuukan], a natural resource that can be exploited (for both recreational and commercial purposes) if reclamation is not pursued. The results of the study were published as the *Sanbanze Guidebook* (see Sanbanze Study Circle 1990).

Meanwhile, Chiba Prefecture had, in 1990, completed a new basic plan that involved the reclamation of 740 hectares of the bay. The Sanbanze Study Circle addressed the plan, arguing that, rather than reclaiming the tidal flats to create parks and beaches, it was far more effective, because environmentally sound, to develop Sanbanze in a way that did not change its natural ecology (see Sanbanze Study Circle 1990). In March 1991, the National Central Port Council [Chuuoo koowan shingikai] approved the plan. However, the Environment Ministry evidently agreed with the study circle's conclusions because, at the meeting of the council at which the project was approved, it made a special request that the environment of Sanbanze not be damaged.

Two study groups grew out of the Sanbanze Study Circle: the Study Circle to Maintain Sanbanze in the 21st Century [Sanbanze wo niju isseiki ni nokosu kai] and the Sanbanze Forum [Sanbanze foramu]. The latter took citizens out on the bay to demonstrate the abundance of Sanbanze. It published *Reclamation Issues at Sanbanze: Resource Book* (see Sanbanze Forum 1991). And, in November 1991, it held an international symposium—funded by a follow-up subsidy to the Sanbanze Study Circle's earlier Toyota grant—titled "Sanbanze, a Natural Sea Area in the Urban Area" ["Sanbanze toshi-no-naka-no shizen-kaiiki kokusai simpogiumu"]. In fact, the Sanbanze Forum's efforts were so successful that it made Sanbanze a topic in development and conservation circles worldwide (Sanbanze Study Circle 1998). At the same time, more groups of citizens concerned about the preservation of Sanbanze sprang up, including (as a direct result of the Sanbanze Forum's symposium) the Tokyo Bay International Cheer Group [Tokyo wan kokusai ouendan] and the Sanbanze Forum Urayasu [Sanbanze foramu Urayasu]. As these groups grew in number, they also grew in size, ever more widely publicizing, and, thus, educating people about, the importance of conserving Sanbanze.

In May 1992, Chiba Prefecture established the Environmental Council [Kankyo kaigi], which had as its charge evaluating the issues raised by the proposed reclamation of Sanbanze. However, the council's meetings were closed to the public. In March 1993, Chiba Prefecture released the revised reclamation plan. One month later, some members of the Environmental Council inspected Sanbanze, a surprising move on the government's part and one showing that some attention at least was finally being paid to environmental issues. This was most likely because, with the growth of grassroots preservation movements, the proposed reclamation of Sanbanze was attracting nationwide attention. For example, the Nature Conservation Society of Japan (NCSJ) submitted a report to both the national and the prefectural governments that, in its opinion, reclamation of Sanbanze was problematic. Other groups brought to public attention the fact that the tidal flats serve as both a nesting and a resting place for migratory birds, making Sanbanze eligible for protection under the Ramsar Convention.

Meanwhile, responding to the fact that the neighboring Yatsu Tideland had, in June 1993, been designated a registered marsh, the Sanbanze Forum held a meeting to explore the possibility of having protected status conferred on Sanbanze under the terms of the Ramsar Convention. In 1994, as a result, 241 parties petitioned the prefectural government, requesting that it seek Ramsar protection. The same group also asked the World Wildlife Fund–Japan (WWFJ) to send a team to inspect Sanbanze, which it did—and then immediately petitioned the prefectural governor, demanding the cancellation of the reclamation plan.

In May 1995, the Sanbanze Forum published a book (Onoo and the Sanbanze Forum 1995) presenting its perspective on

Kenji Yamazaki and Tomoko Yamazaki

nature conservation as well as some concrete methods to achieve that end (e.g., the planting of amamo, a kind of seaweed that purifies water). In another positive development, the Chiba Prefecture Environmental Council, reversing its previous, pro-reclamation position on the election of new, conservation-minded members, issued in November 1995 a report pronouncing Sanbanze a precious natural resource and indicating that further inquiry into the role it played in purifying the ecosystem was called for. Then, early the next year, the Study Circle to Maintain Sanbanze in the 21st Century published *The Thought of the City and the Sea* (Furota 1996), which collected the opinions of researchers and activists across the spectrum about the conservation of Sanbanze, and a lecture/presentation by the novelist Siina Makoto and the underwater photographer Nakamura Masao attracted fifteen hundred people. But still the conservationists and the prefectural government found themselves diametrically opposed, the latter continuing to push for the reclamation of Sanbanze, largely because the venture was profitable.

Nevertheless, the year 1996 marks the end of the second stage of the grassroots movement to save Sanbanze—because that is the point at which the movement had achieved for the general public a voice in policymaking. And it was a July symposium sponsored by the Sanbanze Forum and the WWFJ that truly demonstrated this. The nonpartisan symposium, titled "Japanese Seas Will Change from Sanbanze" [Sanbanze kara nihon no umi wa kawaru], was held at the Makuhari Messe convention center, attracting over a thousand people. Yukawa Reiko, an independent music critic, served as the moderator, and the panelists included the environment minister, the chair of and a representative from the Chiba prefectural parliament, prefectural government administrators, researchers, and environmental activists. Iwadare Sukio, the environment minister,

addressed the panelists and audience, saying: "I think that the preservation of Sanbanze by any means possible is the overwhelming desire of the local people." For recognizing this publicly he received a standing ovation. But even more important is his remark: "It is private citizens who are mainly responsible for the conservation of Sanbanze [i.e., getting the pollution cleaned up]." This signified that policymaking in Japan had entered a new era (WWFJ and Sanbanze Forum 1996).

The Third Stage (1997–2001)

By continuing to conduct investigations, offer counterproposals, and hold symposiums, the Sanbanze Forum managed to turn public opinion solidly in favor of conservation and, thus, continue to check the reclamation forces. So strong was its influence, in fact, that, even though it backed no candidates in elections (fearing that the movement would be jeopardized if candidates it backed lost), no one running for public office at either the local or the national level was willing to support reclamation publicly.

New, pro-conservation mayors were elected in Funabashi City in June 1997 and in Ichikawa City in November 1997, by an electorate favoring the conservation of Sanbanze. In 1998, the supplementary investigation committee of the Chiba Prefecture Environmental Council issued an interim report confirming that the Sanbanze ecosystem supported a wide variety of marine and bird life. In response to this report, the governor announced that the planned reclamation would be scaled down somewhat. Ichikawa and Urayasu city council members who favored the preservation of the tidal flats wrote both the prefectural governor and the prefectural parliament advocating that the plan be even more drastically curtailed. In November 1998, Matsuzaki Hideki, an activist and member of the Sanbanze Forum, was elected mayor of Urayasu City, further proof of the growing influence among local citizens of the environmental movement.

The Sanbanze controversy became linked with the controversy over the Nagoya city government's plan—announced in the 1960s—to reclaim the Fujimae Tideland for use as a garbage dump, a plan strongly criticized by the Environment Ministry. The Sanbanze Forum organized a symposium backing the Environment Ministry's position. In February, the NCSJ, the WWFJ, and the Wild Bird Society of Japan jointly appealed to both the Chiba prefectural government and the Environment Ministry, demanding the preservation of Sanbanze. Environment Minister Manabe then inspected Sanbanze, announcing the following day that it should be preserved by any means possible. In April, elections to the Chiba prefectural parliament were held, and every candidate supported the conservation of Sanbanze. In June, the Chiba prefectural government proposed reducing the area to be reclaimed from 740 to 101 hectares. Minister Manabe responded: "Even if it is as small as 101 hectares, it is still artificial." At the end of the year, plans to reclaim Fujimae Tideland were canceled, giving rise to hope that plans to reclaim Sanbanze would soon be canceled as well.

In December 1999, the Chiba prefectural government made public its estimate of the effect of the scaled-down reclamation plan: it would still affect wildlife such as waterfowl. In February 2000, the Environment Ministry expressed its view that even the scaled-down plan would still significantly adversely effect the marine life and birds that depend on the tidelands and that, if development was, in fact, necessary, other plans should be made.

In March 2000, the Sanbanze Forum held its ninth symposium, "Sanbanze: An Environmental Conference to Open the Door to the 21st Century" ["Sanbanze: 21 seiki-no-tobira-wo hiraku kankyo kaigi"], meant to encourage the participation of the general public in the policymaking process. In the general election campaign held in June, the activist and Sanbanze Forum member Tanaka Kou was elected to the House of Councillors. In Funabashi as well, candidates who favored the conservation of Sanbanze were elected to office. And it was in this political atmosphere that the election for prefectural governor was held in March of 2001. Domoto Akiko, whose platform called for a halt to the Sanbanze reclamation plan, won.

The Fourth Stage (2001–)

At a special session of the prefectural parliament held in April 2001, Governor Domoto announced that plans for the reclamation of Sanbanze were to be withdrawn and work begun on a plan to regenerate the tidal flats as *sato umi* [home sea], a place that all could enjoy.[3] She further announced that, in her opinion, such a plan could succeed only if local residents were involved in both the planning and the implementation stages and that, as a first step, a roundtable discussion was to be convened at which all interested parties (activists, academics, and fishermen as well as the general public) were to be represented. Then, in a regular session of the prefectural parliament in September, she reiterated her call for a collaborative effort—or the "Chiba model"—to save Sanbanze. The Chiba model is conceived as the first step in creating a decentralized society, one in which concerned citizens, not policymakers, identify problems, propose solutions, and then implement the resultant policies.

Summary

The grassroots environmental movement to conserve Sanbanze differs considerably from earlier environmental movements (see figure 12.10). It is not narrowly focused on one solution to the problem, preferring instead to consider as many different alternatives as

Kenji Yamazaki and Tomoko Yamazaki

possible. Its proposals are drawn up in such a way that they can be understood by the general public and, thus, achieve a consensus among interested parties. They are also based on careful fieldwork and are, thus, scientifically valid. The movement also avoids direct involvement in politics, preferring to confine its activities to the education of the general public. These tactics have enabled it to establish a partnership with the government that is rare in Japan, where social lines are rigidly drawn and only recently have activists begun working with government administrators. By acting locally and thinking globally, the Sanbanze activists created a model for preservation that was successfully expanded first to Tokyo Bay and currently to endangered coastlines nationwide.

Notes

1. *Blue tide* is a term used to describe a stream of oxygen-deficient seawater. This condition is the result of decaying organic matter consuming the oxygen in the water and releasing hydrogen sulfide. In Tokyo Bay, strong winds periodically stir this anoxic water up from the depths and push it into the coastal zones, essentially suffocating benthos such as shellfish that cannot escape. It is the hydrogen sulfide that gives the water its distinctive color.

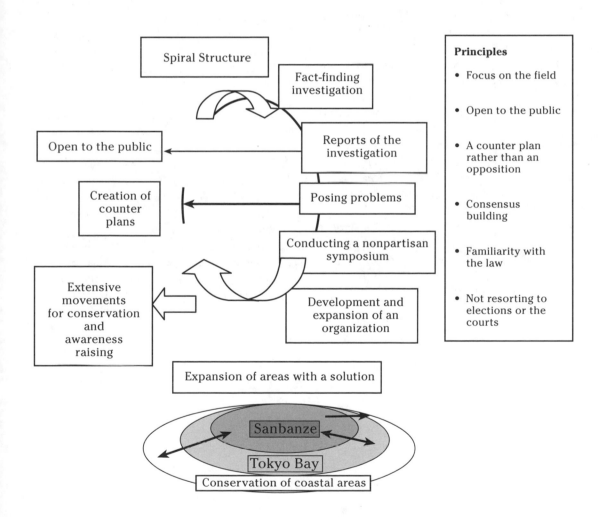

Figure 12.10. Structure of the grassroots movement to preserve Sanbanze.

2. An online version of the original San Francisco Bay Plan can be found at http://www.bcdc.ca.gov/index.php?p=633.

3. The term *sato umi* is derived from *sato yama* [mountains in a hometown]. The idea expressed thereby is that, just as *sato yama* are a bountiful source of nature, so too can *sato umi* be if it is preserved and treasured.

References

Furota, T. 1996. *Sanbanze Bukkuletto* [The Sanbanze booklet]. Vol. 1, *Toshi to umi no siko* [The thought of the city and the sea]. Chiba: Sanbanze wo 21 seiki ni nokosu kai.

Hanayama, Y. 1983. "Tokyo wan umetate shosi" [A brief history of the reclamation of Tokyo Bay]. *Kooai kenkyu* [The study of public pollution], vol. 12, no. 4.

Ishimure, M. 1965. *Kukai-jodo: Waga Minamata-byo* [Harsh life and the pure land: My Minamata disease]. Tokyo: Kodan-sha.

Onoo, S., and Sanbanze Forum. 1995. *Tokyo Wan Sanbanze—Umi wo aruku* [Sanbanze, Tokyo Bay: Walk the sea]. Tokyo: Sanich-shobo.

Ooga, T. 1972. *Watarase-gawa* [Watarase River]. Tokyo: Shinsen-sha.

Oono, K., trans. 1984. *Sanfuransisuko Bei Pulan* [The San Francisco Bay Plan]. N.p.: n.p.

Oono, K., and T. Oono. 1986. *Tokyo Wan de sakana wo ou* [Let's catch fish in Tokyo Bay]. Tokyo: Soushi-sha.

Sanbanze Forum. 1991. *Sanbanze no umetate mondai siryoshu* [Reclamation issues at Sanbanze: Resource book]. Chiba: Sanbanze Forum.

———. 2000. *Sanbanze no hozen to tosi no saisei* [The conservation of Sanbanze and the regeneration of the city]. Chiba: Sanbanze Kenkyusho.

Sanbanze Study Circle. 1990. *Sanbanze gaido bukku* [Sanbanze guidebook]. Chiba: Sanbanze Kenkyusho.

Sanbanze Study Circle, Sanbanze Forum, and Study Circle to Conserve Sanbanze until the Twenty-first Century. 1998. *Sanbanze Bukkuletto* [The Sanbanze booklet]. Vol. 3, *Sanbanze-tosi no nakano sizenkaiiki—kokusai symposium post-waterfront wo sekai kara* [The international symposium on the urban sea area, Sanbanze: Creating a posuto wootaa furonto (postwaterfront) from the world]. Chiba: Sanbanze Kenkyusho.

Shimin-ban: Funabashi no umi wo ikashita machizukuri Canada Grandville no jirei to shimin teigen [The citizens' version: City planning by making good use of the Sea of Funabashi—a case study of Grandville, Canada, and the proposal by its citizens]. 2003. City Planning Series 1. Chiba: *Shimin-ban* Editorial Committee.

Tajiri, M. 1972. *Yokkaichi si no umi to tatakau* [Fight with the dead sea, Yokkaichi]. Tokyo: Iwanami-shoten.

———. 1988. *Tokyo Wan no hozen to saisei* [The conservation and regeneration of Tokyo Bay]. Tokyo: Nihon-Hyoron-sha.

Ui, J. 1971. *Kougai genron* [Principles of public pollution]. Tokyo: Aki-shobo.

World Wildlife Fund–Japan (WWFJ) and Sanbanze Forum. 1996. *Sanbanze Bukkuletto* [The Sanbanze booklet]. Vol. 2, *Sanbanze kara nihon no umi ha kawaru, Neo Tokyo Bay Plan kankyo symposium* [The sea will change from Sanbanze, Neo Tokyo Bay Plan: The environmental symposium]. Chiba: Sanbanze-o-21 seiki-ni-nokosu-kai.

Yamazaki, K. 1988. "Tokyo Wan umetate to toshikinkou gyogyou" [The reclamation of Tokyo Bay and fishing in an urban area]. *Toritsudaigaku-fuzoku koko kiyo* [Annals of Tokyo Metropolitan University High School], no. 10:2–19.

Part IV

Seeking to Preserve the Natural Environment

Chapter 13

Citizens for Saving the Kawabe

An Interplay among Farmers, Fishermen, Environmentalists, and the Ministry of Land, Infrastructure, and Transport

Todd Stradford

The Kawabe River flows south through the Kyushu Sanchi Highlands of Kumamoto Prefecture, in central Kyushu, through the villages of Izumi Village, Itsuki Village, and Sagara Village, joining the Kuma River in Hitoyoshi City (see figure 13.1). The area is noted for its clean waters and is one of the few places where *yamame,* a type of trout, can survive in Japan outside Hokkaido. It is also largely rural, with bottomlands in crops and the surrounding mountains in planted conifers. A single highway, Route 445, traverses the river valley. This has been a remote region throughout Japanese history, and out-migration and *dekasegi,* the practice of leaving the area temporarily for work elsewhere, have long been characteristic. In recent years, these trends have continued, leaving a diminished and elderly population.

The village of Itsuki is used to illustrate. Census data from the period 1986–2001 indicate a decline in the number of households and in the population generally. Both farming and nonfarming numbers are decreasing at the same rate. The economic realities of agriculture in Japan are causing younger people to leave the area for Kumamoto City and Fukuoka City, where they hope to make a better livelihood. The village governments are hard-pressed to find jobs for citizens and a tax base for services (see figure 13.2).

The Kawabe Dam Project Beginnings

In 1962 and 1963, timber planted after the Second World War was harvested using clear-cut methods, stripping large areas of the mountainsides. Under normal weather patterns, the affected area would not have been large enough to cause undue runoff into the streams. However, the years 1963, 1964, and 1965 saw above-average rainfall, causing heavy runoff in the entire Kuma River basin and continuous downstream flooding (see photograph 13.1).

The Ministry of Land, Infrastructure, and Transport (MLIT) saw a need to take steps to prevent future flooding and designed a plan, released in 1966, for a dam on the Kawabe. The proposed dam would be located in the northern part of Sagara Village, be 107.5 meters in height, and create a reservoir of 391 hectares that would back up into Itsuki Village, enough to flood the village. The cost of building the dam, relocating the village and the outlying households, and building new roads above the new reservoir was estimated to be ¥410 billion (US$3.2 billion). In 1968, the plan was amended, irrigation channels

Photograph 13.1. Flooding along the Kuma River in Sakamoto Village. (Courtesy Todd Stradford.)

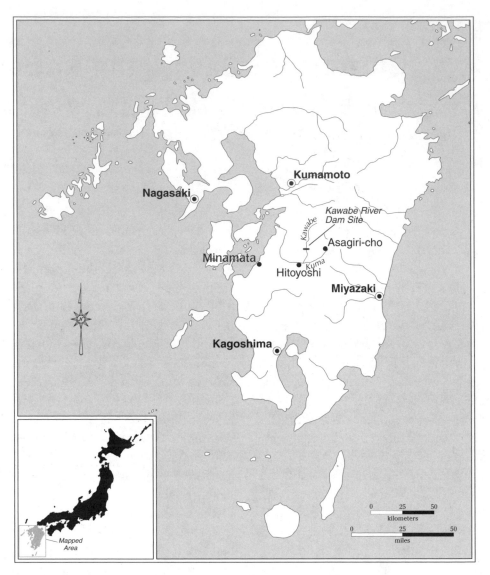

Figure 13.1. Kawabe River Dam site and Kumamoto Prefecture.

and hydroelectric generation being added (see photograph 13.2).

The Formation of Dam Opposition

Villagers

After the announcement of the dam, the ever-present threat of *mura hachibu* [lit., "rejection by the village"; social ostracism] was enough in this conformity-minded society to quell any nascent opposition among the vil-

Photograph 13.2. Artist's rendition of the Kawabe Dam. (Courtesy Old Itsuki Village Web site.)

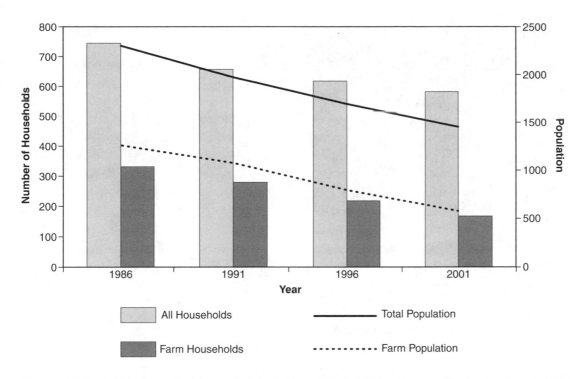

Figure 13.2. Itsuki, households and population, 1986–2001. *Source:* A September 2005 (Heisei 9, Month 9) in-house publication of the Kyushu Government Agricultural Statistics Information Center [Kyushu nohseikyou toukei jouhoubu].

lagers. Residents began accepting payments for lost land and dwellings and agreeing to the relocations proposed under the MLIT's loss-compensation plan.

Farmers

Farmers, however, were not so compliant. In reaction to the high proposed irrigation-service fee, they organized, and twenty-one hundred of the four thousand affected farmers filed suit in the Kumamoto district court against the Ministry of Agriculture, Forestries, and Fisheries (MAFF)—which had informed the MLIT that they were in favor of the plan without actually consulting them—for falsely representing their position. Farmer demonstrations in front of the Kumamoto court made the dam project high-profile news, and the nation began to take an interest in the case. The case was eventually dismissed at the prefectural level, but the farmers appealed to the high court, which, on May 16, 2003, ruled in favor of the farmers, dealing a heavy blow to the state project. Repercussions of this decision are felt to this day (see photograph 13.3).

Fishermen

The members of the Kuma River Fishing Cooperative [Kumagawa kyokyou kumiai], observing the original court case, began to organize and, in group meetings, twice voted to reject the ¥1.65 billion compensation package for fishing losses resulting from the dam's construction. Until the district court's ruling in favor of the farmers, this was the greatest legal obstacle to the construction of the dam.

The Kawabe River is the habitat for the *shaku-ayu* (*Plecoglossus altivelis*), a 30-centimeter-long sweetfish savored by Japanese in restaurants all over the country. The river produces some of the largest of these

Photograph 13.3. Farmers in front of the Kumamoto courthouse protesting water fees from the new dam. (Courtesy Yoko Nishida.)

fish, which is why the prefix *shaku-* (an archaic unit of length measuring 30.3 centimeters) is added to the name. *Shaku-ayu* have a life span of only one year, giving them only one warm season to achieve their size. They are also territorial, each male claiming a portion of some fast-moving river. Already, dam construction in other parts of Japan has seriously affected the species' habitat, reducing its numbers. The fishermen argue that the new dam will diminish its habitat even further—thus affecting their livelihood—even below the dam, where the river will still flow freely, because the temperature of the water there will be 3°C cooler, preventing the fish from growing to full size. The fishermen's right to their livelihood is guaranteed by the Japanese constitution, and, unless they willingly give up this right, it cannot be forcibly taken away. However, the MLIT is attempting to do just that by invoking the principle of eminent domain. In 2000, the ministry suspended the dam project for three years in an attempt to reach a settlement with the fishermen. However, the fishermen refused to budge, and the ministry petitioned the Japanese courts to allow it to proceed. The

decision was due in February 2003, but delays ensued, and, on September 15, 2005, before a decision could be issued, the MLIT withdrew its application to revoke the fishing rights, effectively putting the project back where it was in 1966.

Recreation and Culture

Several companies in Hitoyoshi City, such as the Land Earth Outdoor Sports Club, use the river for rafting trips. They say that the river offers some of the best whitewater in Japan and that they do not wish to see their business options diminished. It is also a weekend destination for Kumamoto kayakers, who would also lose its use.

The villagers of Itsuki have, in the past, sent family members to work elsewhere since the remote village provided few jobs. Many times, young girls were sent to become nannies for the children of wealthy families. One such young woman began singing a song to put her charges to sleep, and it became so popular that it is the best-known lullaby in Japan, the "Itsuki no komoriuta" [Itsuki lullaby]. The dam would submerge the origin of the story and the lullaby.

Citizens for Saving the Kawabe

In 1996, approximately two hundred people organized to become the group Citizens for Saving the Kawabe [Seiryu Kawabegawa wo mamoru kenmin no kai] (CSK), and, under the leadership of Kunitoku Yasuyo and Nishida Yoko, the group began to fight to stop the dam construction. Presently, the group has about four hundred members and has been joined in its endeavor by several other nongovernment organizations (NGOs), some themselves offshoots of the CSK.

Platform

The original goals of the Kawabe River dam project are no longer pertinent. In the years since the project's inception in 1966, the flood control, hydroelectric power generation, and irrigation have been achieved by other means.

Flood control. A newly completed series of levees, combined with a complete forest growth cover, have dramatically reduced the possibility of flooding. In 1995, the amount of rain that the MLIT predicted would cause flooding, 440 millimeters in two days, fell on the watershed, and there was no flooding. A dam would actually increase the chance of flooding if water were released simultaneously by it and the Ichifusa Dam on the Kuma River, a possible scenario in a heavy rainfall. In March 2003, Kaneyuki Nakane, a professor at the Hiroshima Forestry School [Hiroshima daigaku, Daigaku inseibutsukenkagaku kenkyuka], announced to a group in Hitoyoshi City that, owing to "forest reappearance," the dam would be unnecessary ("'Chusuiron wa saido hitsuyou'" 2003). A report published slightly earlier by a private think tank showed that "reinforcing existing banks at the cost of 7 billion yen would

have the same effect as the dam in preventing flooding" ("Fishermen Win Dam Battle" 2001).

Hydroelectricity. There are presently four hydroelectric power generators in operation along the Kawabe River. With the completion of the Kawabe River Dam, all will either be under water or cease operations. The Kawabe River Dam generator that is to replace them will, however, produce less electricity (16,500 kilowatts per second) than the four existing generators do at present (18,900 kilowatts per second).

Irrigation. The MAFF is promoting the State-Run Comprehensive Kawabe River Reclamation Project, in which the water from the Kawabe River Dam reservoir will be used for irrigation.[1] However, farmers in this area have enough water for farming already and do not need any additional irrigation facilities. Furthermore, most cannot afford to pay the proposed irrigation fee. In 1994, there was a drought in the upper reaches of the Kuma Basin and, consequently, almost no harvest in that area despite irrigation water from the Ichifusa Dam. Nonetheless, there was a plentiful harvest in the area where farmers got their water from the Kawabe River.

Legality and Cost-Benefit Analyses

A cost-benefit analysis is standard procedure for public works to ensure that the returns on projects are high. In Shiga Prefecture, with the aid of the Japan Environmental Lawyers Federation, "an administrative lawsuit has been filed to rescind the permit to construct Eigenji dam. While the dam will cause adverse effect upon ecosystem as well as the historic view, the main issue is the illegality of the construction plan as this plan is a typical wasteful public work. A lawsuit focusing on a cost-performance of public expenditure is maybe the first case in Japan" (Kagohashi, n.d.). If the figures for Yatsushiro are not included in the "cost-performance" calculations, the result is only a 73 percent return on the investment, a figure falling far short of present public works standards. The Kawabe Dam Project could generate a similar lawsuit.

Goals

To prevent the building of the dam, the CSK employs the standard tactics of educating people about the issue locally, nationally, and internationally and keeping the issue in the news. Among other projects, it received money from the National Federation of Workers and Consumers Insurance Cooperatives and, in 1999, had six thousand color leaflets printed and distributed on the streets of Kumamoto and Fukuoka (see photograph 13.4). In April 2002, it presented a panel exhibition at the Earth Week conference in Hiroshima. It applied for and received a grant from the World Wildlife Fund and, in 2002, a new pamphlet was printed for distribution to a wide range of citizens' groups. The organizer of the Shakuayu Trust [Shaku-ayu torasuto], the support group for area fishermen, places leaflets in all the boxes of *ayu* it distributes nationwide. The CSK is appealing to other groups to cooperate in the campaign as well in order to further raise public awareness; as of 2003, there were ten NGOs working to

Photograph 13.5. Gathering signatures in Kumamoto City. (Courtesy *Kumanitiniti shimbun.*)

oppose the dam project. In February 2003, the CSK again took to the streets to gather signatures on a petition against the dam that it presented to Governor Shiotani on March 3 (see photograph 13.5).

The group has also held meetings and outings in the Kawabe River area aimed at educating the public about the issues surrounding the proposed dam. One such outing, during which volunteers planted trees in the Kawabe watershed, featured lectures during breaks about the role played by trees in slowing runoff. In the spring of 1999, a bus tour was organized to run from Fukuoka to Minamata to spread the word about the Kawabe River (see photograph 13.6).

Informing the World

Photograph 13.4. Information dispersal on the streets of Fukuoka, March 12, 2000. (Courtesy *Kumanitiniti shimbun.*)

From January 20 to January 23, 2002, Nishida Yoko attended the Rivers Watch East and Southeast Asia (RWESA) conference and

Photograph 13.6. Lecturing during a break while planting trees, March 3, 2000. (Courtesy Nishida Yoko.)

presented the CSK case to save the Kawabe. RWESA is a network of NGOs and citizens' organizations from East and Southeast Asia organized in July 2000 to stop destructive river-development projects and to restore rivers to the communities that depend on them. The network currently consists of around thirty-five organizations. RWESA responded to the presentation by issuing a petition to Prime Minister Koizumi and the MLIT. The Kawabe Dam Project as well as other environmental projects are now followed regularly on the RWESA Web site, which provides links to the latest information.

Keeping It in the News

Since January 2002, articles on the Kawabe Dam have appeared almost daily in the *Kumanichi shimbun* (see figure 13.3).[2] Until March 2003, a fair number covered the actions of citizens' groups. After that time, however, such stories almost completely disap-

peared as the confrontation assumed national proportions. The majority of stories now concern the farmers' court victory and its repercussions and the ongoing attempt to take away the *ayu* fishermen's livelihood.[3] Talks began again in 2006, mainly concerning water rights and flooding concerns. The topic stays in the news of its own momentum without aid of the NGOs.

Appendix

PETITION

To Review Dam Construction in Japan
And To Stop Land Expropriation for the
Tokuyama, Kawabe,
and Tomada Dams
To: Prime Minister Junichiro Koizumi and
Minister of Land, Infrastructure and Transportation Chikage Ogi
February 21, 2002

We, the participants of RWESA's Second International Assembly, call upon the Government of Japan to stop the land expropriation procedures for three dams in Japan and demand that a transparent and democratic review of dam construction is conducted.

Among the members and participants of RWESA* are NGOs and people who have suffered the severe effects of dam projects in Southeast Asia funded with Japanese econom-

---

* RWESA (Rivers Watch East and Southeast Asia) was established in June 2000 at the first East and SE Asia Regional Meeting on Dams, Rivers and People, held in Kong Jiam, Thailand. RWESA was created to develop a strong network in the region to stop destructive river development, restore damaged ecosystems, secure reparations for affected communities, promote alternatives, and broaden public support for affected people. Currently, more than 100 representatives from affected communities, people's organizations, NGOs and universities in 14 countries are members of the network.

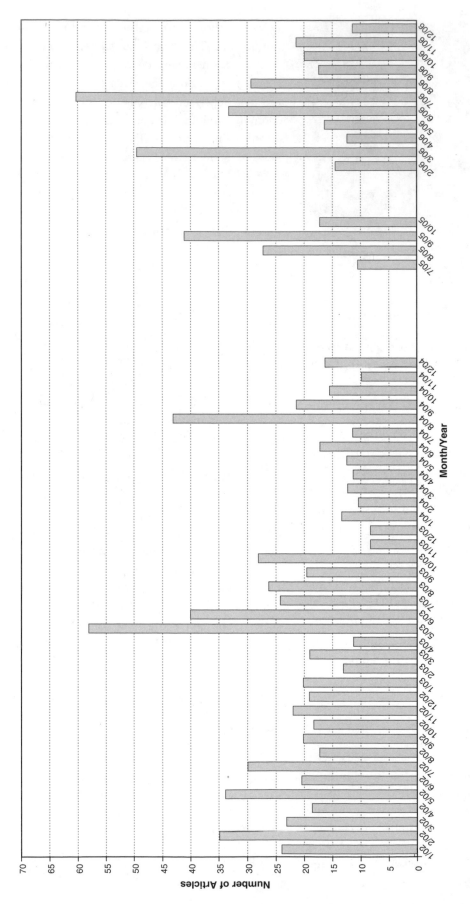

Figure 13.3. Articles on the Kawabe River.

ic aid. Sharing our experiences related to river development projects has highlighted the numerous destructive development projects in East and Southeast Asia. In particular, we are struck by the fact that even in the economically developed nation of Japan, undemocratic processes are propelling the construction of many large dams.

In regard to the land expropriation procedures being used repeatedly in the development of Japan's dams, we demand the following:

1. Reversal of the decision to expropriate land for the Tokuyama Dam

2. Immediate suspension of the land expropriation processes now underway for the Tomada Dam

3. Immediate retraction of the application for the expropriation of fishing rights in relation to the Kawabe Dam

4. A transparent and democratic review of dam construction, including the above mentioned projects, in order to resolve the problems related to dam construction in Japan

Endorsed by Participants of the Second RWESA International Assembly (see separate sheet for names)

For more details, please contact:
National Dam Opposition Network—Japan
Tel: +81 3 5211 5429
Fax: +81 3 5211 5538
Address: 1-7-1 W201 Hirakawa-cho, Chiyoda-ku, Tokyo 102-0093 JAPAN

Expert opinions

It is common practice for environmental movements to bring in experts to support the views of the groups. On many occasions, the groups opposed to the Kawabe Dam have brought in their experts to speak at public meetings. Most recently, a forestry professor from Hiroshima University spoke to a group in Hitoyoshi City just before the Hitoyoshi Mayoral Election. It was his opinion that the reforestation of the hillsides since the 1966 flooding would be sufficient to prevent flooding from occurring again. On November 30, 1997, a Niigata university professor spoke to a Kumamoto City group about the endangered mountain hawk, the *kumataka*. On April 8, 1999, a Meiji Gakuin University Professor was invited to a public meeting in Yatsushiro to discuss "dam construction and fishing rights." The combined NGOs brought 5 professors down from Tokyo to speak at the 7th Open Forum discussions in Kumamoto City. Despite the expense, approximately ¥400,000 for transportation and lodging, the CSK sees this as necessary to win their case. Yoko Nishida states:

"All the members are in a struggle now. They can't afford to do it. We don't have any paid staff member on this issue; we all have [our] own jobs. We need to prepare for the open forum at night after work without sleep for several days. At the same time, we need fundraising for transportation for professors (5–7 people) from another prefecture like Tokyo and fees for photocopies and so on. It costs about 400,000 yen every time. On the other hand, MLIT prepares their case by money from our taxes and it is their jobs, as they are all professionals. It's unfair, I think."

The main goal for bringing in experts is to persuade the Ministry of Land, Infrastructure and Transport to review the Kawabe Dam Project. So far they have refused to do so. Another is to attain an Environmental Impact Assessment for the Kawabe River Project. The EIA law was enacted in June of 1997, well after the proposed project, but an assessment contains consultation with the public and considers the whole ecosystem, and is not limited to solely legal matters. An EIA has never been done on the Kawabe River.

Dam Proponents

The CSK has been fighting well-funded proponents that include MLIT, the construction companies, and some local governments that have benefited by receiving money from the construction companies. "The tradition of pork barrel spending is deeply ingrained in rural Japan, where dying towns get new city halls, soaring bridges, new roads and public museums to give them a facade of prosperity even as their populations dwindle" (*Washington Post*, Monday, September 11, 2000). In the case of the Kawabe Dam Project, the Village of Itsuki has received money that it has used to build an expensive web site, using Macromedia animation, which includes a large portion dedicated to dam construction. The construction companies themselves have given millions of yen to continue the contracts. Construction companies commonly give donations to political parties to secure government contracts. In return, they expect politicians to support the completion of public projects such as dams. According to political-funds reports filed with the authorities, 36 builders operating near the planned Kawabe dam donated a total of 10.4 million yen to the LDP chapter between October and December 2000 alone.

Trends are changing

The Kawabe Dam has become a national concern, through the farmers' case against the Ministry of Agriculture, the fishermen's refusal to budge on their fishing rights, and the efforts of the Citizens to Save the Kawabe. It has now become part of political platforms as the Japanese public slowly becomes aware of their environmental responsibilities.

Naoto Kan, Secretary General of the Democratic Party of Japan, made the dam part of the new DPJ platform. Point 4 reads, "Stop useless public construction. Stop construction project of Isahaya Bay reclamation and Kawabe Dam immediately. Review all the construction projects and put an end to useless ones. Promote [environment] recovery projects" (Yoshida 2003).

Yatsushiro Mayor, Takatoshi Nakashi-

ma, garnered support for his April 9, 2002, election by opposing the dam. He ran as an independent backed by the Democratic Party of Japan, the Liberal Party, and the Social Democratic Party and supported by the Japanese Communist Party.

In February 2001, Nagano Governor Yasuo Tanaka stirred up controversy by declaring that his prefecture will halt construction of concrete dams due to their "heavy burden on the environment." Kumamoto Governor Yoshiko Shiotani announced at a regular meeting of the prefectural assembly on December 10, 2002, that the Arase power dam, run by the prefectural government in Sakamoto-mura [Sakamoto Village], Yatsushiro-gun [Yatsushiro Township], will be completely dismantled. This is a first for Japan removing dams instead of building them. The Kumamoto prefectural government, which had always been an advocate for the dam project, requested the Ministry to account for the necessity of the dam to the general public and held an "Open Forum On Kawabegawa Dam Issue" on December 9, 2001. There have been seven such forums through May 2003. The Governor is now requesting further meetings and discussions between NGOs and local governments. In 2003, Kumamoto Prefecture did not put monies for the dam in the prefectural government budget that it submitted to the Japanese government. This act makes it more difficult for the Japanese government to budget the Kawabegawa Dam project.

Expropriation

"The land ministry announced Tuesday that it will soon seek to expropriate fishing rights" (December 12, 2001). This will deprive the Shakuayu fishing cooperative of its fishing rights in accordance with the Eminent Domain Law. If the application is approved, the ministry will expropriate the fishing rights of the local fishermen's cooperative, which has 1,800 members. This is the first time an agency of the national government has applied for the withdrawal of fishing rights over a dam project. The decision was first due by February 2003, but because of the magnitude of such a decision, as of September 2003 discussions are still ongoing with no end in sight.

Citizens Present Goals

The Citizens for Saving the Kawabe are now working with other environmental NGOs monitoring research on the endangered *kumataka*, the *ayu*, and the water quality of the river. Every month the groups present a signature petition to the Kumamoto governor requesting the discontinuance of the dam to protect the endangered species and preserve the wild nature of the river. The petition presented on 21 September 2003 contained 3800 signatures.

Yoko Nishida sums up the many groups' current efforts: "Now, I'm working for preparing for the 7th Open Forum on the Kawabe issue. The theme is 'The Environment,' especially about the environmental impact on the sea, the endangered species like mountain eagle and so on." The group appears determined to remain active until the dam project is cancelled.

Conclusions

Old projects, such as the Kawabegawa Dam, which have grown into such large pyramids of concessions, contributions, and political promises, have never been reviewed, nor has even holding a review ever been discussed. The construction companies continue to build the peripheral parts of the project, building the diversion tunnel in 1997, new higher roads, and new houses for the people whose homes would be inundated by the new reservoir. The only legality holding back the Ministry and the construction companies was the rights of the *ayu* fisherman, but with the withdrawal

Photograph 13.7. Junior/senior high school being built below the new Itsuki, and above the old Itsuki, June 9, 2006. (Courtesy Todd Stradford.)

of the petition to remove those rights, the Ministry is back to step one, and if they wish to build the dam, they must reformulate its goals and purpose.

The Citizens for Saving the Kawabe began as a single group, uniting the causes of the farmers and fishermen, creating a loose alliance with wildlife preservation groups to further their cause. The Shakuayu Trust and the CSK are now working together to inform the public of their common position. They have worked together with other groups such as the Kawabe River Study Society, the Mountain Hawk Preservation group, and the Kumagawa Shiranui Sea Research Group to combine resources, bring in experts, and reach a greater range of the population. They have brought in topical experts to speak at public meetings to give their cause a foundation based on science. They have garnered support from international organizations and had that support presented to the Prime Minister. They have been very successful in bringing this project to the attention of the Japanese people, and as such, have been extremely successful. The Kawabe Dam Project now is a common occurrence in prefectural and national newspapers. However, only if more politicians are elected who support a free-flowing river or unnecessary public spending, who have the legal ability to stop the dam, will the dam not be built. The CSK and other NGOs actively campaign for candidates against the dam. At the moment, there is no conclusion in sight. Secondary construction continues to take place as if the dam itself were a foregone conclusion. The high profile of the project raised by the NGOs has made all politicians aware that this is now a volatile subject and cannot be rubber-stamped into existence without political fallout. They will continue working as long as others such as Yoko Nishida say, "I believe we still have a way to stop the dam construction." [See photograph 13.7.]

Notes

1. See http://kawabegawa.jp/eng/eng2.html (accessed March 5, 2008).

2. The months in 2005 and 2006 without articles represent missing data, not a lack of articles.

3. Fifty-five articles on Kawabe appeared in the press from January through March 2003. Of those fifty-five, sixteen concerned citizens' groups and their activities, seven covered farmers' appeal to the Fukuoka High Court of a water lawsuit, six covered Kawabe as a campaign issue in the Hitoyoshi mayoral election, five discussed the expropriation of fishing rights, one noted that seven members of the Diet had arrived to review the dam, and the remaining twenty were reports on public meetings and discussion groups.

References

"'Chusuiron wa saido hitsuyou' damu touron shukaikaisai de Shiotani Chiji ["A second discussion on Flood Control is absolutely necessary" says Governor Shiotani at dam debate meeting]. 2003. *Kumanitiniti shimbun,* March 5.

"Fishermen Win Dam Battle, but Gov't to Fight On" (in Japanese). 2001. *Mainichi shimbun,* November 29.

Kagohashi, Takaaki. n.d. "Lawyers Struggle to Protect Nature in Japan." http://www.jelf-justice.org/english/essays/contents/LawyersStrProtectNature.htm (accessed December 12, 2007).

Yoshida, Reiji. 2003. "DPJ Adds Five Pledges to Election Manifesto." *Japan Times,* October 6.

Chapter 14

The Efforts of Japan's Citizens and Nongovernment Organizations to Maintain People-Wildlife Relations in Rural Japan

A Case Study of Monkeys in Mie Prefecture

Kenichi Nonaka

Despite being a small country, Japan is rich in wildlife owing to the diversity of its natural environment. One characteristic of Japanese wildlife is that animals often enter areas where people are living. The human environment often overlaps with animals' habitats, with the result that both humans and animals have become accustomed to sharing the same living environment, though often by necessity rather than by choice. This albeit reluctant coexistence in the marginal areas has allowed familiar relationships to develop between people and wildlife over a long period of time. However, the overlapping of the human environment with animal habitats, resulting from the advance of humans into mountainous areas by means of land development along with changes in people's lifestyles, has also caused conflicts between them.

The relationship between people and wildlife in Japan is diverse. Wildlife conservation is regarded as one of the key environmental issues at the regional level (Akimichi 2003). It is not simply a matter of human-animal conflicts; rather, it is regarded as an environmental issue. Environmental changes affecting animals' habitats have caused them to spread out over a wider area. Their behavior, including feeding, causes problems within the human environment.

This kind of problem occurs in many parts of mountainous rural areas. The kinds of animals that typically are in conflict with humans are deer, boar, serow, and monkeys. Of these, monkeys are the most conspicuous as the damage they cause is both physical and psychological. Both people and their crops are affected by this damage, and monkeys are regarded as being difficult to control as they are clever as well as aggressive. This chapter focuses on the problems caused by Japanese monkeys and how people address them. Described below is how information has been collected and shared and steps taken to alleviate the problem in Mie Prefecture.

Japanese Monkeys and the Damage They Cause

Japan's indigenous monkeys are called *nihonzaru*, or, literally, "Japanese macaques" (*Macaca fuscata*) (see photograph 14.1). They live in troops in mountainous forests from Yakushima Island, a world heritage site in the East China Sea, to the Shimokita Peninsula in Honshu, which is the northern limit of their habitat in Japan. Japanese macaques are around fifty to sixty centimeters in length, with a very short tail and a bright red face. They are omnivorous, eating wild berries, nuts, and small animals. Some groups are designated as protected animals by the Japanese government.

Photograph 14.1. Japanese monkey (*Macaca fuscata*). (Courtesy Kenichi Nonaka.)

Traditionally in Japan, monkeys are familiar to people, and they even appear in folk religions and old tales. In the mountainous areas, monkeys have proved useful as they provide fur, meat, and also medicines (Mito and Watanabe 1999). In addition to performing spiritual roles, monkeys also play a role in entertaining people. A form of entertainment known as *saru-mawashi* [monkey performances] has been a tradition for more than eight hundred years in Japan. These performances are humorous and demonstrate great skill on the part of both the monkeys and their trainers. It can be possible for people to develop friendly relationships with monkeys.

Damage caused by monkeys can be categorized as follows: (1) damage to crops (crops are lost or damaged as monkeys eat, fumble with, or trample on them); (2) damage to people (residents and tourists are harmed or injured, either psychologically or physically); and (3) damage to property (buildings are damaged or looted from) (Muroyama 2003).

Wild monkeys were seldom seen from the postwar era until the 1960s, when there was very little damage due to pressure from hunting and they lived in isolation in the mountains, far from people. During the period between 1950 and the 1970s, natural forests of broad-leaved trees, which were rich in food for the monkeys, were cut down and coniferous trees, which are more useful for timber production, planted in their place. This change of vegetation caused a depletion of food in the monkeys' habitat. Land development, including road construction, also worsened conditions in their surroundings, providing people with easy access to their habitat.

As a result, the distribution of Japanese monkeys became more widespread. In their search for food, they began to approach residential areas in mountainous regions. It became harder to protect the fields from the monkeys' attacks owing to the decline and aging of the population in these areas. As defenses weakened, monkeys began to make more frequent visits to fields and villages, becoming more and more accustomed to people. Since the 1970s, damage caused by monkeys has been reported in many parts of Japan. These monkeys have been regarded as harmful animals, and their extermination has been permitted under the Control of Wildlife Conservation Law. Extermination using guns is also practiced by licensed hunters. The number of monkeys killed annually in this way has increased over the past forty years. Extermination, however, is not regarded as an adequate solution. Even if all the monkeys of an entire group were exterminated, their territory would merely be taken over by another group. But, because people don't like to have to kill the monkeys and conservationists especially object to the practice, extermination is a last resort.

Another solution is to drive the monkeys away. Several methods have been developed, such as netting them and scaring them with loud noises. Japanese monkeys, however, are quite dexterous, very agile, and able to adapt easily to changes in their surroundings. (The latter characteristic explains why they are easily tamed.) So they are hard to net, and they readily become accustomed to certain loud noises (e.g., gunshots). Greater success has been had with rocket-type fireworks (see photograph 14.2). Equipment for shooting

Photograph 14.2. Rocket-type fireworks driving monkeys away. (Courtesy Kenichi Nonaka.)

the fireworks has been developed by the Station for Agricultural Experiments in Nara and is provided by the local authorities. It can be operated easily by local residents, even the elderly.

Whatever countermeasures are used, however, it is necessary to collect such information about monkey groups as their demography, location, and movements. Various other local conditions, such as the impact made by humans on the monkeys' habitats and social factors, should also be considered.

A Case Study in Mie Prefecture

The Surveyed Area

Mie Prefecture is located in central Japan, facing the Pacific Ocean (see figure 14.1). It encompasses a vast mountainous area inhabited by monkeys and, consequently, suffers a great deal of damage at the animals' hands.

Examples of crop damage caused by monkeys are shown in photograph 14.3. Monkeys go into fields that are ready for harvesting and eat the crops. They sometimes even pull up crops, seemingly just for fun. It is difficult to predict when the monkeys will make an appearance, so people often realize what has

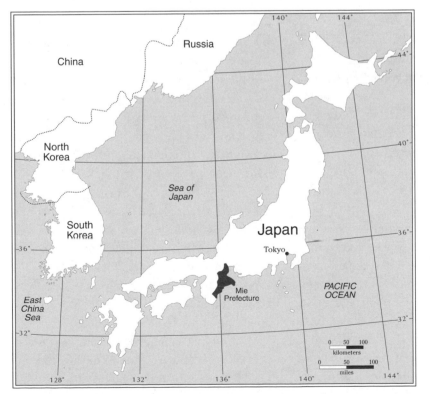

Figure 14.1. Location of Mie Prefecture.

Photograph 14.3. Examples of crop damage caused by monkeys. (Courtesy Kenichi Nonaka.)

happened only after they have gone. It is also difficult to estimate the extent of the damage the monkeys cause in terms of economic value. With crops such as rice, wheat, beans, and mandarin oranges, which are sold on the open market, the cost of the damage can be calculated and the government appealed to for compensation. However, values for crops that farmers grow for their own use are not so easily determined.

Still, as difficult as the situation is, people are unwilling to give up on finding a solution. To do so would mean that the monkeys would be free to feed when and where they choose. And, with a stable supply of food, they would greatly increase in number, causing even greater damage.

One effective means of protecting crops is with electric fencing, but the cost is pro-

hibitive, and careful maintenance is required. Most effective is a cage. Monkeys cannot get in (unless they get hold of the key); they can only stand by and watch the farmers work. A rigid iron cage, however, is also expensive. The cage shown in photograph 14.4, covering an area of fifty square meters, cost ¥500,000—equivalent to approximately US$4,000 (in 2003 terms)—and still covers only half of the crop. Although the farmer wanted to surround the whole area, he could not afford to, the extra cost being equivalent to what he would earn by selling four thousand packs of vegetables. One might think that the cost of a cage too would be prohibitive. But this example just goes to show that, however small a patch of field might be, its yield is precious to its owners. Clearly, quality-of-life, and not just economic, issues are involved here.

Photograph 14.4. An antimonkey cage. (Courtesy Kenichi Nonaka.)

The Development of a System for the Collection and Distribution of Monkey Data

In Mie Prefecture, a specialist in environmental conservation spent the years 2001–2003 working on a project commissioned by the prefectural administration and aimed at developing a system for the radio tracking of monkeys. Adult females monkeys are trapped, fitted with radio transmitters, and then released and tracked. Their position is then charted on a "third mesh map," a map divided into one-kilometer squares, making it easy to locate and identify the monkeys quickly.

Under the recent government drive to create jobs in remote areas, the Mie Forestry Union [Mieken shinrin kumiai] has been commissioned to carry out wildlife surveys, and eighteen people have been employed to observe and report on the behavior and movements of monkeys. After training in

the use of radio telemetry, they patrol their neighborhoods and, after determining the whereabouts of monkeys, transmit the appropriate grid references daily to the Society for Monkeys [Saru no kai], a nongovernment organization (NGO) staffed mainly by university students, which processes the data. Of the 333 monkeys captured by spring 2003, transmitters were attached to 109, allowing the positions of seventy-four of the ninety-seven groups identified in the prefecture to be pinpointed.

From 2002 to 2003, this information was made available to the public on an experimental basis in the form of a GIS (geographic information system) map posted on the prefecture's official Web site. The map was, however, difficult to interpret as the monkeys' actual positions were indicated simply as red squares each representing one square kilometer in a grid overlaid on an aerial photograph. The map also took a great deal of time to create, owing to the large volume of data involved. It thus became clear to those of us working on the project that a new system—one more easily and efficiently operated, even by nonspecialists, in terms of both the inputting and the retrieval of data—was required.

In collaboration with Dr. Tani, and working with his shareware software Mandara, I developed a new geographic information system named SIS, or Saru Information System. Figure 14.2 shows how SIS is used to display the distribution of monkey groups in Mie Prefecture. Identified groups are marked on the map. Their movements can be seen on any scale in any area simply by zooming in or out. Various maps—for example, vegetation, land use, and geomorphologic—are used as base maps, depending on the purpose of the study in question. This new system enables the analysis and display of such tracking data as time and location of observation, feeding conditions, and demographic/social factors.

Building on this technology, in 2004 Saru-

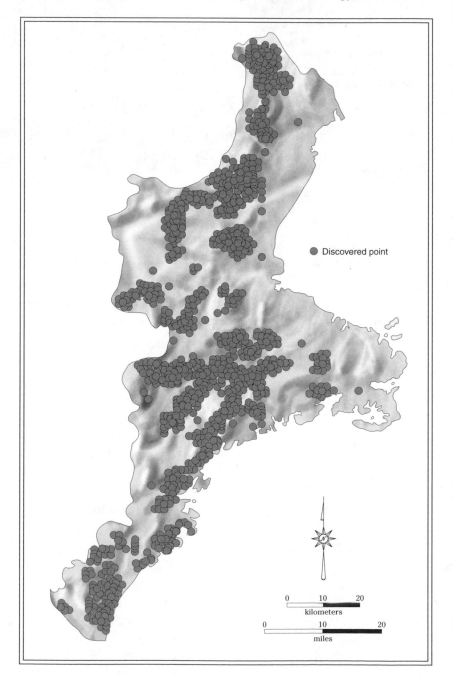

Figure 14.2. Distribution of monkey groups in 2002 shown by SIS.

Photograph 14.5. An example of information displayed on a cellular phone. (Courtesy Kenichi Nonaka.)

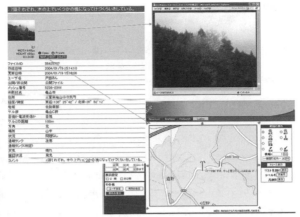

Photograph 14.6. An example of Web site information displayed on a computer. (Courtesy Kenichi Nonaka.)

Doko-Net [lit. "Monkey-Search Network"], a newly formed nonprofit group (see below), the Society for Monkeys, and the Mie prefectural government together used GPS (global positioning system)–enabled cellular phones together with Web-GIS to build an automated system for sending, capturing, and viewing information about the monkey groups that could be used without prior training (see table 14.1). The system was customized using Map Extreme by Map Info as the Web-GIS engine, Eco Mapics by Vertex System for the application software, and Proto Atlas Raster by Alps Mapping K.K. for the maps. All information collected is made available on Saru-Doko-Net's Web site, and subscribers to a mailing list receive updated information about the monkeys' movements as soon as it becomes available (see photograph 14.5). Users are able to find out whether any monkey groups are active in their area (see photograph 14.6). They are also able to obtain information about groups' territories and seasonal movements.

Table 14.1. Information on Discovered Monkeys		
	Content	**Data**
1	Current location	Select from list of place names
2	Monkey group detected	Select from list of monkey groups
3	Means of detections	Radio signal, sighting, encounter, capture
4	Distance from current position	Select from list
5	Direction to monkeys	Select from list
6	Strength of radio signal	Strong, medium, weak
7	Location type	Settlement, crop or rice field, forest, N.A.
8	Situation	No problem, crop damage, humans harmed, N.A.
9	Warning level	Danger, advisory, safety
10	Weather conditions	Sunny, cloudy, rainy, snow, showers, foggy
11	Comments	Input freely

Note: Name of sender, the date and time, and mesh code number on the map are automatically added by the host computer.

The Continuing Effort

There has, in fact, been a rapid escalation in recent years in the damage caused by monkeys. As raids by the animals were previously less frequent, many local residents and even some local authorities were unfamiliar with their behavior and how to deal with it. In order to address this problem, in 2003 the Society for Monkeys launched an initiate to assess awareness of monkeys and their behavior, the extent of any damage they had caused, and the effectiveness of any countermeasures taken, the information thus obtained being made freely available to all interested parties (Toda 2007).

Even after the autumn 2003 completion date of the Society for Monkeys survey project, many of the researchers involved continued gathering information on a voluntary basis. Their activity lead to the formation of the previously mentioned Saru-Doko-Net, which was registered with the government as an official nonprofit organization in July 2006. Besides collecting and distributing information on monkey groups, Saru-Doko-Net has also taken on projects involving wild birds and animals and regularly holds workshops.

Conclusion

In this chapter, I have attempted to show an integrated network made up local residents, local authorities, and NGOs working cooperatively to solve a particular wildlife-management problem, the damage caused by monkey groups. More specifically, information is obtained and technology developed and made available to local residents as necessary by local authorities and NGOs (see figure 14.3). It is hoped that this model can be effectively applied to such other situations as will inevitably arise, people and animals more and more frequently these days becoming reluctant neighbors (see figure 14.4). Especially important to note is the way in which human interests and wildlife protection have both been taken into account. But the most important lesson to be taken away is the necessity of grassroots activism, the involvement of people with a clear personal interest and an integral understanding of the problematic situation.

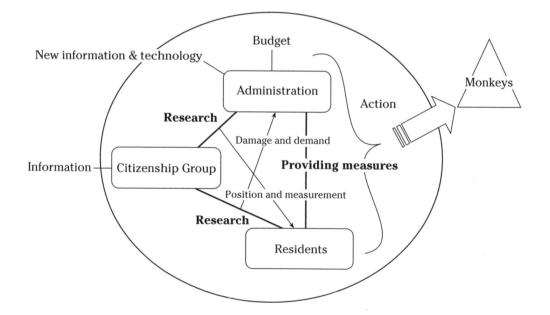

Figure 14.3. The tripartite collaboration.

Kenichi Nonaka

Figure 14.4. Cartoon depicting relations among monkeys and people.

Note

I am grateful to Mr. Yoshihisa Suzuki (Mie prefectural government), Ms. Haruka Toda (the Society for Monkeys), and Dr. Kenji Tani (Saitama University) for providing valuable information on their activities in Mie and helping develop the Saru Information System (SIS) and to Ms. Nozomi Yanahara for preparing figure 14.4.

References

Akimichi, T., ed. 2003. *Yaseiseibutsu to chiiki-shakai* [Wildlife and regional society]. Kyoto: Showado.

Mito, Y., and K. Watanabe. 1999. *Hito to saru no shakaishi* [Social history of human beings and monkeys]. Tokai: Tokai University Press.

Muroyama, Y. 2003. *Sato no saru to tsukiauniwa* [How to live with monkeys in villages]. Kyoto: Kyoto University Press.

Toda, H. 2007. "Crop Damages Caused by Spatial Movements of Japanese Macaques in the Kameyama Hills of Mie Prefecture, Japan." *Geographical Review of Japan* 80, no. 11:614–34.

Chapter 15

The Grassroots Movement to Preserve Tidal Flats in Urban Coastal Regions in Japan

The Case of the Fujimae Tidal Flats, Nagoya

Akiko Ikeguchi and Kohei Okamoto

The preservation of biodiversity has become one of the major objectives of environmental movements in various countries. Among ecosystems, tidal flats, along with tropical forests and coral reefs, are major targets for preservation. In Japan, tidal flats are common geomorphic features along the Pacific Coast. The action of river and tide continuously reshapes them, creating diverse habitats supporting a wide variety of living organisms. Since the 1990s, the preservation of biodiversity on the tidal flats has motivated Japanese local environmental movements. Recently, the national government and local administrations have become increasingly involved in environmental conservation or restoration projects.

The value of tidal flats in Japan depends on the way in which they are viewed. For example, they can be seen as necessary to small-scale coastal fisheries, much marine life either depending to some extent on them as habitat or preying on species that do. They can be seen as relatively easily reclaimed land that can be devoted to farming or industry. And they can be seen as possessing scenic or aesthetic value. Coastal development can focus on any of these values, and over the years the environmental movement has worked hard to bring its alternative values to public attention, realizing some degree of success in affecting government policy.

The early, postwar conservation movement in Japan was mainly concerned about the value of tidal flats as natural resources supporting the coastal fishing villages. It had little success in mobilizing the public, however, and, thus, made little headway until the 1970s, when the impact of pollution was brought home by the emergence of Minamata disease. Several lawsuits brought at that time against chemical companies and the national government on behalf of Minamata patients contested the development of the coastal environment and did lead to change in environmental policy, most notably the establishment of the national environmental agency.

Since the 1980s, another type of environmental movement, one contesting the value of biodiversity and favoring development, has arisen, influenced by similar developments in the West. This movement mobilized people nationwide but was especially influential in urban coastal areas, which favored reclamation because of the business that would be attracted. With reclamation, fishing villages declined and, with them, grassroots support for the fishing industry. There are, however, urban dwellers who value biodiversity over development. And it was such people who spurred the environmental movement to save the Fujimae Tidal Flats in Nagoya (see figure 15.1). The movement was the first case in Japan in which a reclamation project was contested on the grounds of biodiversity. It is a useful subject of study in that it provides insights into how Japanese environmental

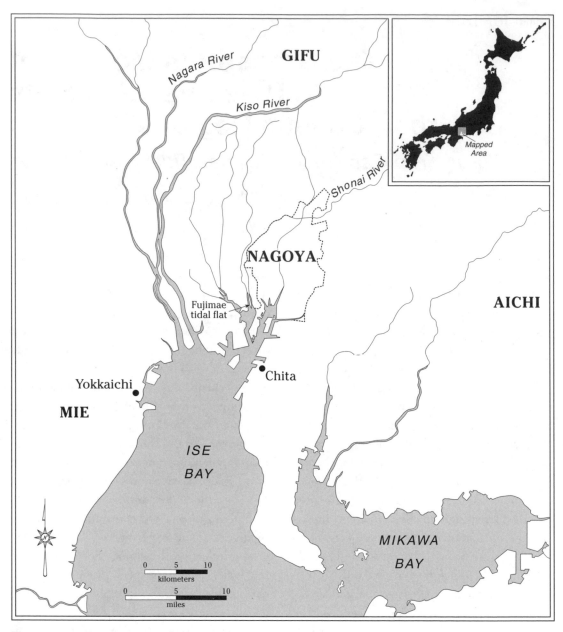

Figure 15.1. Fujimae Tidal Flat.

policies are changing at both the local and the national levels. We begin our discussion with some historical background.

The Development of Tidal Flats Historically

Tidal flats in Japan are distributed mostly along the Pacific Coast and on Kyushu and Okinawa islands. They have historically been important fishing grounds, remaining so even in the nineteenth century as the commercial economy of Japan developed and fishermen increasingly turned to offshore fishing. Beginning in the Meiji period (1868–1912), a series of legal reforms facilitated the reclamation process. For example, an 1886 law establishing a cooperative fishermen's union—intended to solve a dispute over fishing grounds

Akiko Ikeguchi and Kohei Okamoto

among local villages—had the unintended consequence of making commercial fishermen the only authorized users of coastal fishing grounds. Similarly, a 1921 law gave the union, in consultation with the governor and the Ministry of Transportation, the power to authorize reclamation projects and excluded users other than union members from compensation. Thus, until 1997, when the Environmental Assessment Law gave local communities and scientific experts a stake in the decisionmaking process, tidal flats were easy prey for developers and development-minded local governments. The years after the Second World War in particular saw a significant loss of these wetlands.

The decrease in tidal-flat area (by 35 percent) in the thirty-three years from 1945 to 1978 is shown in figure 15.2. Ten major tidal flats are highlighted. Reclamation proceeded by means of both drainage and landfill, and the most significant loss occurred in indus-

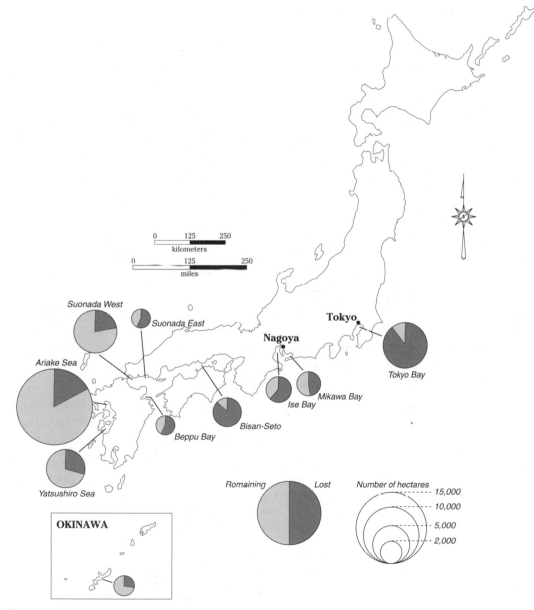

Figure 15.2. Change tidal-flat area, 1945–1978. *Source:* Nature Conservation Bureau (1980).

trial areas: Tokyo Bay, Ise-Mikawa Bay (near Nagoya), and the Seto Inland Sea. The 1962 First National Comprehensive Plan, which promoted the development of the heavy-industry and export sectors of the economy, mostly allocated the reclaimed land to the industrial complex, although some in the Seto Inland Sea area was devoted to rice production.

During this period, fishermen and other local residents were the major actors in the various grassroots movements against reclamation. One such movement was that of fishermen in the Tokyo Bay area who were trying both to prevent further reclamation and to stop the pollution of Tokyo Bay by factories built on already reclaimed land, both causes of damage to their fishing grounds (Wakabayashi 2000). Another was that of fishermen and local residents protesting the restriction of coastal access by firms that surrounded their factories with fences and seawalls (Takasaki and Takakuwa 1976; Honma 1977).

The decrease in tidal-flat area (by 7 percent) from 1978 to 1992 is shown in figure 15.3. Highlighted areas constitute 80 percent of all tidal flats in 1978. During this period, the rate of reclamation decreased sharply, particularly in industrial areas, discouraged by the oil crises of 1973 and 1979. However, some reclamation projects were carried on, particularly in urban areas, where they were spurred by population growth and, during the economic boom of the 1980s, the sharp rise in land prices. The reclaimed land was devoted to the construction of housing, waste-treatment plants, and parks. A typical example is Urayasu City on Tokyo Bay. Urayasu was once a small village where the main economic activities were fishing and gathering seaweed and shellfish. Large-scale reclamation projects since the 1960s have paved the way for apartment buildings, most famously, and Tokyo Disneyland, which opened in 1984.

Rural projects continued as well. In the Ariake Sea of Kyushu, which has the largest tidal flats in Japan, a project to reclaim land in Isahaya Bay for agricultural production was first proposed in 1953, only to fail because agreement could not be reached on how to compensate fishermen. The plan was revived periodically, and unsuccessfully, through the 1980s, finally gaining approval in 1988 when the rationale was shifted to disaster prevention. The subsequent campaign against the project was initially organized by fishermen and other local citizens but, because it received wide media coverage, soon attracted a larger base of citizens concerned about preserving the ecosystem. When the environmental impact of the project became apparent, most noticeably in the form of wretched seaweed harvests, criticism of this and similar public works projects grew. Still, the campaign failed, and the project was completed in 1997.

Also attracting criticism for its role in the Isahaya and other projects was the national Environmental Agency. The agency had been established in 1971, staffed initially by temporary personnel. However, as Japan became a signatory to various international environmental conventions, as environmental laws were established, and as public concern for environmental issues increased, the agency was expected to take an ever greater role in environmental protection. Because of its failure in Isahaya, its dedication to its mission was called into question. But it was presented with an opportunity to redeem itself in the Fujimae Tidal Flats project. And redeem itself it did, officially opposing the project and leading the opposition movement, a development that marked a critical turning point for environmental policy in Japan.

Another new development is the trend toward the creation of artificial tidal flats. Such projects were seen first in Tokyo, then in Osaka, Hiroshima, and Aichi, where most of the naturally occurring coastal wetlands had long since disappeared owing to reclamation. In Tokyo, 30 hectares of artificial tidal

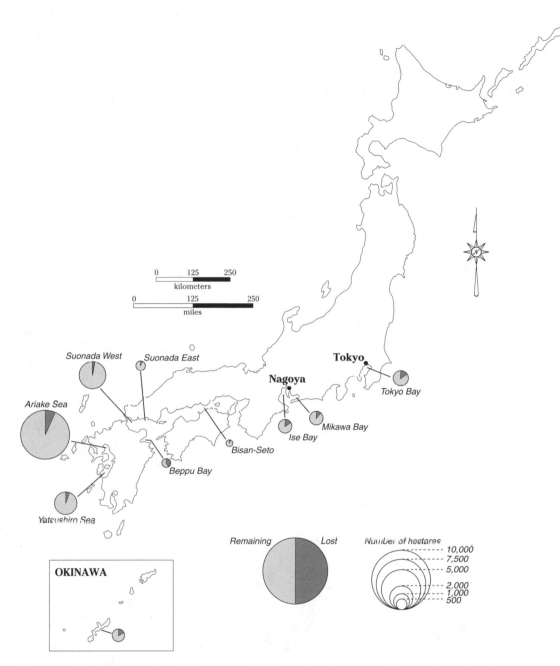

Figure 15.3. Change tidal-flat area, 1978–1992. *Source:* Nature Conservation Bureau (1994).

flats were created in 1983 and 38 hectares in 1988, the combined area being equivalent to 80 percent of the rest of Tokyo. The first of these two wetland areas became a bird sanctuary, the second a waterfront marine park. In Aichi, nearly 350 hectares of tidal flats had been created by 2001 (Suzuki 2003).

This trend in coastal planning follows directly from national environmental policy. In response to the burgeoning environmental consciousness, the law was revised in 1999 to guarantee the preservation of the coastal environment and citizens' access to it. The 2002 Nature Restoration Act strengthened the initiative by establishing public works projects aimed at the regeneration of the natural environment.

The reaction to such public works projects on the part of grassroots conservation groups has been mixed, some in favor and some opposed. On the one hand, the designation of coastal areas as public spaces was welcomed by those who insisted on access. On the other hand, the creation/restoration of wetlands was called into question by those advocating the preservation of biodiversity. The history of such environmental interventions in Japan is too short yet to allow the collection and evaluation of evidence as to their effects, positive or negative. Still, the homogenization of microhabitats (and, thus, of benthos and birds), secondary environmental destruction caused by the dredging of sand for tidal-flat construction, and the cost of maintenance remain hotly debated (Furota 1996). In areas where reclaimed land is allocated for development, the possibility of such environmental intervention often justifies proceeding with reclamation without conducting the appropriate environmental assessments. Nevertheless, such creation/restoration projects are being enthusiastically promoted.

In order to further the debate of such issues, we turn next to an analysis of a specific case, that of the Fujimae movement.

The Grassroots Movement to Preserve the Fujimae Tidal Flat

The Fujimae Tidal Flat Reclamation Project

The Fujimae Tidal Flat is located at the head of Ise Bay (see photographs 15.1 and 15.2). Several rivers have deposited sediment there since prehistoric times, forming an extensive delta. From the seventeenth century to 1890, about 10,000 hectares of the bay were reclaimed, the land devoted to rice production. The remaining wetlands went largely undisturbed throughout the first half of the twentieth century, but during the period 1955–1975 a total of 6,000 hectares was reclaimed (about 200–400 hectares each year), mostly in the Nagoya and Yokkaichi industrial areas. Underwater exploitation and decreased sedimen-

Photograph 15.1. The Fujimae Tidal Flat is barely protected within the industrial complex of the Nagoya port. (Courtesy Akiko Ikeguchi.)

Akiko Ikeguchi and Kohei Okamoto

tation also caused subsidence in some areas, and a further 500 hectares was reclaimed after 1980 (Mizuno 2003), leaving only 105 hectares of tidal flats, an important feeding ground for migratory birds in Japan (Tsuji 1999).

The Fujimae Tidal Flat reclamation project was formally announced in 1984, with work to begin by 1990. The project was part of a larger Nagoya city government and port-management-union plan to develop the harbor to serve as an international trading hub as well as a waste-treatment area. The latter need was particularly pressing because, owing to population pressures (the population of Nagoya City in 1981 was 2.17 million), the amount of waste to be disposed kept increasing. Even though a program to recycle paper, glass, and aluminum had been instituted by the city in 1980, only seven of sixteen wards participated, and disposal sites were expected to reach capacity by 2001.

In 1987, a survey commissioned by the port-management union revealed that the tidal flat was an important feeding area for migratory birds. Also in that year, the conservation group the Nagoya Port Tidal Flat Association [Nagoya-ko hozen group]—later the Save the Fujimae Association [Fujimae higata o mamoru kai] (SFA)—was organized and began its campaign against the project. In 1989, the city government reduced the planned reclamation area from 105 to 70 hectares (owing to concerns about flood control) and then again, in 1990, to 52 hectares (in response to a petition drive that had collected 100,000 signatures opposing the project). After another reduction to 46.5 hectares (owing to land-purchase problems), the city decided to start the environmental impact assessment (EIA) process.

As data for the EIA were collected, the anti–reclamation project movement grew stronger. When a public hearing on the EIA was held in 1996, the SFA, as well as other nongovernment organizations (NGOs) and academics, criticized the report as inaccurate,

Photograph 15.2. Yamatoshijimi (*Corbicula japonica*), one of the life forms inhabiting the Fujimae Tidal Flat. (Courtesy Akiko Ikeguchi.)

citing as evidence their own, independently collected data. In 1998, the city organized a committee to explore the effects of the construction of an artificial tidal flat as compensation for proposed reclamation. That same year, however, at a symposium sponsored by the SFA and other NGOs, the Environmental Agency made clear its position that such a project should not be pursued. In October 1998, the Fujimae Neighborhood Association [Fujimae jichikai] polled the residents of Fujimae, and 223 of 231 votes expressed opposition to the dumping of waste in the coastal wetlands. That same month, a petition demanding a public referendum on the project was presented to the Nagoya city council. The petition contained 108,155 signatures, more than three times the required 34,000 signatures. In the face of such profound opposition, the city abandoned the reclamation project in January 1999 and began to revise its waste-management policy.

The opposition to the Fujimae Tidal Flat reclamation project offers us a representative case study of a grassroots environmental movement valuing biodiversity. Characterizing the movement's success is that it conducted its own independent research (instead of relying on the official EIA), that on the basis of that research it suggested a policy alternative, and that it mobilized an extensive network of concerned citizens. While it is not uncommon for grassroots movements to conduct their own independent research, it is uncommon for the information thus obtained to have such a marked effect on public opinion. We now detail the characteristics of the movement and its sociopolitical background, drawing on Matsuura (1999) and a newspaper database for the period 1989–1999.

The Assessment of the Project

The EIA process in Japan was established by a 1972 cabinet decision. Even so, assessments of large-scale public works projects were conducted only at the discretion of the government. Although some local governments did come to require EIAs, especially after the process was recommended by the 1993 Environment Basic Law, it was not until 1997 that the process became legally mandated nationwide. In the case of the Fujimae Tidal Flat reclamation project, the SFA did identify and bring to public attention certain defects in the plan and the associated evaluation process, defects that had also been seen in earlier evaluations. One significant defect was that no account was taken of the fact that the plan might have to be revised once the findings of the EIA were revealed. Another is that the EIA was overseen by the project director, not by an independent third party. Also, although the final EIA report was made available to the public and public hearings were held, the evaluation procedure was not entirely transparent.

More specifically, the Fujimae project assessment proceeded as follows. First, a preliminary report was compiled by Nagoya City. That report was then submitted for review to a committee composed of local residents and scholarly experts. When completed, that committee's assessment report was submitted to the Ministry of Transportation along with the reclamation project application. Prior to the preliminary report being made available to the review committee, however, several NGOs, including the SFA, made available independent research covering the effects on migratory birds and benthos, the status of artificial tidal flats, the availability of alternative waste sites, the legal issue raised by the land-purchase procedure, and an economic evaluation of Fujimae employing contingent valuation methodology. The most influential was SFA research on tidal-flat organisms.

For example, the preliminary report concluded that, owing to the low (1.3 percent) tidal-flat use rate, the effect on migratory birds would be minor. (The use rate was calculated as the percentage of a four-day survey period during which birds of any species were sighted on the tidal flats.) The SFA was critical of this conclusion on the grounds that the role of tidal flats in birds' lives differs by species and that birds use different parts of the tidal flats at different times and for different purposes. It was also critical of the report's data, which were collected by people who had little local experience and employed generalized techniques, whereas the SFA data were obtained by people who had long observed bird behavior in the area and were familiar with tidal-flat use patterns.

The preliminary report also concluded that, owing to their low biomass on the tidal flats (only twenty to forty grams per square meter), benthos did not play a significant role in the purification of seawater. The SFA was critical of this conclusion because it was based on a generalized technique employed by environmental research companies, the collection by dredge of animals in the tidal-flat sediment to a depth of only ten centimeters. The SFA data, however, collected by hand-dredging to a depth of more than one meter, showed that the actual biomass was twenty times higher than the report indicated. In a public hearing, an SFA representative reported that "each

 Akiko Ikeguchi and Kohei Okamoto

tidal flat has unique characteristics that any survey must evaluate" (Matsuura 1999, 65).

The independently compiled data—and especially that coming from the SFA—was considered by the review committee to be so credible that, in its report, it concluded that the project would have a negative impact on the tidal flat (Hayakawa 1999). Such an admission by a review committee being rare, the media took great interest in the story and publicized the assessment report widely. In response, Nagoya City revised the project plans, the final reclamation project application being for permission to construct an artificial tidal flat by adding sand to thirty-eight hectares of shallow water in order to raise the seabed.

But the SFA was again critical, responding as follows: "The plan of artificial tidal flat is human insolence. . . . Even one burrow of a Japanese mud shrimp (*Upogebia major*) is part of a pattern of life that has existed since ancient times. Now is the time to reconsider the significance of the tidal flat. . . . This plan was exclusively an engineer's idea. The Fujimae Tidal Flat was formed over a period of more than 150 years, beginning at the mouth of the old Kiso River, where it was known formerly as *Chidori gata* [tidal flat of a plover], and long fostered life. They [Nagoya City] close their eyes to the criminal attempt to bury and, thus, destroy this place with waste and, in the name of compensation, plan another massacre [by constructing artificial tidal flats], although perhaps not knowingly" (Matsuura 1999, 65, 151).

In refuting the claims made in the preliminary report, opponents of the reclamation project stressed both the ecological importance of the tidal flats and their historical significance. And it was from such a perspective that they came to the conclusion that the project was inappropriate, a conclusion compelling enough that the project was eventually abandoned. Thus, we see the importance of independent research, media exposure,

and the pressure of public opinion in a grass-roots movement. I now go on to describe how the network that saved the tidal flats was assembled.

The Development of the NGO Network to Save Fujimae

The core of the Fujimae movement was the SFA, which had been organized by bird-watchers in Nagoya City. Among conservation groups in Japan, those devoted to bird-watching were among the earliest to establish both national and international networks. The Wild Bird Society of Japan [Nihon yacho no kai] (WBSJ) is the largest of the bird-watching groups, having a branch in every region of Japan. And among the international networks in which it participated was one including groups in China and Korea that was established to exchange information about tidal flats, which are important relay points for birds migrating from Australia to Siberia. Recognizing the importance of tidal flats, the network has held symposia about and proposed international conventions promoting their conservation. The Ramsar Convention—which Japan joined in 1980—was one such convention; it obliges member countries to establish wetland sanctuaries.

The initial motivation for the development of a national network promoting tidal-flat conservation was the Ramsar Convention symposium held in Kushiro in 1993 (Tsuji 1999). Spurred by the campaign against the Isahaya Bay reclamation project, the Japan Wetland Action Network [Nihon shitchi network] (JAWAN) was organized in 1991 and, at the symposium, proposed stricter Environmental Agency enforcement of wetland-conservation policy. In 1992, the WBSJ proposed that the Environmental Agency register nineteen wetlands as sanctuaries. The International Wetland Symposium, organized by JAWAN, was held in Tokyo that same year, and tidal-flat-conservation groups from Fujimae and the Isahaya

Bay, Hakata Bay, Tokyo Bay, and Nagara River areas attended. The Environmental Agency also organized the Asia Wetland Symposium to be held prior to the Kushiro congress and invited representatives from a variety of Asian countries. At the symposium, the agency did, in fact, propose the establishment of new sanctuaries, including some on the WBSJ's list, but tidal flats already slated for development were not among them. Protesting this decision, the World Wildlife Fund–Japan [Sekai shizen hogo kikin Nihon shibu], the Nature Conservation Society–Japan [Nihon shizen hogo kyokai], JAWAN, and the WBSJ organized the '93 Wetland Alliance [93 wetland kaigi] and started a campaign to register, and, thus, protect, wetlands slated for development under the Ramsar Convention. This network of NGOs became the basis from which the movement against the Fujimae project grew.

Also important in the development of the national network was the establishment in 1997 of a listserv managed by a representative of the Forum on Environmental Administration Reform [Kankyo gyosei kaikaku forum], a nongovernment think tank (Matsuura 1999). The activities of the SFA and the problems with the Fujimae project EIA process were already widely known to activists concerned with environmental policy, and, therefore, a representative of the SFA was invited to join the mailing list, which currently includes about two hundred activists, government officials, lawyers, and journalists. The SFA has used the mailing list both to publicize its activities and to exchange information with other organizations. The advantages that it gained were several.

First, the effort to keep journalists informed about the Fujimae project garnered such media attention that what was once a local environmental issue became a national political one. This is especially evident in the spike in newspaper articles about the project around March 1998, when the negative EIA committee report was issued (see figure 15.4). But attention to the project had been growing

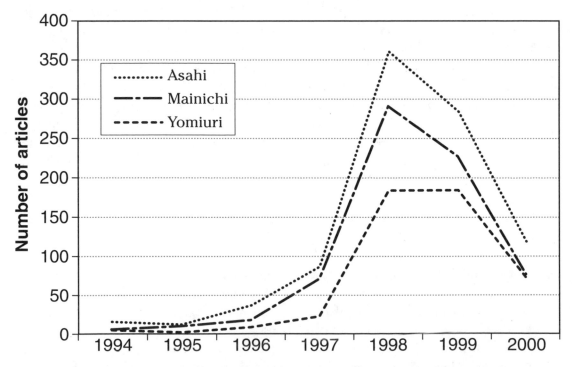

Figure 15.4. Number of articles about Fujimae Tidal Flat reclamation in three major Japanese newspapers.

even before that, prompting an invitation to the SRA from the Forum on Environmental Administration Reform to send a representative to a January 1998 meeting of its executive council at which members of the National Diet in charge of environmental policy would be present. The idea was that, by presenting a case that the issues surrounding the Fujimae project extended beyond migratory bird conservation to such other aspects of environmental policy as waste management and international law, the SRA could affect policymaking at the national level (Matsuura 1999, 76). The move was successful, and, in September 1998, over one hundred Diet members from different parties organized a workshop on the tidal-flat issue (*Asahi shimbun*, September 15, 1998).

Second, when in 1998 the focus of SFA activity shifted from the reliability of the EIA process to waste-management policy generally, the mailing list offered activists a much-needed resource for researching alternatives to reclamation. Because local governments in rural areas were increasingly refusing to accept waste from nearby urban areas, Nagoya City had no choice but to construct the needed waste-treatment site inside city limits. But it had to be convinced that there were alternatives to reclamation. And it was up to NGOs like the SFA to come up with a feasible alternative, a task greatly aided by members of the mailing list familiar with environmental policy.

Still, cooperation with other NGOs in the city beyond the sharing of information was needed if a workable strategy was to be formulated and implemented. I turn now to the process by which the local network that constructed the alternative waste policy developed.

Development of Local Network and Construction of an Alternative Policy

Local networks among NGOs with different political goals but a common commitment to public involvement in policy were first developed in Japan in the early 1990s in response to large public works projects in Aichi Prefecture, such as the world's fair and the new international airport, that were pushed through rapidly with little public input (Shimazu 2001). In April 1994, nineteen NGOs, including both conservation groups and consumers' networks, held a symposium in Aichi. The next month, the group Save Our Health and Environment! Aichi Residents' Action [Kenko to kankyo a mamore! Aichi no jumin issei kodo] was organized in the name of sixty NGOs, and representatives of the group met with the mayor of Nagoya City.

As we have seen, the issue prompting the SFA and other NGOs to join forces was waste-management policy. Since the 1980s, the rapid increase of waste as a result of rapidly increasing consumption has been recognized as a critical urban problem in Japan. Although smaller cities tried to solve the problem by promoting recycling, such an approach was thought to be unrealistic in the larger urban areas, where the waste generated by business activity exceeds that generated in domestic settings. Nagoya City's response to the problem was to construct disposal sites for untreated waste outside Aichi Prefecture, in Gifu Prefecture, but few municipalities could be found to host such sites owing to concern with dioxin pollution. Also, Nagoya City was criticized by other local governments for dumping its waste on others instead of reforming its waste policy.

There did, however, exist some grassroots groups promoting recycling. The Chubu Recycle Action Network [Chubu recycle undo shimin no kai], the largest such group, was established in 1980. It established recycling stations that collected newspapers and plastic containers, using volunteer workers to promote its efforts. From its inception in 1987, the SFA held the yearly collaborative "Waste-Bird Symposium" [Gomi-tori symposium] at which alternative waste policies were discussed. Other grassroots recycling activities with the ultimate aim of tidal-flat conservation also emerged in Nagoya. For example, in April 1995, two housewives from Meito Ward—one a member of the Chubu Recycle Action Network—published a book about the ecology of Fujimae (*Asahi shimbun,* April 22, 1995). And about a hundred housewives from the same ward organized the Society for Zero Emission Nagoya [Zero-emi Nagoya no kai] in August 1998, establishing stations for the collection of recyclable waste, which was to be transported from the stations to the treatment site by lorry (*Asahi shimbun,* August 4, 1998). The society also helped the neighboring ward, Higashi, launch a similar project the next month. The success of all these projects suggests that it was easier to spur the conservation minded to action when the focus shifted from the more esoteric goal of tidal-flat ecology to the more familiar one of recycling everyday waste.

The efforts of such local networks were unexpectedly aided by the 1990 recession and the 1991 Law for the Promotion of Utilization of Recycled Resources. While the amount of waste generated by business activity did continue to increase, it did not increase nearly as much as was expected. In 1992, for example, only 8.2 percent of the expected amount of waste was generated in the city, meaning that the Chita disposal site could easily handle it (*Asahi shimbun,* November 20, 1992), and calling into question the city's claim that metropolitan waste generation would continue to increase almost exponentially.

The growing concern of the residents of Aichi Prefecture with the establishment of an alternative waste policy was demonstrated at the ballot box. In the 1997 election for the House of Councillors, the upper house of the Japanese Diet, for example, the Democratic Party, running on a platform of the conservation of Fujimae and the revision of the city's waste-management policy, defeated the Liberal Democratic Party by gaining two seats out of three in the Aichi Prefecture proportional representation block. Even the Communist Party, also calling for the revision of waste-management policy, tripled its total from the previous election, earning one in three seats in the process. Subsequently, some members of the Diet from Nagoya who had previously supported the mayor's position began calling for a revision of waste-management policy (*Asahi shimbun,* August 4, 1998).

The growing concern with waste policy extended to the Fujimae neighborhood as well. For much of the 1990s, the Fujimae neighborhood association had made no express objection to the reclamation project, which allowed the city to claim that it was outsiders fueling the opposition. This state of affairs can be attributed to the fact that Fujimae was a relatively closed community—presumably because of the long-standing ban on the construction of new residential housing under the Nagoya City industrial land use plan—and, thus, local residents had little contact with the SFA (Matsuura 1999, 252). Things changed, however, when one resident organized a study group on dioxin pollution, after which a number of housewives organized the Association to Save the Children from Dioxin [Daiokisin kara kodomo o mamoru kai] in August 1998 and started meeting with the SFA. These activities prompted the Fujimae neighborhood association to sponsor a referendum—ultimately successful—calling for the abandonment of the Fujimae reclamation project.

An Evaluation of the Fujimae Tidal-Flat Conservation Movement

The previous section describes the development of the Fujimae Tidal Flat conservation movement, with special attention paid to the process of network building. This remarkable movement was a turning point in the history of coastal development in Japan. It showed that the tidal flat has value other than as a resource to be exploited by industrial, agricultural, and government interests. Nevertheless, the means by which the end of stopping the reclamation project was achieved had unintended consequences, affecting local waste-management policy, national environmental policy, and the perceived value of tidal flats.

The Impact on Local Waste Policy

After canceling the Fujimae reclamation project, Nagoya City first attempted to find an alternative waste-disposal site—with the assistance of Aichi Prefecture, the Ministry of Transportation, the Ministry of Health and Welfare, and the Environmental Agency— without success. The city therefore had no choice but to change its waste policy drastically. In February 1999, it issued an emergency declaration that waste was to be reduced by 20 percent (about 200,000 tons) over the next two years. In August, it made recycling mandatory, with five categories of recyclables identified—the strictest recycling program ever attempted in a metropolitan area. To aid those citizens—especially the elderly and ethnic minorities—who found the rules for separating recyclables confusing, the city made available a booklet explaining the procedure. It also established sanitary committees in all school districts to ensure that recyclables had been properly separated. The NGOs aided the process by establishing collection sites in vacant lots or supermarkets. The program proved successful, the city's waste being reduced by 23 percent in two years, and the amount of recyclables increasing from 13 percent in 1998, to 18 percent in 1999, to 28 percent in 2000 (Matsubara 2001; Nagoya City 2003).

National Environmental Policy

The debate over the Fujimae reclamation project brought to public attention several problems with existing environmental policy at the national level, including the lack of public involvement in the decisionmaking process and the limits to the authority of the Environmental Agency director. The result was that, when, in 1997, the EIA process was legally mandated nationwide, these problems were directly addressed. First, in addition to the submission of proposals to review committees, more general surveys of public opinion about environmentally sensitive public works plans were required. Second, whereas, before, the director of the Environmental Agency could comment publicly on projects only if specifically requested to do so, now he was free to act on his own initiative, allowing him to participate more effectively in the policymaking process.

Recently, the Environmental Agency and the Ministry of Land, Infrastructure, and Transport (an amalgam of the former Ministry of Transportation and Ministry of Construction) have been considering the introduction of a more direct form of public involvement: the strategic environmental assessment. Under such a plan, local residents would be asked their opinion of alternative plans as they are being developed. Of particular concern here are establishing lines of communication and facilitating cooperation among the various parties with an interest in environmental policy.

Regeneration of the Tidal Flat?

In opposing the Fujimae reclamation project, the SFA harbored a different notion of the value of the tidal flat than did the government, seeing it not simply for its potential to be exploited, but as a valuable natural resource in its own right and a unique historical heritage. Such a view became widespread, and, in response, Prime Minister Koizumi set up a national environmental commission that called for public works that had as their aim the regeneration of natural environments.

Tidal flats were among the ecosystems targeted by the commission, and the Ministry of Land, Infrastructure, and Transport is currently conducting conservation projects in forty ports. Some of the projects are, unfortunately, problematic. That in the Ise-Mikawa Bay, for example, aims at regenerating a twelve-hundred-hectare area by raising the seabed with sand dredged from the bay, a process that, as we saw earlier, can have dire consequences for marine life.

The Ise-Mikawa Bay project reveals that disconnects still exist in the networks established to date by environmental activists. As with the Fujimae reclamation project as initially conceived, the value that local activists place on tidal flats is not necessarily the value that national policymakers place on them. It is to be hoped, however, that, as public involvement in environmental policymaking grows, all involved will make their views more widely known and that better-functioning partnerships will be developed. Here, the initiative clearly must be taken by the grassroots movements.

Conclusion

This chapter is a case study of a grassroots conservation movement in Japan that successfully opposed the Fujimae Tidal Flats reclamation project. It details the differing values that different actors placed on the tidal flats and how the influence of the various actors changed, environmental concerns increasingly being incorporated in public policy at both the local and the national levels over time. It also details how major components of the movement's success were the mobilization of a large-scale network, a process made possible largely by sociopolitical developments, and an insistence on the need for public input in the decisionmaking process. Despite the movement's success, however, the value of the tidal flat remains contested. The

challenge for this grassroots movement—and others throughout Japan—in the twenty-first century remains the education of both the general public and policymakers and the fight to incorporate local knowledge in public policymaking.

References

Furota, T. 1996. "Touron: Seitaikei shuhuku toshi-teno jinkokaihinzousei no mondaiten" [Comments on construction of artificial beaches as a method for restoration of ecosystem in developed shorelines]. *Engan kaiyo kenkyu* [Bulletin on coastal oceanography] 33, no. 2:163–67.

Hayakawa, Y. 1999. "Kankyo yusen no 'zenrei': Fujimae higata, umetate kaihi o shuzaishite" [A precedent of priority in environment: From covering case of Fujimae Tidal Flat conservation]. *Shinbun kenkyu* [Newspaper studies] 572:86–87.

Honma, Y. 1977. *Irihamaken no shisou to koudou: Umi wa minna no mono Nagisa wo kaese* [Thought and action on coastal access rights: The sea is everyone's; return Nagisa to us]. Tokyo: Ochanomizu Shobo.

Matsubara, T. 2001. *Isshu okure no toppuranna: Nagoya shimin no gomi kaikaku* [A leading runner of the round delay: Waste renovation of Nagoya citizens]. Nagoya: KTC Chuo Shuppan.

Matsuura, S. 1999. *Soshite higata wa nokotta: Internet to NPO* [Then, the tidal flat was saved: Internet and NPO]. Tokyo: Riberuta Shuppan.

Mizuno, T. 2003. "Isewan ni okeru higata, moba, kakouiki no hensen to busshitsu junkan no genjo" [The change of tidal flat, sea grass bed, estuary and current status of material recycling in Ise Bay]. In "Suisankaiyo chiiki kenkyukai 'Ise-Mikawawan no kankyo to gyogyo o kangaeru' Kouen youshishu" [Proceedings of symposium on environment and fisheries in Ise-Mikawa Bay, March 8, 2003], 13–17. Nagoya: Nagoya Nogakudo.

Nagoya City. 2003. "Nagoya gomi report '03" [Nagoya city waste report '03]. Nagoya City.

Nature Conservation Bureau. Environmental Agency. 1980. *Dai nikai shizen kankyo hozen kiso chosa* [The second survey report for natural environment conservation]. Tokyo.

———. 1994. *Dai yonkai shizen kankyo hozen*

kiso chosa [The fourth survey report for natural environment conservation]. Tokyo.

Shimazu, Y. 2001. "Kyodai koukyo jigyo asesumento karano kyokun" [Lessons learned at assessment of megascale public works: The cases of Chubu International Airport, Aichi World Exposition, and Fujimae Tidal Flat]. *Kankyo to kogai* [Research on environmental disruption] 30, no. 4:34–40.

Suzuki, T. 2003. "Higata moba zosei no genjo to gijutsuteki kadai" [Current status of tidal-flat creation and technical problems]. In "Suisankaiyo chiiki kenkyukai 'Ise-Mikawawan no kankyo to gyogyo o kangaeru' Kouen youshi-shu" [Proceedings of symposium on environment and fisheries in Ise-Mikawa Bay, March 8, 2003], 35–40. Nagoya: Nagoya Nogakudo.

Takasaki, Y., and M. Takakuwa. 1976. *Nagisa to Nihonjin* [Nagisa and Japanese]. Tokyo: Nihon Housou Shuppannkyoukai.

Tsuji, A. 1999. "Kankyo hakai to shizen tono kyosei: Fujimae higata ga mamorarete" [Environmental destruction and symbiosis with nature: On the preservation of Fujimae Tidal Flat]. *Gunshuku mondai shiryo* 223:14–19.

Wakabayashi, K. 2000. *Tokyo wan no kankyo mondaishi* [The history of environmental issues in Tokyo Bay]. Tokyo: Yuikaku.

Chapter 16

The Protection of the Shiraho Sea at Ishigaki Island

The Grassroots Anti–Ishigaki Airport Construction Movement

Unryu Suganuma

The Ishigaki Island airport construction controversy has been a matter of public debate for more than thirty years, going back to February 1972. In February 1974, the mayor of Ishigaki City officially petitioned the Okinawa prefectural government, asking for new airport construction. In May 1976, in response, the prefectural government identified (but did not immediately make public) six potential plans: (*a*) the Expansion of Current Airport Plan; (*b*) the Shiraho Sea Plan; (*c*) the Fusakino Plan; (*d*) the Karadake East-Side (or Shiraho Land) Plan; (*e*) the Miyara Plan; and (*f*) the Karadake Land (or Karadake) Plan (see figure 16.1). In June 1992, it narrowed the potential plans to four—the Karadake East-Side Plan, the Miyara Plan, the Karadake Land Plan, and the Fusakino Plan. In April 2000, it decided to construct the airport at Mt. Karadake in the Shiraho area (the Karadake Land Plan). Construction began on October 20, 2006, and is expected to be completed by 2012 (*Ryukyu shimpo*, October 21, 2006). However, residents of the Shiraho area have strongly resisted the Karadake Land Plan. There is a long history surrounding the controversy. This chapter will deal only with current developments, examining the pros and cons of the plan as well as the grassroots movement to protect the Shiraho Coral Reef. The Shiraho Reef has often been called Japan's last ocean paradise. The ten- by one-kilometer stretch of reef adjacent to the seashore is especially valuable, supporting as it does an amazing 50 genera and 120 species of coral (see photographs 16.1–16.2). This is more than two-thirds of the number found on Australia's Great Barrier Reef of Australia, an ecosystem about two thousand kilometers long. Critically, the Shiraho Reef contains a large community of blue coral that is more than a thousand years old ("World's Coral Reefs" 2003).

Geographically, Ishigaki Island is located about 411 kilometers southwest of Naha (the capital of Okinawa Prefecture), 1,019 kilometers from Kagoshima, 1,957 kilometers from Tokyo, and about 277 kilometers from Taiwan. As part of the Yaeyama District, Ishigaki Island is currently administered as a city and is the largest islet of the district in terms of both area and population. In 1972, along with the Okinawa Islands, Ishigaki Island was returned to Japan under the U.S.-Japanese Reversion Treaty. By May 1975, the existing Ishigaki Island airport was fully equipped, but its fifteen-hundred-meter runway could handle only YS-11-type airplanes (*Yaeyama yoran* 2001)—hence the need for new airport construction.

Grassroots Mobilization

In July 1979, surprising news spread around Shiraho—it was one of six sites being considered for airport expansion. Most island residents were totally astonished. The Ishigaki city government and the Okinawa prefectural government met with a small group of leaders from the Shiraho area to explain that the new airport would be built on the Shiraho Sea (Ikushima 1989). In September

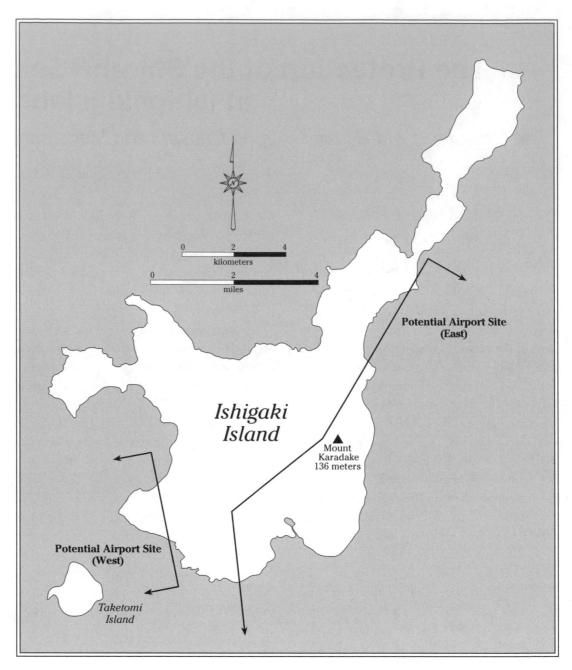

Figure 16.1. Potential new airport sites, Ishigaki Island.

1982, the Okinawa prefectural government, over objections from local leaders, surveyed the seabed so that construction plans could be formulated. A citizen protest, led by Mukazato Kiyoshi, opposing the survey project was easily broken up by the police (Ikushima 1989). This was because, since its inception in the late 1970s, the protest movement worked in isolation, without any support from out-side organizations. This changed, however, shortly after the September 1982 protest.

In December 1983, a meeting was held in Tokyo at which the journalist Shimojima Tetsuro and the environmentalist Ikushima Toru showed photographs of the Shiraho Coral Reef, generating lively discussion among the environmentally conscious audience (see photograph 16.3). Interest in reef preserva-

Photograph 16.1. Shiraho coral reefs. (Courtesy Takashi Kobayashi.)

Photograph 16.2. Shiraho Sea floor with rich marine life. (Courtesy Takashi Kobayashi.)

tion was so great, in fact, that several leading conservationists, including Shimojima and Ikushima, were induced to form the nonprofit Association to Protect the Yaeyama-Shiraho Coral Reef [Yaeyama-Shiraho no umi wo mamoru kai] (APYSCR) (Ikushima 2003a). The original twenty members initially worked mainly distributing petitions opposed to the Shiraho Sea Plan, ultimately collecting more than forty-five thousand signatures (Ikushima 2003a). Following the founding of the APYSCR in Tokyo, similar groups sprang up in other cities, for example, Naha and Osaka. And scholars from Ryukyu University also joined the movement.

Photograph 16.3. An anti–airport construction meeting in the 1980s. (Courtesy Toru Ikushima.)

Grassroots Tactics

The grassroots tactics of the APYSCR involve two fundamental strategies: external support and internal effort (see figure 16.2). At the first layer of external support, the APYSCR enlisted experts, both foreign and domestic, in its fight to conserve the Shiraho Coral Reef. Among the foreign scholars was Katherine Muzik, a specialist in corals from the Smithsonian Institution who conducted extensive research on the Shiraho Sea. She would later describe Shiraho as follows: "Food from the sea is a well-known gift. Shiraho lagoon provides an endless supply of fishes, lobsters, crabs, clams, topshells, squids, octopuses, urchins, seaweeds. . . . There are other foods too, less well known, but just as delicious: sea-water itself, for example! Sea-water is added to soymilk, to make it into nutritious tofu. And traditionally, sea-water was the source of precious salt. Other sea-treats? A little powdered coral skeleton was added to sugar-cane molasses, to make it harden into candy" (Muzik 1991, 18).

At the first layer of internal effort are appeals to the public: petitions and public demonstrations (see photograph 16.4). As we have seen, one of the APYSCR's first organized activities was the circulation of petitions. Then, in September 1984, it organized a demonstration against another government-conducted environmental survey of Shiraho. A police riot squad confronted the protesters, and three of the demonstration's leaders were arrested, one suffering a concussion in the process. Mukazato Kiyoshi was imprisoned for twenty-two days (Ikushima 1989). The momentum of the grassroots movement did not stop there, however. In April 1985, local citizens organized by the APYSCR began a sit-in/hunger strike in front of the Okinawa prefectural government building that lasted for eight days. On July 7, 1985, the APYSCR organized a nationwide day of demonstrations protesting the proposed new airport construction at Shiraho. Cities where demonstrations were held included Tokyo, Ishigaki, Naha, and Osaka. The most crucial date in the protest movement, however, was

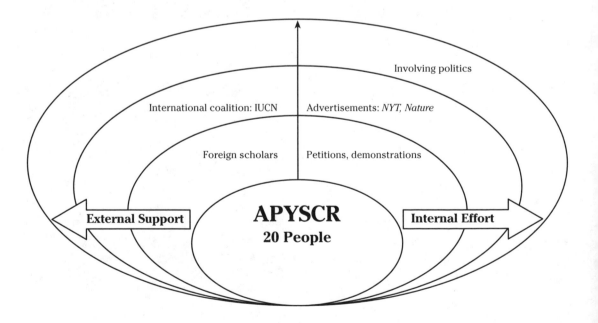

Figure 16.2. Analysis of the APYSCR's activities.

Photograph 16.4. The APYSCR appeals to the public to save the Shiraho Sea on April 11, 1992. (Courtesy Toru Ikushima.)

August 1, 1986, when the APYSCR bought a small piece of land at the proposed construction site. The news spread, and a year later more than eighteen hundred people had registered as the owners of one-*tsubo* (roughly three-square-meter) plots there, paying the APYSCR a minimum ¥5,000 fee (Ikushima 1989). Nevertheless, the central government has continued to devote about ¥3 million annually to researching the construction of the new airport.[1]

At the second layer of external support, and building on the first level, the APYSCR worked to create a broad-based international coalition. Toward that end, it dispatched three representatives to the seventeenth conference of the International Union for the Conservation of Nature (IUCN), held February 1–10, 1988, in San José, Costa Rica, seeking its continued support for the movement to conserve the Shiraho Coral Reef. (The World Wildlife Fund–Japan and other Japanese nongovernment organizations [NGOs] had arranged for the IUCN to conduct an ecological survey of

the reef in 1987.) They were successful, and, on the last day of the conference, the general assembly unanimously approved a resolution urging the protection of the coral reef: "The Government of Japan [should] take immediate steps to reconsider with [sic] the construction of an airport facility at the Shiraho Reef site in view of the serious environmental consequences on the reef of such an activity" (IUCN 1991, 21; see also Kobashigawa and Mezaki 1989). Furthermore, an IUCN research team issued a statement based on its 1987 ecological survey of the reef expressing concern that "construct[ing] a jet airport facility at and on the Shiraho Reef site [would] result in irreparable damage to the ecological processes and the biological diversity of the reef" and that "continued soil position in the Todoroki River watershed [would] degrade and stress the biological communities of the Shiraho Reef" ("World's Coral Reefs" 2003).

As the news of the IUCN resolution spread, the Okinawa prefectural government began receiving letters from private citizens expressing their opinion on the proposed airport construction. Of the thirty-seven hundred letters ultimately received on the subject, the vast majority—thirty-five hundred—were against construction. And the politicians began to get nervous (Ikushima 1989). On August 22, 1987, the APYSCR, with the help of several other NGOs, gathered more than seventy-eight thousand signatures on petitions calling for the Environment Agency to halt the proposed project. In addition, the head of the IUCN sent letters to Prime Minister Takeshita Noboru and Nishime Junji, the Okinawa prefectural governor, asking them to protect the Shiraho Sea (Ikushima 1989). Meanwhile, the national newspapers published editorials opposed to the proposed airport construction (e.g., *Asahi shimbun,* August 17, 1988). Faced with such determined opposition, in late 1988 the Okinawa prefectural government decided to withdraw the Shiraho Sea Plan from consideration.

This success was short-lived, however. On April 26, 1989, Governor Nishime announced his support for the Karadake East-Side Plan, which proposed a site for the airport four kilometers north from that proposed by the Shiraho Sea Plan.[2] This was not much of a change as the distance between the northern part of the Shiraho Sea Plan and the southern part of the Karadake East-Side Plan is only about 1.5 kilometers (Ikushima 1989). The APYSCR called a meeting of the people of Shiraho at the community center in Ishigaki City to explain this. And, on May 25, 1989, the head of the IUCN once again sent letters to the prime minister and the governor, this time pointing out the danger to the coal reef if the Karadake East-Side Plan were to be carried out. The IUCN followed up on these letters by conducting, on October 25, a second ecological survey of the Shiraho Sea, a survey that confirmed its concerns.

Efforts to prevent construction at Shiraho did not end there. For example, the APYSCR placed advertisements in, among other places, the November 30, 1989, issue of *Nature* and the April 17, 1990, issue of *Greenpeace USA,* seeking support for the grassroots movement (the second layer of internal effort).[3] More than six hundred letters were received in response (Ikushima 1989). The APYSCR also dispatched four representatives, including Katherine Muzik, to the eighteenth annual IUCN conference, held November 28–December 5, 1990, in Perth, Australia, again seeking support for the movement. Again, they were successful, the IUCN adopting another resolution in support of the cause. Also at the conference, the Shiraho Community Center received the Sir Peter Scott Award for conservation merit, a first for a Japanese community center (Ikushima 1989).

At the same time as the APYSCR was promoting its cause in the international arena, it was also doing so in the local political arena, supporting politicians who shared its views (the third layer of internal effort), for example, Ota Masahide, who was challenging the incumbent, Nishime Junji, in the November 18, 1990, gubernatorial race. Ota, who opposed the Shiraho Sea Plan, won and immediately began promoting the selection of alternative construction sites. His advocacy was successful, the airport site selection committee narrowing the field to the Fusakino Plan, the Miyara Plan, and the Karadake East-Side Plan. Since the site to be utilized under the Karadake East-Side Plan is far from downtown Ishigaki (ten kilometers, about twenty minutes by a car), local politicians preferred the Fusakino or Miyara plans. And, on November 26, 1992, Governor Ota announced the selection of the Miyara Plan, which would locate the new airport at Miyara (Ikushima 1989).

While this decision represented a victory for the Shiraho activists, it was not without its opponents. Only those owning land near the new site supported construction of the airport (K. Motomura 2001). When Governor Ota flew to the Ishigaki airport on August 13, 1992, to talk to the residents of Miyara in an attempt to defuse the situation, protests prevented him from leaving the airport terminal. The confrontation lasted eight hours, ending only when he returned to Naha (Ikushima 1989). The governor's popularity continued to decline, owing to his stance on airport construction, and, in the November 1998 election, he was defeated by Inamine Keiichi, who was supported by the Liberal Democratic Party. After winning the election, Governor Inamine announced a change in plans, the adoption of the previously rejected Karadake Land Plan, which meant locating the new airport in the area of Mt. Karadake.

In the meantime, and capitalizing on the momentum it had obtained, the APYSCR had, in February 1992, begun to mobilize local citizens to push for the recognition of the Shiraho Sea as a World Heritage Site.[4] But that momentum was interrupted by Governor Ota's announcement in 1992 that the Miyara Plan had been selected, and it was difficult for the APYSCR to jump-start it in 1998 when Governor Inamine adopted the Karadake Land Plan (Ikushima 2003a). Nevertheless, the grassroots movement is not completely moribund. When the prefectural government was ready to start construction at Mt. Karadake, it was surprised to learn that the APYSCR, along with other NGOs in Osaka, Tokyo, and Naha, had mobilized 24 people who registered as the owners of a 1,560-square-meter plot of land in the center of what was to be the terminal at the new airport (*Ryukyu shimpo*, October 20, 2001). This "trust movement" continued, and, just six months later, 150 people had registered, and 300 more were willing to register, as the owners of parcels of land at the airport site (*APYSCR Newsletter*, April 30, 2002), with the APYSCR's goal being 1,000 or more (*Mainichi shimbun* [Yaeyama ed.], December 26, 2001). This placed the government in an awkward position, and, publicly, the governor promised to negotiate with the "landowners" (*Mainichi shimbun* [Yaeyama ed.], December 8, 2001; February 16, 2002). But, ultimately, he refused to do so, and the government was able to expropriate the land under the right of eminent domain.

The APYSCR has subsequently been unable to regain its former strength. On August 22, 2002, the APYSCR wrote both the governor of Okinawa and the mayor of Ishigaki City proposing an international academic symposium at which both pro- and antiairport views would be presented (*APYSCR Newsletter*, November 20, 2002). The mayor agreed to attend (*Mainichi shimbun* [Yaeyama ed.], August 26, 2002), but the governor rejected the idea. The same day, the organization also proposed the creation of a national park at Mt. Karadake, but both the mayor (*APYSCR Newsletter*, November 20, 2002) and the governor (Ikushima 2003b) rejected the idea. The APYSCR's popular support is also waning. Whereas in its heyday the movement had over forty-five hundred members, it currently has only eleven hundred. Consequently, the APYSCR has limited financial support and can employ only one full-time employee, Ikushima Toru, who manages a staff of ten to fifteen paid part-timers and volunteers (Ikushima 2003b). The major newspapers— for example, the *Asahi shimbun*, the *Yomiuri shimbun*, and the *Mainichi shimbun*—no longer cover the airport issue (Ikushima 2003a).

Issues Related to Airport Construction

Views regarding the construction of a new airport on Ishigaki Island have been shaped, not only by the long history of local opposition, but also by the fundamental debate between economic growth and environmental conservation. Those in favor of airport construction emphasize the importance of economic growth in Ishigaki City, those opposed the importance of preserving the area around Mt. Karadake.

The Pro–Airport Construction Position

Those in favor of new airport construction include politicians, high-level bureaucrats, local businessmen (particularly those involved in construction), and scholars whose focus is on economic growth. This group offers four arguments in support of its position.

First, the new airport is necessary to meet the increased demand for air transportation and to revitalize the local economy (Ishigaki 2001). The construction phase alone (expected to cost ¥63.3 billion) is expected to create forty-eight hundred jobs, more than thirty-one hundred of which will be with construction companies. The impetus that the operational airport will give to the service and manufacturing sectors will in ten years' time, it is argued, have pumped ¥403 billion into the local economy and generated sixty-seven hundred jobs. Also, exports from the agriculture and fishing industries during that same time period are expected to push another ¥3.6 billion into the local economy (see table 16.1).

Second, the new airport is necessary because cargo transportation by jet plane is crucial to island life (Okinawa Prefectural Government 1998). Because the current airport is temporary, it has only a fifteen-hundred-meter runway, which means that it can handle only smaller airplanes, such as the 130-seat Boeing 737-200 and the 150-seat Boeing 737-400, and no large cargo planes. The new airport will have a two-thousand-meter runway, which means that it can handle larger airplanes, such as the 234-seat Boeing 767-200 as well as large cargo planes (Ukai 1992). It will also relieve cargo and passenger limitations on flights (Ohama 2001). It will even allow Ishigaki Island to become a resort destination (Ishigaki 2001).

Third, the new airport is necessary to relieve flight-related noise. Because the current airport is only temporary, residential construction (homes and apartment buildings) has boomed in its vicinity since the 1970s. At the same time, increased flight demands have increased noise levels. Politicians in particular argue that moving the airport will bring much-needed relief to local residents (Ohama 2001).

Finally, the new airport is necessary on the grounds of safety. The short runway with which the current airport is equipped has already been the cause of a landing accident. The new airport's longer runaway will prevent such accidents (Ohama 2001).

Thus, it can be seen that economic growth is the top priority behind the push to build the new airport. The pro–airport construction group expresses little interest in the concerns of local citizens, especially those living in the Shiraho area. It believes that the economy of Ishigaki Island will come to a standstill if plans for the new airport do not proceed.

The Anti–Airport Construction Position

All the arguments of the pro–airport construction group—which is seen as motivated only by self-interest—are rejected by the anti–airport construction group, which includes the APYSCR.

First, the argument that the new airport is necessary to revitalize the local economy is flawed. That is because, in this scenario, revitalization is based on new construction jobs. But the new airport is classified as third rank.[5] And that means that only large construction companies from the Japanese mainland will be able to bid on the job, leaving only the occasional subcontract to be handled by local companies (Washio 2000).

Second, the argument that the new airport is necessary because cargo transportation by jet plane is crucial to island life is likewise flawed. Currently, no airline company plans to fly cargo planes in to Ishigaki Island (Washio 2001). And few farmers in rural prefectures

Table 16.1. Economic Effects in Yaeyama County				
A. Construction Investment Effects during the Building Period				
	Effect Created by Production (¥100 million)		Effect Created by Employment (jobs)	
Total	633		4,754	
Agriculture, stock breeding	2		4	
Manufacturing	82		340	
Construction	430		3,228	
Commerce	21		453	
Transportation/information	20		124	
Service	45		517	
Other	33		88	
B. Increase in Visitors after Airport Opening				
	Total 10 years	Used 1 year	Used 5 years	Used 10 years
Total effect created by production (¥100 million)	4,030	224	386	577
Agriculture, stock breeding	21	1	2	3
Manufacturing	113	6	11	16
Construction	35	2	3	5
Commerce	878	49	84	126
Transportation/information	831	46	80	119
Service	1,784	99	171	255
Other	368	21	35	53
Total effect created by employment (visitors)		2,613	4,514	6,736
Agriculture, stock breeding		3	6	8
Manufacturing		19	32	48
Construction		14	25	37
Commerce		1,077	1,860	2.776
Transportation/information		285	492	735
Service		1,135	1,961	2,926
Other		80	138	206
C. Other Potential Effects after Airport Opening (¥100 million)				
		Used 1 Year	Used 5 Years	Used 10 Years
Agriculture		0	14	29
Fishing		3	5	7
Source: Okinawa Prefectural Government (1998).				

succeed in selling their produce in big-city markets; in fact, big-city products dominate rural markets. The effect of the new airport on Ishigaki Island will, therefore, be to facilitate the importation of products from the mainland. This will be exacerbated if Ishigaki does, in fact, become a resort destination because mainland tourists will prefer products from the mainland. The end result will be,

not the stimulation, but the weakening of the local economy. Even increased tourism itself is unlikely to help much. For example, new and renovated hotels are likely to be mainland owned and to bring most of their employees with them, and few visitors are likely to leave the downtown resort area (Gushiken 2001a; Ui 2001).

Third, the argument that the new airport

is necessary to relieve flight-related noise is also flawed. Yes, residents living near the current, temporary airport will experience relief. But the problem of noise pollution will simply be transferred to the Mt. Karadake area (Washio 2000).

Finally, the argument that the new airport is necessary on the grounds of safety is no more than a smoke screen. The fact of the matter is that many people, including local and mainland politicians, high-ranking bureaucrats, and prominent businessmen, have invested heavily in land in the area around Mt. Karadake. The expectation is that they will be able to sell the land to the government at an inflated price (Y. Motomura 2001b). Indeed, since the mid-1980s, some real estate company employees have been arrested and charged with land fraud (Ukai 1992).

The main concern of those opposed to the new airport construction (see photograph 16.5) is, of course, the preservation of the natural environment. Of particular concern is the survival of the coral reef. Coral requires clean water to thrive. But the leveling of portions of Mt. Karadake to build the airport threatens to release large amounts of red clay into the Shiraho Sea. The local government has yet to find a way to control the naturally occurring red clay runoff in the island's rivers (Gushiken 2001b) and, thus, cannot be expected to deal with that generated by the airport construction (K. Motomura 2001).

Photograph 16.5. An anti–Karadake Plan meeting on April 9, 2001. (Courtesy Ikushima Toru.)

Fundamental Questions Regarding Airport Construction

Fundamental questions remain to be raised. For example, is the capacity of the current airport so limited as to call for the construction of a new airport? According to statistical data provided by the Okinawa prefectural government, the number of landings has steadily increased over the period 1995–2005—from 8,989 in 1995 to 12,317 in 2005—as has the number of passengers carried—from 1.14 million in 1995 to 1.86 million in 2005—but not enough to overburden the current airport (see tables 16.2–16.3). The local government does claim that the number of visitors to the island will increase to 2.09 million by 2020 (*Yaeyama yoran* 2001), but no evidence has been presented to support this claim—hardly a compelling case for new airport construction. More telling is an analysis of the Ishigaki–Naha route, which accounts for most of the island's air traffic. According to government statistics from 2004, this route accounts for 62.2 percent of passenger and 82.5 percent of cargo traffic. Of the other routes connecting Ishigaki to major cities on the mainland, only three others (those to Tokyo, Osaka, and Miyako) break out of the single digits (10.4, 15.6, and 15.6 percent, respectively) for even *combined* passenger and cargo traffic (see table 16.4). These data suggest that Ishigaki Island is only a stopover for most travelers.

What *is* a compelling case is the one *against* new airport construction based on a national debt that, as of December 2000, stood at more than ¥660 trillion (or 130 percent of GDP) and was expected to reach ¥700 trillion by 2006 (Schoppa 2001). Government projections for public works projects (highway, railroad, and airport construction) always maintain that there will be tremendous public demand in the near future, but most such projects (e.g., the Honshu-Shikoku Highway, the Tokyo Bay Bridge, and the Kansai Airport), once completed, are not even

Table 16.2. Ishigaki Airport Status

Year	Number of Landings	Number of Passengers			Tons of Cargo		
		On	Off	Total	Load	Unload	Total
1995	8,989	572,048	569,360	1,141,408	4,345	3,816	8,161
1996	8,976	577,951	574,516	1,152,467	4,278	4,039	8,317
1997	8,980	637,548	633,171	1,270,719	4,547	4,219	8,766
1998	9,796	659,835	655,889	1,315,724	4,246	4,498	8,742
1999	9,198	699,995	697,108	1,397,103	4,536	4,739	9,302
2000	9,737	713,767	708,968	1,422,735	4,429	5,064	9,493
2001	9,931	722,168	716,494	1,438,662	4,703	5,305	10,008
2002	9,970	754,417	750,048	1,504,465	4,710	5,298	10,008
2003	11,662	879,457	875,855	1,755,312	4,864	5,483	10,347
2004	12,487	895,991	885,913	1,781,904	4,857	6,159	11,016
2005	12,317	934,304	929,304	1,863,608	4,589	6,437	11,026

Source: Yaeyama yoran (2001, 89); Yaeyama yoran (2006, 88).

Table 16.3. Cargo Freight at Ishigaki Harbor (tons)

Year	Total	International Trade			Domestic Trade		
		Total	Import	Export	Total	In	Out
1995	607,762 (100)	49,918 (8.2)	39,834 (6.6)	10,084 (1.7)	557,844 (91.8)	388,956 (64.0)	168,888 (27.8)
1996	1,150,174 (100)	98,692 (7.4)	98,692 (7.4)	0 (0)	1,051,482 (79.3)	793,840 (59.9)	257,642 (19.4)
1997	1,325,186 (100)	300,975 (23.9)	152,130 (12.1)	148,845 (11.8)	1,024,211 (81.3)	857,904 (68.1)	166,307 (13.2)
1998	1,259,455 (100)	295,834 (23.9)	146,211 (11.8)	149,623 (12.1)	963,621 (77.9)	721,123 (58.3)	242,498 (19.6)
1999	1,236,008 (100)	252,200 (20.6)	169,032 (13.8)	83,168 (6.8)	983,808 (80.5)	707,9063 (57.9)	275,902 (22.6)
2000	1,222,495 (100)	174,869 (14.3)	98,673 (8.1)	76,196 (6.2)	1,047,626 (85.7)	772,559 (63.2)	275,067 (22.5)
2001	1,274,738 (100)	233,229 (18.3)	103,651 (8.1)	129,578 (10.2)	1,041,509 (81.7)	771,642 (60.5)	269,867 (21.2)
2002	1,312,985 (100)	266,771 (20.3)	88,279 (6.7)	178,492 (13.6)	1,046,214 (79.7)	778,443 (59.3)	267,781 (20.4)
2003	1,327,092 (100)	159,866 (11.0)	68,825 (4.7)	91,041 (6.3)	1,167,226 (80.3)	827,336 (56.9)	339,890 (23.4)
2004	1,453,073 (100)	314,149 (21.6)	68,015 (4.7)	246,134 (16.9)	1,138,924 (78.4)	862,819 (59.4)	276,105 (19.0)

Sources: Yaeyama yoran (2000, 91); Yaeyama yoran (2001, 91); Yaeyama yoran (2006, 91).
Note: Percentages are given in parentheses. Because the data given in the various editions of Yaeyama yoran differ slightly, rows do not always total to 100 precisely.

Table 16.4. Passenger and Cargo Flights from Different Routes, 2004		
	Passenger Flights	**Cargo Flights**
Ishigaki–Tokyo	7.3	3.1
Ishigaki–Osaka	10.8	4.8
Ishigaki–Fukuoka	3.6	1.8
Ishigaki–Nagoya	3.6	1.3
Ishigaki–Naha	62.2	82.5
Ishigaki–Miyako	7.3	3.1
Ishigaki–Yonaguni	3.8	3.1
Ishigaki– Hateruma	0.1	0.1
Ishigaki–Tarama	0.1	0.1
Ishigaki–Tohoku	1.2	
Ishigaki–Kumejima	0.1	

Note: Figures do not total to 100 owing to rounding. There are no cargo flights between Ishigaki and Tohoku or between Ishigaki and Kumejima.

self-supporting, let alone in a position to repay the loans required to build them. It seems unlikely that the new airport on Ishigaki Island will be any different.

Another compelling case against new airport construction is that those visitors who *do* come to Ishigaki come to enjoy the natural environment. Without the coral reefs in particular, few tourists would come to Ishigaki. But the reefs will be endangered by red clay runoff from construction. Construction will also endanger the animals—including a number on the endangered species list—that make their home on Mt. Karadake. Not to mention the fact that ignoring the natural environment at Shiraho compromises the environmental assessment law and goes against the wishes of the local citizens (Kobayashi 2001a).

Other unanswered questions remain. For example, according to the rules of the site selection committee, its decision had to be unanimous, yet the representative of the World Wide Fund for Nature–Japan was opposed (Washio 2000). So how could the committee have reached a final decision? There is also the question of why all the scholars on the committee were from the Ryukyu University (and none from Okinawa University). And

why did the committee not choose to expand the current airport? Certainly, huge amounts of taxpayers' money would have been saved. (Admittedly, land owned by the Pacific Golf Club would have had to have been purchased for runway expansion. And of course many important people had already bought land in the area of Mt. Karadake.) What will be done with the current airport when the new airport is built? It is hardly likely that it will be abandoned (despite the government's claims). Will it be used by the Japanese air force—or by the American forces in Okinawa?

In sum, too many questions remain unanswered. The official estimate of the project's environmental and ecological costs has not even been made public. The project seems to be controlled by self-interested investors out for nothing but the profit.

Future Prospects

With little public debate and scant input about the environmental impact, the government decided to build a new airport at Mt. Karadake. Sadly, few local citizens really understand what the consequences for Ishigaki

Island's ecosystem will be. Indeed, few expressed any interest whatsoever in protecting the environment (Washio 2000). Why is this? One of many reasons is the role of the local media. The majority of the people living in the Yaeyama District subscribe to the Yaeyama edition of the *Mainichi shimbun,* which does not report the views of the opposition movement (Kobayashi 2001a; Y. Motomura 2001a). The newspaper's editor even refused an interview with me, stating (when I called on July 2, 2001): "You can write whatever you want; I do not care." The generation that most strenuously opposed the new airport construction is aging, and the younger generation seems to have little interest in having the torch passed to it (Miyara 2001). Time is running out for the coral reefs.

Notes

This research was supported by the 2001 Hokuriku University Special Research Fund.

1. Cutting off promised funding for public works projects is unthinkable in Japan, unlike in the West.

2. The government claimed that, under the Karadake East-Side Plan, more space would be available to build the airport, meaning that less of the sea would need to be reclaimed and, thus, less of the coral reef destroyed. It also claimed that the new location would mean reduced noise levels for surrounding communities.

3. For reproductions of the ads, see Muzik and Makishi (1991).

4. More than 440 sites, including the Great Barrier Reef, the Galapagos Islands, and Yosemite National Park, have been registered as World Heritage Sites. In 2003, the Ryukyu Islands became one among three candidates for the Japanese government to recommend to the Unesco World Heritage Programme.

5. In Japan, airports are classified as follows: the two largest airports, Haneda (formerly Tokyo International) and Narita (which now handles international traffic), are in the first rank; city airports, including Naha, Nagoya, and Asahikawa, are in the second rank; and regional/local airports, such as Miyako, Yanakuni, and Ishigaki, are in the third rank.

References

Gushiken, Minoru. 2001a. Interview with the author. July 1. Shiraho District, Ishigaki.
———. 2001b. Interview with the author. July 2. Shiraho District, Ishigaki.
Ikushima, Toru. 1989. "Shiraho no sango no umi wo mamoru tatakai" [The fight to protect the Shiraho Coral Reef]. In *Shizen hogo jiten* [Encyclopedia of nature conservation], ed. Association for Japanese Nature Conservation. Tokyo: Ryofu Shuppan.
———. 2003a. Interview with the author. January 12. Shinjuku, Tokyo.
———. 2003b. Interview with the author. March 25. United Nations University, Shibuya, Tokyo.
International Union for Conservation of Nature (IUCN). 1991. *Shiraho no umi wo shizen kankyo hozen chiiki ni sitei saseyo!* [Let's declare the Shiraho Sea as a natural monument!]. In *Let the Sea Stain Our Hearts,* ed. and trans. Katherine Muzik and Yoshikazu Makishi. Naha: Association to Protect the Shiraho Coral Reef. Originally published in 1988.
Ishigaki, Sosei. 2001. Interview with the author July 2. Shiraho District, Ishigaki.
Kobashigawa, Tomoo, and Shigekazu Mezaki. 1989. *Ishigakijima, Shiraho Sango no umi* [The Ishigaki Island, Shiraho Coral of Sea]. Tokyo: Kobunken.
Kobayashi, Takashi. 2001a. Interview with the author. June 2. Shiraho District, Ishigaki.
———. 2001b. Interview with the author. June 3. Shiraho District, Ishigaki.
Miyara, Chochin. 2001. Interview with the author. July 4. Shiraho District, Ishigaki.
Motomura, Keiwa. 2001. Interview with the author. July 3. Shiraho District, Ishigaki.
Motomura, Yoshiko. 2001a. Interview with the author. July 2. Shiraho District, Ishigaki.
———. 2001b. Interview with the author. July 3. Shiraho District, Ishigaki.
Muzik, Katherine. 1991. "Blessing from the Sea." In *Let the Sea Stain Our Hearts,* ed. and trans. Katherine Muzik and Yoshikazu Makishi. Naha: Association to Protect the Shiraho Coral Reef.
Muzik, Katherine, and Yoshikazu Makishi, eds. 1991. *Let the Sea Stain Our Hearts.* Naha: Association to Protect the Shiraho Coral Reef. Includes material in both English and Japanese.
Ohama, Nagateru. 2001. Interview with the author. July 5. Shiraho District, Ishigaki.
Okinawa Environment Network. 2001. *Okinawa*

kara sekai e: Heiwa, kankyo, fukushi no 21 seiki wo [From Okinawa to the world: Peace, environment, and social welfare in the twenty-first century]. Naha: Okinawa Kankyo Nettowaku.

Okinawa Prefectural Government. 1998. *Shin Ishigaki Kuko keikaku* [Plan of the new Ishigaki Airport]. Okinawa: Shin Ishigaki Kuko Kensetsuka.

Schoppa, Leonard J. 2001. "Japan, the Reluctant Reformer." *Foreign Affairs,* September/October, 76–90.

Ui, Jun. 2001. Interview with the author. June 30. Okinawa University, Naha City, Okinawa Prefecture.

Ukai, Teruyoshi. 1992. *Okinawa: Kyodai kaihatsu no riron to hihan* [Okinawa: The theory and criticism of large development projects]. Tokyo: Shakai Hyoronsha.

Washio, Masahisa. 2000. "Yaeyama ni kurasu: Shin Ishigaki kuko kensetsu ni handai sitsutsu" [Living in Yaeyama: To continue opposing the new Ishigaki Airport construction]. *Higashi ajia shakai kyoiku kenkyu* [Studies of social education in the East Asia], no. 5 (September): 171–80.

———. 2001. Interview with the author. July 4. Shiraho District, Ishigaki.

"The World's Coral Reefs Are in Danger!" 2003. http://www8.cds.ne.jp/~nature/shirahonokai/ntimese.html (accessed November 8, 2007).

Yaeyama yoran [Handbook of Yaeyama]. 2000. Ishigaki: Okinawa Yaeyama Shicho Shinko Somuka.

———. 2001. Ishigaki: Okinawa Yaeyama Shicho Shinko Somuka.

———. 2006. Ishigaki: Okinawa Yaeyama Shicho Shinko Somuka.

Chapter 17

The Management of Mountain Natural Parks by Local Communities in Japan

Teiji Watanabe

High mountainous areas in Japan have not historically been targeted for intensive use by animal herders, having been considered sacred since at least the sixth or seventh century (Koizumi 2001). Use patterns have begun to change, however. With the construction of roads and the development of recreational facilities, the mountains have become a resort destination. And with this change also has come natural resource deterioration, such as rapid soil erosion on trails, the accumulation of human waste and trash, and the loss of alpine vegetation by trampling and by illegal collection.

Most high mountainous areas in Japan are designated as natural parks, whether national, quasi-national,[1] or prefectural. The highest areas—those above sixteen hundred meters on Hokkaido Island in northern Japan and above twenty-six hundred meters on Honshu Island in central Japan—are alpine zones characterized by extremely strong winter winds and an uneven distribution of snow cover that ranges from zero to more than twenty meters. The result is beautiful, high-altitude landscapes, complex, patchy patterns of dwarf pine (*Pinus pumila*), alpine meadows, permanent snow, and active periglacial landforms. Nineteen of twenty-eight national parks are in high mountainous areas (Norihisa and Suzuki 2006).

The national park system of Japan is described in Hiwasaki (2005, 2006) and Norihisa and Suzuki (2006). One of its distinctive features is that, unlike the system prevailing in the United States, the government can, un-

der the Natural Parks Law (which regulates national, quasi-national, and prefectural natural parks), establish national parks without first procuring the land (Norihisa and Suzuki 2006).

This chapter first discusses trail deterioration, describing the present state of park trails and the causes of deterioration; provides examples of grassroots movements attempting to protect parkland; considers the role that grassroots movements played in the amending of the Natural Parks Law in 2002; and discusses the importance of the continued involvement of grassroots movements in protecting mountain natural parks.

Nature Deterioration and Its Management

A trail is the basis of the management and use of mountain national parks, which normally have roads (paved and unpaved, for both vehicles and hikers) only in buffer zones. The deterioration of natural conditions usually starts on and near trails. It can be observed in the form of soil erosion, sedimentation of the eroded soils, and damage to surrounding vegetation. Many of the major trails in Japanese mountains are among the most severely deteriorated in the world.

Soil erosion on trails has been studied most in Daisetsuzan National Park (DNP), located in the central part of Hokkaido Island, northern Japan (see figure 17.1). Determining the extent of soil erosion is difficult,

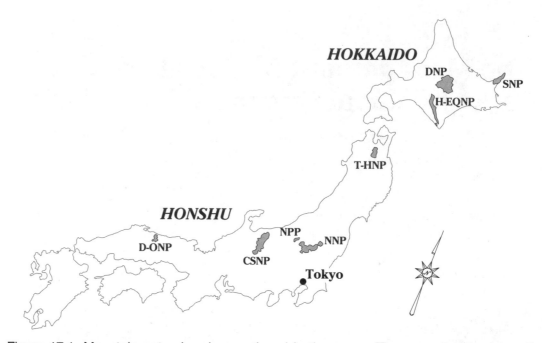

Figure 17.1. Mountain natural parks mentioned in the paper. There are twenty-nine national parks, fifty-five quasi-national parks, and more than three hundred prefectural natural parks. DNP = Daisetsuzan National Park; SNP = Shiretoko National Park; T-HNP = Towada-Hachimantai National Park; NNP = Nikko National Park (the Oze area of Nikko National Park became Oze National Park in July 2007); CSNP = Chubu Sangaku National Park; D-ONP = Daisen-Oki National Park; H-EQNP = Hidaka-Erimo Quasi-National Park; NPP = Niigata (Uonuma-Renpou) Prefectural Natural Park.

so previous studies conducted in Japan (Yoda 1991; Gotoh 1993; Watanabe and Ono 1996; Watanabe and Fukasawa 1998; Yoda and Watanabe 2000; Oki 2001; Watanabe, Ota, and Gotoh 2004) have adopted the method outlined in Cole (1983) whereby cross sections are taken at two different dates and the difference between them determined (in centimeters squared) (see figure 17.2a).[2] Yoda and Watanabe (2000), measuring the erosion rates from 1990 to 1997 on the south side of Mount Kurodake (in the northern part of DNP), found that they ranged from 100 to 1,557 centimeters squared per year. They also suggested that the amount of erosion is larger and the rate faster at snowy sites than that at windy sites. Watanabe and Fukasawa (1998) estimated the trail erosion rates on the northern slope of Mount Kurodake to range from 57 to 557 centimeters squared per year. In the area they studied, 70 percent of the wooden steps and water-drainage bars—installed in

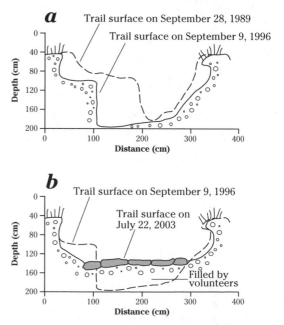

Figure 17.2. A cross section of a deteriorated trail in DNP. Graph a shows change in the trail surface owing to severe soil erosion, 1989–1996. Graph b shows the trail surface after the addition of sand and gravel by volunteers in 2003.

1989, seven years before they conducted their study—were damaged, indicating that regular maintenance is essential.

Because the deterioration of mountain trails is one of the most serious problems facing the national park management system, the Environment Ministry (and its predecessor, the Environment Agency) has, in recent years, been spending a large portion of its budget on so-called public works—large-scale trail-maintenance projects. Despite the application of advanced engineering methods, however, the problem still remains. One reason is that the process of constructing boardwalk and steps can itself cause deterioration, and for such a heavy-handed approach the ministry has been criticized. Another reason is that there is no money in the ministry's budget for day-to-day maintenance once the large-scale construction projects are completed.[3] The same problem is experience by quasi-national and prefectural natural parks throughout the country.

Overuse of mountain national park areas in recent years has also accelerated deterioration, especially soil erosion. This overuse is directly related to improved accessibility.

Road construction in and near the mountain national park areas intended to facilitate forestry management and timber logging has at the same time facilitated recreational visits. It has also led to massive landslides and weekend/holiday traffic congestion.

Grassroots movements have actively engaged such issues. The problems they tackle range from trail maintenance (e.g., counteracting or at least mitigating deterioration) to development (e.g., opposing road construction and deforestation).

Examples of Grassroots Activism

Trail Maintenance

Trail maintenance has always been a matter of treating the symptoms, as it were, not the underlying disease, a matter of correction rather than prevention (Watanabe 2008). The national park management has long considered boardwalk construction to be the best method of dealing with erosion (see photograph 17.1). It also sometimes employs rock-filled gabions

Photograph 17.1. An example of boardwalk constructed as a big-budget public works project in DNP. (Courtesy T. Watanabe)

Photograph 17.2. An example of boardwalk constructed by local volunteers in DNP. (Courtesy S. Saito)

(wire cylinders) to fill gullies. (Helicopters are used to carry materials and equipment to mountain areas.) Such projects are undertaken only infrequently, however, and deterioration still plagues trails in many park areas.

The problem of deterioration is so widespread, in fact, that citizens' groups are often prompted to attempt their own solutions. One such initiative has, since 1977, targeted the prefectural natural park of Mount Makihata in central Japan (Kurita and Aso 1995), a group of volunteers—mostly university students under the guidance of their professors—undertaking a revegetation project. Similar projects have been successful in other mountain areas, such as those mounted in the Mount Shirouma region of Chubu Sangaku National Park (Tsuchida 2002) and the Mount Daisen region of Daisen-Oki National Park.

DNP, as described above, is faced with the problem of rapid soil erosion, to remedy which volunteers started a trail-maintenance program in 1998 (see figure 17.2b). Personnel manning the Kogen-Onsen Brown Bear Information Center in the park also started a voluntary trail-maintenance program. They have even built a boardwalk, using discarded lumber and donated materials (see photograph 17.2).

The most important effect of these movements, however, has been that they have caused the government to reconsider its approach to trail maintenance. An Environment Ministry task force was created in 1999 and charged with finding new methods that employ fewer workers and are less expensive (Watanabe 2008). The task force began testing some of those methods in DNP in 2001. Whatever the results of those tests, grassroots

Teiji Watanabe

involvement will clearly be integral to successful future trail maintenance.

Road Construction

Road construction in mountain natural parks has occasioned an even stronger grassroots response than the problem of trail maintenance. For example, the construction of Shihoro Kogen Road in the eastern part of DNP (a project initiated in 1964 by the town of Shihoro and taken over in 1969 by Hokkaido Prefecture) was highly controversial from the start. Opposition in the 1960s was conducted mainly on an individual basis. Then, in the early 1970s, with support from university professors, some activists began forming various grassroots groups with the aim of building a more powerful and systematic opposition movement. These groups remained active through the 1980s and into the 1990s. Some hosted symposia spotlighting the importance of the area's ecosystem. One collected more than 200,000 signatures in opposition to the project and, in 1995, filed a lawsuit to halt construction. The prefectural government finally decided to halt construction in 1999. At that point, only 2.7 of 21.6 kilometers remained to be built. Although $15 million had been wasted, $67–$80 million (earmarked mainly for tunnel construction) was saved.

One other road-construction project in the alpine area of DNP was brought to an end by citizen opposition. So was one in Hidaka-Erimo Quasi-National Park (also in Hokkaido Prefecture). The Hidaka-Erimo construction began in 1984 and cost $430 million before the prefectural government called a halt to it—on the face-saving grounds that the expected return was not worth the continued investment.

One other important factor in the success of such grassroots activism must here be mentioned—the voluntary contributions of university researchers. Local researchers, especially from Hokkaido University and the Obihiro University of Agriculture and Veterinary Medicine, collected, and made available to opposition groups, data detailing the harmful effects of road construction in DNP and Hidaka-Erimo Quasi-National Park. Similar partnerships have been established between activists involved with Chubu Sangaku National Park and researchers from Toyama, Kyoto, and Shinshu universities; activists involved with Towada-Hachimantai National Park and researchers from Hirosaki University; activists involved with Niigata Prefectural Natural Park; and researchers from the Tokyo University of Agriculture and Engineering.

Deforestation

Deforestation too has occasioned grassroots protest. For example, the initial attempts of an anti-deforestation group to stop the April 1996 clear-cutting by the Forestry Agency of certain areas of Shiretoko National Park in eastern Hokkaido were frustrated by the police. Immediately after the incident, however, the group turned to the mass media to publicize the fact that much old growth—including many trees older than four hundred years—had been cut and animal habitat destroyed. The resulting furor led the Forestry Agency to abandon its plans to conduct further logging in the park, having cut only 5 percent of what it had originally planned to cut over a ten-year period. This success encouraged further protests against deforestation in Towada-Hachimantai National Park, and the Forestry Agency had no choice but to change its policy from logging to conserving forests.

At the same time, reforestation projects have gained grassroots attention. A good example is the Shiretoko 100 Square Meters Movement [Shiretoko 100 heihou metoru undou] (Nakagawa 2006). Over the roughly twenty-year period from 1979 to 1997, some fifty thousand people nationwide donated some $4.3 million to purchase a 450-hectare parcel of land in Shiretoko National Park for reforestation. Renamed the Shiretoko 100 Square Meters Movement Forest Trust [Shiretoko 100 heihou metoru undou no mori torasuto], the movement continues today, involving roughly ninety-one hundred people who have donated a further $1 million to the project.

Grassroots Movements at the National Level

Grassroots movements have met with both failure and success at the national level. Most notable is their failure to address an important structural problem plaguing all, and not just mountain, national parks in Japan: they have historically been managed by multiple agencies (some combination of the Environment Ministry, the Forestry Agency, the Agency for Cultural Affairs, and various local governments), making it difficult to tell where management responsibility lies. Although in January 2001 the national government implemented systemwide structural reforms, for example, upgrading (as we have seen) the Environment Agency to the Environment Ministry, the plural-management structure remained in place. Grassroots attention to this problem is sorely needed.

Grassroots movements can, however, be credited with the amendment in 2002 of the Natural Parks Law, albeit indirectly. The amended Natural Parks Law mandates the assumption of partial responsibility for park management by nongovernment organizations (NGOs), the recognition of the importance of ecosystem conservation, and the introduction of use-limitation areas within natural parks. The involvement of NGOs in park management in particular can be regarded as the result of long years of grassroots activism, a recognition by the Environment Ministry of its past policy mistakes and its limited human and financial resources in the face of government cutbacks (Hiwasaki 2005, 2006). A possible management framework with grassroots involvement is presented in the following section.

The Future of Grassroots Movements

In many countries, various stakeholders—including local communities, central and local governments, educational and research institutions, nature-conservation groups, volunteers, and users—are involved in park management (Hall and McArthur 1996). In fact, the Nature Conservation Society of Japan stressed the importance of collaboration when proposing a partnership program meant to bring the management of national parks in Japan into the twenty-first century (NACS-J 2000). But, as we have seen, only recently has the amended Natural Parks Law made this possible in Japan. Also, there is no tradition of communication among these various groups in Japan, and few have a history of involvement with the park system. While, under the Natural Parks Law, the nongovernment body assuming management responsibility need have no prior experience, it seems likely that seasoned nature-conservation groups will take the lead. And, if this experiment is to succeed, the involvement of all interested parties, and especially local communities, is essential. It is, therefore, incumbent on the Environment Ministry to ensure that no stakeholders are excluded from the process and that those new to it are involved from the very beginning (Brandon 1993; Watanabe 2003; Hiwasaki 2005).

This assertion is borne out by two case studies of similar initiatives reported in Brandon (1993). The success of that in the Annapurna Conservation Area in Nepal can be attributed to the involvement of local residents from the beginning; the failure of that in the Monarch Nature Reserve in Mexico, to their involvement only later. Ali and Butz (2003) report similar findings in a case study of Khunjerab National Park in northern Pakistan. In establishing the Khunjerab Villages Development Organization, the national park authority left out one local community, Shimshal, which now refuses to acknowledge the existence of the park.

Teiji Watanabe

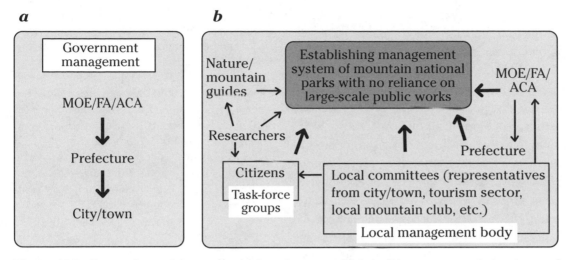

Figure 17.3. Comparison of the present (*a*) and proposed future (*b*) management structures of mountain national parks in Japan. MOE = Environment Ministry; FA = Forestry Agency; ACA = Agency for Cultural Affairs.

Since the new management structure has not yet been established, there is still time to ensure that the experiment is successful. In what follows, I present one possible new management framework. Whereas, up to now, Japanese national parks have been managed exclusively by the central and local governments (see figure 17.3*a*) in a top-down approach that allows little room for local community participation (Hiwasaki 2005), I suggest a system in which grassroots task force groups operate under the umbrella of the local management body (see figure 17.3*b*).

The local management body would be composed of one or more committees whose membership is representative of all interested stakeholders. Several such local management mechanisms already in existence in Japan provide useful reference points here. Kamikawa, for example, one of the ten municipalities located in DNP, has established committees to handle specific aspects of park management, such as trails, toilets, and brown bears. The drawback to this particular approach is its inefficiency. Although some individuals serve on more than one committee, the committees rarely coordinate their efforts (Hiwasaki 2006) and are controlled by powerful

bureaucrats uninterested in views that differ from theirs. Also, the committees remain in existence for only one to three years (depending on their purpose) and, thus, can have no significant long-term effect. A similar example is the Shiretoko Regional Liaison Committee, which manages the Shiretoko World Natural Heritage Site (also known as the Shiretoko National Park), serving to build a consensus between the national government and local communities (Yoshinaka 2006). A slightly different type of local management body is observed in the Oze National Park (in central Japan), where an incorporated foundation has, with support from the prefectural government and the Environment Ministry, assumed some management functions.

Such a management structure will not necessarily ensure the desired grassroots involvement and, thus, the desired budget reduction. These ends can be most effectively achieved by the creation of task force groups that operate under the guidance of the local management body. This arrangement maximizes volunteer involvement by allowing the participation of novices and optimizes performance by promoting specialization. Researchers also play an important role in such a scheme, making available professional knowledge and skills to the local management body, the task force groups, and nature/mountain guides alike. And guides should, in turn, be permitted to play a role in park management themselves.

Ideally, all stakeholders should function with some degree of autonomy (Cernea 1991; Brandon 1993). Still, inexperienced grassroots activists cannot be allowed to proceed with their work without at least some guidance from more experienced stakeholders (Watanabe 2003, 2008). Such systematic control will, likely, be different from the operating procedures with which most volunteers are familiar. But all parties must work to ensure their smooth incorporation in the new system to whose success they are essential.

Conclusion

Grassroots environmental movements in Japan have succeeded in demonstrating that the successful management of the country's high mountain natural parks requires input from, and the continued involvement of, local communities. Such an arrangement is relatively common worldwide, for example, in New Zealand (Hall and McArthur 1996) and Nepal (Lama 2000; Sharma 2000). It has even begun to be employed in the U.S. national park system (Brown, Mitchell, and Tuxill 2003) despite the important differences between its management style and the plural-management style of the Japanese system.

Notes

1. A quasi-national park is one so designated by the minister of the environment on the recommendation of the prefectural governor.
2. Soil erosion is usually measured cross-sectionally, not volumetrically, because of the difficulty involved in taking accurate volumetric measurements, a practical method for doing so not yet having been developed (see Hammitt and Cole 1998). The method outlined in Cole (1983) is the

most frequently used because it saves time in the field and provides reliable results.
3. Another symptom of current budget constraints is the relatively small number of park rangers the system is able to employ. DNP, e.g., which has an area of 216,000 hectares, has only seven rangers. Overall, some 250 people (including park rangers and other staff members) are employed to administer the 2,065,156 hectares of parkland found nationwide, a figure that works out to one staff member per every 8,261 hectares (Norihisa and Suzuki 2006). Parks cannot be properly managed by such small numbers.

References

Ali, I., and D. Butz. 2003. "Report on Shimshal Nature Trust (SNT), Ghojal, Northern Areas, Pakistan." Paper presented at the Fifth World Parks Congress, Durban, South Africa, September 11–13, 2003.

Brandon, K. 1993. "Basic Steps toward Encouraging Local Participation in Nature Tourism Projects." In *Ecotourism: A Guide for Planners and Managers,* ed. K. Lindberg and D. Hawkins, 1:134–51. North Bennington, VT: Ecotourism Society.

Brown, J., N. Mitchell, and J. Tuxill. 2003. "Partnerships and Lived-in Landscapes: An Evolving Use System of Parks and Protected Areas." *Parks* 13, no. 2:31–41.

Cernea, M. 1991. *Putting People First: Sociological Variables in Rural Development.* New York: Oxford University Press.

Cole, D. N. 1983. *Assessing and Monitoring Backcountry Trail Conditions.* Research Paper INT-303. Washington, DC: U.S. Department of Agriculture Forest Service.

Gotoh, T. 1993. "Daisetsuzan Kita-Hakkoudasan ni okeru tozandou no shinshoku" [Erosion of mountain hiking trails in the Daisetsuzan Mountains and Kita-Hakkoudasan Mountains, Japan]. Master's thesis, Hokkaido University.

Hall, C. M., and S. McArthur. 1996. "Strategic Planning." In *Heritage Management in Australia and New Zealand,* ed. C. M. Hall and S. McArthur, 22–36. London: Oxford University Press.

Hammitt, W. E., and D. N. Cole. 1998. *Wildland Recreation: Ecology and Management.* 2nd ed. New York: Wiley.

Hiwasaki, L. 2005. "Toward Sustainable Management of National Parks in Japan: Securing Long Community and Stakeholder Participation." *Environmental Management* 35:753–64.

———. 2006. "Community-Based Tourism: A Pathway to Sustainability for Japan's Protected Areas." *Society and Natural Resources* 19:675–92.

Koizumi, T. 2001. *Tozandou no tanjo—hito wa naze yama ni noboruyouni nattanoka* [Cultural history of mountaineering—why people began to climb mountains]. Tokyo: Chuko-shinsho.

Kurita, K., and M. Aso. 1995. "Tasetsu sangakuchi ni okeru setsuden shokusei no hukugen houhou ni kansuru kenkyu" [Studies on the revegetation method of snow-patch vegetation at heavy snow montane region]. *Nihon ryokuka kougakkaishi* [Journal of the Japanese Society of Revegetation Technology] 20:223–33.

Lama, W. 2000. "Community-Based Tourism for Conservation and Women's Development." In *Tourism and Development in Mountain Regions,* ed. P. M. Godde, M. F. Price, and F. M. Zimmermann, 221–38. Oxford: CABI.

Nakagawa, H. 2006. "Historical View of Shiretoko National Park." In *Wildlife in Shiretoko and Yellowstone National Parks,* ed. D. R. McCullough, K. Kaji, and M. Yamanaka, 221–24. Shari: Shiretoko Nature Foundation.

Nature Conservation Society of Japan (NACS-J). 2000. *Yutakana shizen hukai hureai patonashippu—21 seiki no kokuritsukouen no arikatawo kangaeru* [Diverse nature, deep experience and partnership—a proposal for national parks in the twenty-first century from the National Parks Subcommittee of the Nature Conservation Society of Japan]. Tokyo: Nature Conservation Society of Japan.

Norihisa, M., and W. Suzuki. 2006. "Mountainous Area Management in Japanese National Parks: Current Status and Challenges for the Future." *Global Environmental Research* 10, no. 1:125–34.

Oki, K. 2001. "Daisetuzan kokuritsukouen Kurodake ishimuro shuuhen ni okeru tozandou no hozen notameno kenkyu" [Study for the mountain hiking trail conservation near the Kurodake hut, Daisetsuzan National Park, central Hokkaido, Japan]. Master's thesis, Hokkaido University.

Sharma, U. R. 2000. "People's Participation in Sagarmatha Park." In *Himalayas, Ecology and Environment,* ed. M. S. Kohli and Y. Bali, 164–66. New Delhi: Har-Anand.

Tsuchida, K. 2002. "Tozandou shuuhen no kou-

haishita kouzanshokusei no hukugen" [Rehabilitation of deteriorated alpine vegetation around trails]. *Kokusaisangakunen kinen gyouji yama to shizenno shinpojiumu yama to shizen wo aisuru watashitachino ayumu michi shiryoushu* [Proceedings of the mountains and nature, International Year of Mountains], 14–15. Tokyo and Matsumoto: Kankyoushou and Yama to Shizen no Shinpojiumu Jikkou Iinkai.

Watanabe, T. 2002. "Deterioration of Mountain Trails and Their Management in Mountains of Japan." In *International Seminar on the Utilization and Conservation of Mountains,* 34–38. Seoul: Korean Geographical Association.

———. 2003. "Nihon no sangaku kokuritsukouen ni okeru turizumu to shizenkankyou hozen: Himaraya kara manabu koto" [Tourism and natural environmental conservation in mountain national parks of Japan: Lessons learnt from the Himalaya]. *Chiri Kagaku* [Geographic sciences] 58:146–56.

———, ed. 2008. *Tozandou no hozen to kanri* [Mountain trail studies]. Tokyo: Kokon Shoin.

Watanabe, T., and K. Fukasawa. 1998. "Daisetsuzan kokuritukouen Kurodake nanagoume kara sanchou kukan ni okeru kako nananenkan no tozandou kouhai to sono keigen no tameno taisaku" [Seven-year deterioration of a hiking trail and measures to mitigate soil erosion, Mount Kurodake, Daisetsuzan National Park, Hokkaido, northern Japan]. *Chirigakuhyouron* [Geographic review of Japan] 71A:753–64.

Watanabe, T., and Y. Ono. 1996. "Human Impact on High Mountains of Japan." In *Mountains of East Asia and the Pacific,* ed. M. Ralston, K. Hughey, and K. O'Connor, 70–78. Lincoln: Lincoln University, Centre for Mountain Studies.

Watanabe, T., K. Ota, and T. Gotoh. 2004. "Daisetsuzan kokuritukouen Susoaidaira shuuhen ni okeru tozandou shinshoku no chouki monitaringu" [Long-term monitoring of trail erosion around the Susoaidaira Area, Daisetsuzan National Park, northern Japan]. *Kikan chirigaku* [Quarterly journal of geography] 56:254–64.

Yoda, A. 1991. "The Erosion of Mountain Hiking Trails near Mt. Kurodake Hut in the Daisetsuzan Mountains National Park, Central Hokkaido, Japan." Master's thesis, Hokkaido University.

Yoda, A., and T. Watanabe. 2000. "Erosion of Mountain Hiking Trail over a Seven-Year Period in Daisetsuzan National Park, Central Hokkaido, Japan." In *Wilderness Science in a Time of*

Change, ed. D. N. Cole and S. F. McCool, 172–78. Ogden: U.S. Department of Agriculture Forest Service, Rocky Mountain Research Station.

Yoshinaka, A. 2006. "Conservation and Management Policy in Shiretoko National Park." In *Wildlife in Shiretoko and Yellowstone National Parks,* ed. D. R. McCullough, K. Kaji, and M. Yamanaka, 284–86. Shari: Shiretoko Nature Foundation.

Teiji Watanabe

Part V

Protesting the Effect
of Military Activity

Chapter 18

Antimilitary and Environmental Movements in Okinawa

Jonathan Taylor

The subtropical island of Okinawa, Japan, is a major hub of the U.S. armed forces in the Pacific. More than 75 percent of the land area of U.S. bases in Japan is in Okinawa (see figure 18.1), and more than 60 percent of the U.S. forces deployed to Japan are stationed there. The estimated number of American forces and dependents currently in Okinawa is around fifty thousand. The main military bases include Kadena Air Base, the largest U.S. air base outside the United States and home of the air force's Eighteenth Wing, and a number of U.S. Marine Corps bases and training facilities, including Camp Hansen and the adjoining Central Training Area, the controversial Marine Corps Futenma Air Station, and the Jungle Warfare Training Center (JWTC, formerly the Northern Training Area).

Historically, the people of Okinawa have been reluctant hosts to the bases but have had little say in the matter. The bases on Okinawa date to World War II, the first ones being Japanese airstrips and garrisons in places such as Yomitan, in Central Okinawa, and all over the southern part of the island. U.S. control of the island and the presence of the U.S. military there date to the Battle of Okinawa. This battle, waged from March to June 1945, was the bloodiest of the Pacific war. More Okinawan civilians were killed during this period than in either the Hiroshima or the Nagasaki atomic bomb attacks. While estimates cannot be precise, most use benchmarks of around 150,000 civilian deaths, some one-quarter of Okinawa's population at the time. The damage done to the natural environment was also

staggering. During the eighty-day battle, an estimated 7.5 million howitzer rounds, more than 60,000 naval shells, 20,000 rockets, and almost 400,000 hand grenades were fired just by the American side (Feifer 1992, 533). Beyond the horrors this caused to civilians and combatants alike, the effect was the utter environmental transformation of large sections of the island, especially the southern part, where the most intense fighting and bombardments took place. Thus, from at least this time on, the effects of war and the presence of military forces are associated in Okinawa with, not only tremendous human suffering, but also severe environmental degradation.

It is important to recognize the impacts this battle has had on Okinawan views of war and the military and the influence the influence these views have, in turn, had on recent and contemporary opposition to the U.S. bases, for, as one U.S. naval officer was quoted as saying, "Of the many places on this globe that were touched by the withering blast of war I doubt if in any the life of the people has been more completely changed than on Okinawa" (Karasik 1948, 254). Antiwar sentiments have been further exacerbated by the postwar revelations by Japanese military historians that Okinawa was deliberately sacrificed. The leaders of its defense were aware of the impossibility of their victory but viewed their task as necessary to weaken the enemy before the dreaded invasion of "home soil," that is, the principal islands of Japan. For, despite being a part of Japan politically since the Meiji era, Okinawa was still viewed as

OKINAWA

Jungle Warfare Training Center

Ie Jima Auxiliary Airfield   *Ie Island*

Okuma Rest Center

Katena Bay

Nago Bay

Camp Schwab

Henoko Ordnance Depot

Kin Red Beach   Gimbaru Training Area
Kin Bay   Central Training Area
    Kin Blue Beach

Sobe Communication Site   Tengan Pier
Yomitan Auxiliary Airfield   Camp Courtney
Torii Station   Camp McTureous

Kadena AFB
Camp Kuwae

Camp Zukeran   Awase Communication Site
Futenma Marine Air Station   White Beach
Makiminato Service Area   Futenma Airport

Naha Military Port Facility   Naha   *Nakagusuku Bay*

0   5   10
kilometers

0   5   10
miles

Figure 18.1. Military bases in Okinawa.

something quite apart from the main islands. Its people were racially, linguistically, and culturally different. This difference in the eyes of the mainlanders ultimately rendered Okinawa expendable in the eyes of the Japanese government, a pattern that has, some would argue, continued from World War II until the present day. To the continuing anger of many Okinawans, Japan has never issued anything even resembling an apology for the suffering and sacrifices endured by the civilian population of Okinawa during the Pacific War. The Ministry of Education banned from textbooks the mention of Japanese murders of Okinawan civilians during the battle. An understanding of the suffering Okinawans endured and the attitude of the Japanese government toward them is integral to understanding the enduring strength of Okinawan antibase and peace movements.

After the battle, the United States began constructing bases in Okinawa, both on the sites of former bases and on plots of flatland, which more often than not coincided with good agricultural land. Construction projects, whether roads, arsenals, camps, or airfields, were designed with no consideration of previous patterns of land use or ownership. This sowed seeds of resentment toward U.S. land-confiscation policies, and tension continued as U.S. confiscations of land intensified throughout the 1950s and early 1960s.

At the war's end, it was decided that Okinawa and all the Ryukyu Islands south of 30° north latitude would be indefinitely administered by the United States. As the Communists achieved victory in China in 1949,

Okinawa became a new strategic locale for U.S. troops. Renewed base construction started in 1950 and quickly expanded as the Korean War broke out in the same year. By the late 1950s, all U.S. ground troops in Japan had been moved to Okinawa.

Base construction, often facilitated by land seizure, continued through the 1960s. Eventually, by 1967, the United States had 51,586.3 acres of private/municipal property under lease and was using 24,147.72 acres of public domain land free of charge (High Commissioner 1967). This area was 13.8 percent of the total land area of the Ryukyu Islands and 10.9 percent of the arable land. At the height of its presence on Okinawa, the United States had bases on over 25 percent of its land area. This caused tremendous changes to the lifestyles and livelihoods of Okinawans, most of whom had been able to own land privately only since 1888. Conflicts over land confiscation were the key issue in increasing public opposition to the U.S. bases, although environmental degradation, crime, and accidents by military personnel were also factors. Okinawan and Japanese pacifism was also a major issue, especially in light of opposition to U.S. foreign policy in Southeast Asia. Increased public opposition eventually helped result in the politically strong reversion movement, leading to the U.S. decision to grant Okinawa's reversion back to Japanese control in 1972.

After reversion, the U.S. bases on Okinawa remained, and, although some have been consolidated and others returned, about 18 percent of Okinawa's land area is still occupied by U.S. bases. This was generally prime agricultural land, so the bases have directly affected Okinawan land use and economic patterns. Agriculture was forced into hillside areas as the United States confiscated flatland for bases, and the economy became by stages highly base dependent, although, since reversion, this has largely changed to a dependency on largesse from the central government.

Local opposition to the bases has ebbed and flowed. Opposition has tended to coalesce around key events, such as the rape of a twelve-year-old girl by three servicemen in 1995. This galvanized the anti–U.S. base movement and led to protest assemblies of up to eighty-five thousand. Local opposition is also partially dependent on the stance of the current political leadership of the prefecture. During the rape incident and trial in 1995–1996, the opposition movement was strongly supported by the then governor, Ota Masahide, despite his prior probase stance. The election of the probusiness governor Inamine Kenichi in 1998 and his reelection in 2002 dampened public opposition to the bases, although protests have since regathered steam over the environmental effects of the pending relocation of Futenma, about which more below. However, even more status quo–oriented governors than Inamine have requested a reduction of U.S. forces in Okinawa, the most recent request being by Inamine in a November 2003 face-to-face meeting with Donald Rumsfeld—and, as usual, to no avail (AFIS 2003). In fact, the United States seems more determined than ever to hold on to the Okinawan bases in the wake of September 11 and the War against Terror being waged by the Bush administration ("Okinawa's Strategic Role" 2002).

Environmental Impacts of the U.S. Bases

It goes without saying that military bases produce environmental degradation of various kinds. In the case of Okinawa, where a large proportion of the surface area of the main island is covered with bases, the bases have had a number of direct environmental impacts. However, they have also indirectly affected Okinawa's environment through their effects on its political economy and thus, its political ecology.

Direct impacts have included toxic dumping, water pollution, noise pollution, land degradation, and soil erosion. Indirectly, the bases have contributed to Okinawa's increasing economic dependence on construction and public works, the damming of Okinawan rivers and the building of large-scale dams, and other actions linked to the peculiar political economy that Okinawa first developed under U.S. occupation. Moreover the bases have taken up valuable farm- and urban land, increasing pressures both on marginal uplands and on the densely populated urban sprawl of southern Okinawa (Taylor 2001).

Land degradation is an especially serious problem. On one of the large bases, the Central Training Area, in Okinawa's north, there has been extensive firing of artillery shells. This has led to deforestation, frequent fires, vegetative denudation, and soil erosion (see photograph 18.1). In addition, the area is now covered with unexploded ordnance. This may well be the worst case of direct environmental degradation caused by the U.S. military in Okinawa; however, cleanup of the area is not the responsibility of the United States, and the future status of the area has not been discussed by the Japanese and U.S. governments.

On the opposite extreme however is the other large military base in the northern part of the island, now called the JWTC. This area is extremely undeveloped, with only a few facilities, one main road, and a few small helipads. The nature of the training conducted there (jungle combat training) has led to minimal environmental impacts. In essence, this base is the de facto largest semiwilderness area in Okinawa and certainly by far the largest contiguous protected area in the Ryukyu Islands. Recent surveys have found scores of endangered species found *only* in this area (Ito 1997; Ito, Miyagi, et al. 2000). This phenomenon of bases as accidental wilderness sanctuaries mirrors the situation at military bases within the United States as well; for ex-

Photograph 18.1. The effects of artillery bombardment on upland areas of Okinawa. (Courtesy Jonathan Taylor.)

ample, Camp Pendleton in southern California is now a last resort for some seventeen endangered or threatened species ("Military Chafes" 2001).

Meanwhile, in similar ecological areas outside the base, the forests are being cut and the land cleared. This so-called land improvement is supposed to result in increased agricultural land, but, in reality, most fields lie barren. A cynic might consider these social-welfare projects for Okinawan contractors. In addition, roads with various steep cuts have been built into the forest, expanding human penetration and use of the forests. The roads remain largely unused, but the damage remains. These activities have now imperiled the Yanbaru region to the extent that numerous Okinawan nongovernment organizations (NGOs) have lobbied for its benefit. However, the best remaining examples of this ecosystem now exist on the U.S. bases.

Overall, Okinawa's environmental problems are significant. They include deforestation, soil erosion, coral reef degradation, the depletion of fisheries, ground water depletion, groundwater pollution, hazardous waste and toxic dumping, riverine pollution and siltation, land salinization, the destruction of wetlands and mangroves, coastal erosion, marine pollution, and the depletion of biological diversity. Okinawa suffers especially from a shortage of water, surface water pollution, soil erosion, deforestation, and the degradation of coral environments. All its terrestrial and aquatic environments are endangered, some catastrophically so. In addition, environmental problems associated with quality-of-life issues include noise pollution and destruction of the scenic environment. Despite this, Okinawa has never had an environmental movement to rival that of the home islands. This can be attributed to a number of causes, but one is the fact that environmental problems in Okinawa have seldom been associated with immediate hazards to human health. This is, ultimately, the case

because the main culprit in Japan's toxic crises, industry, has a disproportionately small presence in Okinawa. Thus, the main catalyst for environmental grassroots movements in Japan has largely been missing from Okinawa.

However, Okinawa does have an obvious target of protest that is less common in the rest of Japan—the U.S. bases. Because of this, antimilitary and environmental groups in Okinawa have a symbiotic relationship unusual in Japan. Regardless of the reality of which factors are the most important in causing environmental degradation in Okinawa, for many Okinawans struggles to protect the environment and struggles to remove the bases and prevent their expansion are inextricably intertwined.

Okinawan Peace and Environmental Groups

While peace and environmental groups in Okinawa have a long and complex history, more pertinent at present is the coalescence of groups currently occurring. It has recently become difficult to divorce the peace and antibase movements in Okinawa from the environmental movement. While this is not a completely new trend, it has been greatly amplified in the last five to ten years as protests against the bases have increasingly turned to environmentalism for their rationale and local groups have reached out internationally to large environmental organizations for support.

The last few years have seen a flurry of new environmental groups take root in Okinawa while some of the previously most active groups seem to be dormant. Newer groups include the Dugong Network Okinawa, the People's Network against Construction and Strengthening of Military Base, the Okinawa-Yaeyama-Shiraho Association for the Protection of Sea and Life, the Dugong

Protection Fund Committee, the Association to Walk in the Nature of Yanbaru, No to Heliport Association of 10 Districts North of Futami, the No to Heliport 10,000 Voices Movement, the Nago Citizen's Network, the Okinawa International Forum for People's Security, the Save the Dugong Foundation (SDF), and many others. Longer established groups include the Okinawa Environmental Network (OEN), the Okinawa Clean Beach Club and its associated group, the Okinawa Ocean Culture and Environment Action Network, and the Okinawa Citizen's Recycle Movement, among others.

Two of the main groups exemplifying the coalescence of struggles against the bases and for environmental protection are the OEN and the SDF. One of the most well-known Okinawan environmental groups currently is the OEN, headed by a number of professors associated with Okinawa University, especially Dr. Ui Jun, the "godfather" of environmental studies in Japan. The OEN was formed in 1997 after the Japan Environmental Conference was organized by Ui in Okinawa ("NGO Can Help" 2003). The OEN has been involved with a number of controversial environmental issues in Okinawa: protesting a major land reclamation project at the Awase Tidal Flats, investigating water pollution caused by livestock breeding in southern Okinawa, and investigating lead pollution from shooting on U.S. bases. As with most Okinawan environmental groups, currently the main issue with which the OEN is involved is the pending relocation of the air station at Futenma to Henoko.

The OEN has organized two international conferences on environmental issues, the first coinciding with the G-8 Summit held in Okinawa in 2000, the second in March 2003. These conferences are efforts to both hear the views of other organizations working with military/environment issues from around the world and further publicize Okinawa's military/environment issues. Participants thus represent a range of organizations and nations, with an emphasis on Okinawa, Japan, and the Asia-Pacific region. Much attention is focused on the effects of military bases on the environment. Nonetheless, the OEN does not take a completely adversarial stance toward the U.S. military presence on Okinawa itself. In fact, it clearly relishes the opportunity to work with officials from the U.S. military, as is evident by the presentation by the deputy environmental officer of the Environmental Branch, Marine Corps, as part of the OEN's 2003 conference. Other environmental officers from the U.S military have also told me that they have had cordial relationships with the OEN.

Despite this, the level of antimilitary sentiment expressed by OEN members is significant. In interviews that I conducted with Ui and other main members in June 2002, it was clear that the military bases were viewed as being a large component of the environmental problems in Okinawa but also that, regardless of the bases' environmental implications, there were social and political rationales for opposing them. Interestingly, however, the OEN demonstrates an understanding of the role the military plays in Okinawa's political economy and political ecology and the relationship between military bases and the national government's largesse to construction projects in Okinawa. In fact, many of the scholars associated with the OEN were some of the first to make clear the ties between the bases and the government's "bribes" to Okinawa Prefecture in the form of public works.

The SDF was established in October 1999. Although ostensibly directed at the preservation of the dugong, an endangered sea mammal similar to the manatee whose northernmost breeding grounds are thought to be in the Pacific east of the administrative district of Nago, the movement was formed around the struggle to prevent the construction of a U.S. Marine Corps heliport near Henoko Village in eastern Okinawa. This group is part of a larger movement formed in opposition to the proposed base, which, if constructed, would also have significant economic and quality-of-life impacts (Taylor 2000). Attracting the most international attention, however, have been environmental issues, especially the fate of the dugong.

In a fashion typical of Japanese grassroots organizations, the SDF and a number of other groups with similar goals are working together in networks of convenience. In Okinawa, local groups often interface with Japanese groups in the home islands as well. During the 2000 Nago G-8 Summit, a statement to the press about the impacts of the proposed Henoko heliport was signed not only by the SDF but also by the Dugong Network Okinawa, the Association to Save the Dugongs of the Northern-Most Habitat (from Yokohama), the We Mammal (from Kyoto), the Supporting Fund for Movements of Saving Dugongs (from Kanagawa), the World Wide Fund for Nature–Japan (WWFN-Japan, based in Tokyo), and twenty other NGOs. In a more recent statement, the SDF, in association with the WWFN-Japan and a variety of other organizations, addressed both the Henoko heliport issue and the military and environmental issues raised by the JWTC. As discussed above, the JWTC is a large base in the lushly forested low mountains of Yanbaru. With only jungle warfare training there, no live fire, few roads, and sparse facilities, environmental impacts and development in the area have been minimal. As a consequence, this is a highly protected area, at least when compared to the rest of Okinawa. Controversy arose in 1999, however, over the U.S. Marine Corps' plan to replace helipads lost when land was returned by building seven new, and considerably larger, helipads in areas of pristine forest that serve as habitat for a number of endangered species, including the Okinawa rail and the Okinawa woodpecker.

Affiliated with the SDF is the Save the Dugong Campaign Center (SDCC), a Tokyo-based group petitioning the Japanese government to take a number of legal measures to protect dugong in Japanese waters. In April 2002, the SDCC began a campaign to collect 200,000 signatures on a petition demanding that the Japanese government implement three measures to protect the endangered dugong from the dangers posed by the construction of a new U.S. military air base off the shores of Okinawa. The first measure is to establish conservation areas for the protection of the dugong under the Law for the Conservation of Endangered Species of Wild Fauna and Flora. Protected-species status would also help protect the dugong's natural habitat, which in Japan is only around the northeastern section of Okinawa's main island. The second measure is to perform an environmental impact assessment (EIA) using internationally accepted standards for the Henoko heliport construction project. The SDCC argues that, in Japan, EIAs are conducted under the assumption that the plan will go forward despite some environmental damage. It calls for the possibility of a "zero option"—that the plan itself be shelved should the environmental damage be considered too great. It also calls for public input into the EIA process. The third measure is to establish means by which the government can control fishing nets. The SDCC's petition has been endorsed by the WWW-Japan and the Nature Conservation Society of Japan. In late 2003, it had more than fifty thousand signatures. A subsequent petition drive was undertaken in 2004 to demand that the Japanese government designate a protected area for the dugong.

Henoko and the Coalescence of Okinawa's Antimilitary and Environmental Movements

The tendency of the Okinawan peace and antimilitary groups and environmental organizations to coalesce is being bolstered by one particularly controversial issue, a plan to build a U.S. military heliport at Henoko. As originally stated, the plan called for the relocation of Futenma Air Station to a large heliport planned offshore of the village of Henoko, near the current location of Camp Schwab (see photograph 18.2). The plan and the location have subsequently been amended, but not to the satisfaction of the local communities or the peace and environmental movements. While approved by all levels of the Japanese and Okinawan governments, the plan engendered great opposition from its announcement, especially from environmentalists, largely because the base would be constructed in an area that is home to the dugong. It is also host to one of the few thriving live coral areas on the eastern coast of Okinawa. Given the state of Okinawa's coastline and the existing damage to its coral reefs, protecting and preserving this marine environment is viewed by many as crucial, and the plans to relocate the heliport here have run into substantial local, national, and even international opposition.

While the Japanese government has promised enough cash to northern Okinawa to buy some local support, it has been hard-pressed to paint the picture of the development of the base at Henoko as anything other than an unambiguous environmental disaster. Initial plans called for the heliport to be constructed directly on top of the coral reef. This would have endangered a large section of relatively pristine coastline, destroyed the peace and harmony of the small fishing village of Henoko, drastically affected the ecology of the area, and, in all probability, hastened the demise of the dugong, whose habitat's northernmost extent coincides almost exactly with the location for the proposed base. The dugong feeds on sea grass that lies inland of the five coral reefs that compose the reef structure around Henoko. The breaks between the five reefs allow the dugong access to this feed-

Photograph 18.2. Village of Henoko near the proposed site of the heliport. (Courtesy Jonathan Taylor.)

ing area. The proposed heliport would have closed off the dugongs' access to their food supply (Urashima 2003).

Since the Henoko plan has been proposed, the battle lines between environmental and peace activists, on the one hand, and the prefectural and national governments and the U.S. military, on the other, have been dramatically drawn. Whatever valid claims the military can make about its overall environmental footprint and its positive effects on Okinawa's environment, the environmental disaster of building the Henoko heliport will overshadow them, and the United States will continue to receive negative press. The protest against the Henoko base has already grown from a local, Okinawan movement to a national, Japanese one and has also picked up support in the international arena. In September 2003, six Okinawan, Japanese and U.S.-based environmental groups filed a lawsuit in U.S. court against the U.S. Department of Defense over environmental issues related to the construction of the heliport, while others have run full-page ads in the *New York Times* and other major U.S. newspapers urging Americans to support the removal of troops from Okinawa. This movement continues to grow even as the United States and Japan agreed to make changes in the location of the heliport to the nearby Camp Schwab.

Since 2002, the decision about Futenma's relocation moved to the hands of the Council for the Construction of the Relocated Facility, a group of nine representatives of the Japanese government, the Okinawa prefectural government, and Okinawan municipalities. This group included the then governor, Inamine, Mayor Kishimoto of Nago, the Japanese foreign minister, Kawaguchi Yoriko, and Ishiba Shigeru, the director general of the National Defense Agency. At the group's first meeting, it became apparent that the dispute over the length of time the base will be used as a military airport was unresolved ("Final Agreement Reached" 2002). Inamine had said from the outset that he wanted a fifteen-year limit on the military airport (this position was a major campaign promise when he ran for reelection in 2002 ["Inamine Re-Elected" 2002]), but the United States refused to agree to this, and the Japanese government has sided with the United States. This development, coupled with opposition to the U.S. invasion of Iraq, revitalized the Okinawan peace movement. And, if construction actually begins on the new base, heating up the movement to preserve the dugong, public opinion could turn even more harshly against the U.S. bases. Whether the newly reenergized antimilitary and environmental movements will then be able to duplicate some of the successes of their counterparts in the home islands and, perhaps, in the process, hasten the removal or reduction of the U.S. military presence in Okinawa remains to be seen.

References

American Forces Information Service (AFIS). 2003. "Rumsfeld Visits Okinawa; Meets with Troops, Local Officials." November 16. http://www.defenselink.mil/news/Nov2003/n11162003_200311161.html (accessed November 8, 2007).

Feifer, George. 1992. *Tennozan: The Battle of Okinawa and the Atomic Bomb*. New York: Ticknor & Fields.

"Final Agreement Reached on Futenma Alternative." 2002. *Okinawa Times*, August 3. Available at http://www.okinawatimes.co.jp/eng/20020803.html#no_1.

High Commissioner. 1967. *Civil Administration of the Ryukyu Islands*. Naha: GRI.

"Inamine Re-Elected, Pushes 15 Years." 2002. *Okinawa Times*, November 30. Available at http://www.okinawatimes.co.jp/eng/20021130.html#no_1.

Ito, Y. 1997. "Diversity of Forest Tree Species in Yanbaru, the Northern Part of Okinawa Island." *Plant Ecology* 133:125–33.

Ito, Y., K. Miyagi, et al. 2000. "Imminent Extinction Crisis among the Endemic Species of the Forests of Yanbaru, Okinawa, Japan." *Oryx* 34, no. 4:305–18.

Karasik, Daniel. 1948. "Okinawa: A Problem in

Administration and Reconstruction." *Far East-ern Quarterly* 7, no. 3.

"Military Chafes at Wildlife Rules." 2001. *Los Angeles Times,* August 23.

"NGO Can Help Save Okinawa's Environment." 2003. *Asahi shimbun,* March 3.

"Okinawa's Strategic Role Rules out Change: Baker." 2002. *Japan Times,* January 1.

Taylor, J. 2000. "Okinawa on the Eve of the G-8 Summit." *Geographical Review* 90, no. 1 (January): 123–30.

———. 2001. "Environmental Change in Okinawa: A Geographic Assessment of the Role of the U.S. Military." Ph.D. diss., University of Kentucky, Department of Geography.

Urashima, Etsuko. 2003. "Okinawa Base Dooms Dugong." Trans. Steve Rabson. *Zmag,* February 6. http://www.zmag.org/content/showarticle .cfm? SectionID=17&ItemID=2988 (accessed November 8, 2007).

Chapter 19

Grassroots Participation in Hawaiian Biodiversity Protection and Alien-Species Control

Christopher Jasparro

The Hawaiian island chain is one of the most isolated places on the planet, more than two thousand miles away from the nearest landmass (Jasparro and Shibuya 2002). However, it now faces some of the highest extinction rates of indigenous and endemic flora and fauna in the world. By the early 1990s, it had become apparent that existing biodiversity-protection measures were woefully inadequate (Jasparro and Shibuya 2002), but, over the past ten years, there has been a significant increase in grassroots attention to biodiversity issues and, in particular, invasive-species threats. Hawaii now has arguably the best invasive-species-control program in the United States after California. The state has become a laboratory for technological and organizational innovations with the potential to become a national and global model (Loope and Canfield 2000). Much of this progress is the result of strong grassroots efforts by environmental professionals, local and national environmental groups, local businesses, and dedicated individuals, often in partnership with government agencies.[1]

Four features characterize these efforts:

- the role of midlevel environmental professionals, land managers, and scientists as initiators, claimsmakers, and leaders;
- voluntary and participatory institutional partnerships;
- the importance of volunteer labor; and
- the extensive application of geographic and information technologies.

After some general background comments on Hawaiian biodiversity and alien-species issues, this chapter will examine these four characteristics. First, they will be discussed in the context of general biodiversity protection (with a focus on alien-species control); then detailed comments on island invasive-species committees (ISCs) from Maui and Oahu will be presented as case studies.

Background

Before humans arrived roughly fifteen hundred years ago, plant species arrived and established themselves in the Hawaiian Islands at the rate of about one species every twenty to thirty-five thousand years (Loope and Canfield 2000; Bisignani 1999). The offspring of many who survived the journey produced a spectacular legacy of adaptive radiation and one of the planet's most unique biota (Mlot 1995). In the course of 70 million years, two thousand species arrivals produced eleven thousand or so new species (Loope and Canfield 2000).

The geographic situation that once facilitated the development of Hawaii's unique biota proved, with the arrival of man, ultimately to be devastating to the island's biodiversity. Pigs, rats, and numerous horticultural and medicinal plants arrived with Polynesian immigrants between A.D. 400 and 500. These first human residents also began a long history of forest clearance. The arrival of Europeans in 1778 then heralded a rapid acceleration in habitat conversion; alien-species introduction and deforestation became particularly acute with the birth of the sandalwood industry in the 1790s. Land clearance for agriculture and grazing (see photograph 19.1) furthered the process of deforestation, while feral cattle and goats devoured and trampled undergrowth and young seedlings. Meanwhile, introduced plants and animals began displacing native species.

Urbanization, resort development, and sprawl exploded after World War II, further assaulting both terrestrial and marine biodiversity. Hawaii's central location in the Pacific and emergence as a major tourist destination transformed the island into a regional transportation hub. The Honolulu airport is now the seventeenth busiest airport in the world, with a flight arriving every 1.3 minutes. Nineteen million tons of cargo from Asia, the U.S. mainland, and other Pacific Rim countries pass through Hawaiian ports annually (CGAPS 1996). Today, the greatest threat facing Hawaii's 297 endangered species (a quarter of the entire U.S. total) is alien species, now arriving at a rate 2 million times more rapid than would naturally occur (CGAPS 1996). Continued habitat destruction, meanwhile, remains a persistent threat (Stone 1999; Mlot 1995).

Biodiversity loss and invasive-species arrivals have serious economic as well as potentially negative public health and cultural impacts. For example, Formosan ground termites exact a toll of US$150 million per year, fruit flies up to US$300 million a year (CGAPS 1996). The papaya ringspot virus threatens a US$16 million industry plus twelve hundred jobs. The potential establishment of brown snakes, biting sand flies, piranhas, and various disease-vectoring insects threatens Hawaii's carefully cultivated image as a tourist destination free from tropical diseases, dangerous pests, and snakes. Recent outbreaks of dengue fever and typhus could be harbingers of things to come. Root aphids, snails, insects,

Photograph 19.1. Cattle damage to native koa trees can scar for decades. This photograph taken in 2004 clearly shows scarring from the 1960s. (Courtesy Christopher Jasparro.)

and viruses threaten the resurgence of taro cultivation, which not only has economic impacts but also hurts native Hawaiian cultural revitalization efforts (CGAPS 1996). Hawaii has faced recurrent water shortages since the 1980s. Invasive alien species and forest clearance seriously threaten Hawaiian forested watersheds, on which the islands' water supply depends. Oahu's Koolau Watershed Reserve, for example, is thought to provide a net present value of US$7.4 and US$14 billion in water and other ecosystem services, respectively (Jasparro and Shibuya 2002). Deepwater surveys in 2001–2004 determined that over 50 percent of all black coral colonies below 230 feet had alien snowflake coral overgrowth, with most being completely smothered and dead (Kahng 2005). This threatened widespread black coral die-off poses a significant economic threat because black coral (the state gemstone) accounts for several million dollars worth of annual jewelry sales (Kahng 2005).

Characteristics of Contemporary Grassroots Biodiversity Protection

Hawaii's modern conservation movement originated in the late nineteenth century and the early twentieth, just as it did in Europe and North America. Watershed and forest restoration and protection were the foremost concerns. As in Europe, early conservation efforts were largely an elite preserve (Dalton 1994). In Hawaii, much of the impetus for this movement was utilitarian and driven by economic elites, particularly sugar planters, who in the late nineteenth century became worried about the deteriorating state of Hawaiian forests, particularly in regards to water supply (Woodcock 2003, 624–45). Forest conservation and restoration were promoted by wealthy planters but implemented by professional scientists and foresters (Woodcock 2003).[2] To some degree, the importance of professional scientific leadership, organizational partnerships, and even volunteer labor today has its antecedents at the turn of the last century.

A territorial Division of Forestry was set up in 1903 owing to pressure exerted by the influential sugar industry. Once established, the driving force behind reforestation efforts was individual professional foresters such as R. S. Hosmer (the first superintendent of forestry), Harold Lyon (Hawaii Sugar Planters Association), and Charles Judd (Hosmer's successor). In order to overcome limited resources and funding constraints, partnerships and cooperation between government and private organizations as well as the use of volunteers were salient characteristics of early reforestation efforts. For much of the first half of the twentieth century, there was a high degree of cooperation between the Hawaii Sugar Planters Association (which established its own forestry division in 1918) and the territorial forestry department on a variety of fronts, including funding for salaries, research, and tree planting (Woodcock 2003,

627). The U.S. Army Air Service assisted with aerial seeding, local residents and groups such as the Boy Scouts participated in hunts designed to control feral ungulates, while the Civilian Conservation Corps planted 12.5 million trees and eradicated more than 100,000 animals between 1932 and 1940 (Woodcock 2003, 629–30).

The instigation of contemporary biodiversity-protection efforts has also been an elite activity, but the drivers have been midlevel environmental professionals (versus economic-political elites). Today, partnerships and volunteers are arguably even more important and incorporate a greater range and number of actors. The development of GPS (global positioning system) technology, GIS (geographic information system) technology, and the Internet have also revolutionized the surveying, mapping, and analysis of biodiversity and alien species as well as the exchange and dissemination of information and data.

Leadership

One of the more significant features of the Hawaii experience is that most of the innovation and initiatives has been spawned by midlevel professional applied scientists and land managers. These are people who are, on the one hand, personally passionate and dedicated to protecting the state's biodiversity while, on the other, professionally convinced that existing professional resources are insignificant and that greater public awareness and political will are needed to produce the necessary resources and policies. Remarks by two prominent figures in Hawaiian biodiversity protection are illustrative. According to Randy Bartlett (Maui Invasive Species Committee [MISC] chairman and Pu'u Kukui Watershed supervisor, Maui Pineapple Company): "Scientists and conservation professionals must make a conscious effort to inform and educate both the general public and political leaders at all levels if any progress is to be made" (Jasparro and Shibuya 2002, 12). Lloyd Loope (the U.S. Geological Survey [USGS] Haleakala field station leader) has said: "My job (which I've been at for nearly 23 years) involves research and vision toward how to protect the Park [Haleakala National Park] and Hawaii's native plants and animals longer than the next few years. Over the years I've become increasingly struck with the fact that we on Maui are always on the verge of losing it because of new [invasive-species] introductions. That is how I've gotten involved in this 'prevention stuff'" (Loope 2003).

The leadership dimension of grassroots biodiversity efforts in Hawaii in some respects mirrors that of grassroots environmentalism in other parts of the world, such as Western Europe, where "the leadership of environmental groups has obvious origins in a distinctive stratum of the middle class. Environmental elites disproportionately come from professions that specialize in the creation and application of symbols: scientists, teachers, journalists, and public administrators" (Dalton 1994, 118). Local environmental professionals seem to regard inconsistent funding and high-level institutional and political support plus limited public awareness as the primary two obstacles to the conservation and restoration of Hawaiian biodiversity. Therefore, it is not surprising that, in Hawaii (as in Europe), those who are adept at the "creation and application of symbols" play an important leadership role. In a recent talk to the Hawaii Native Plant Society, Loope (2003) expressed this commonly held sentiment by concluding: "What is now required to achieve significantly better protection seems to be active public support and political will." This professional leadership cadre (particularly scientists and educators) acts as "environmental claims-makers, in other words people who play an influential role in shaping how the public perceives and interacts with the environment" (Meindel, Alderman, and Waylen 2002, 682).

Volunteers

Volunteer labor is particularly critical in activities ranging from alien-species eradication, to fence building, to biodiversity surveying, to tree planting, to public awareness, to political advocacy. Park and land managers rely on volunteers as a reserve force that can bolster ongoing operations as well as provide surge capacity. For example, two-thirds of the fence construction undertaken in the Honouliuli Preserve of the Nature Conservancy of Hawaii (TNCH) in 2001 was done with the help of volunteers (LaPierre 2001, 4). The ability to mobilize large numbers of volunteers to tackle acute problems is illustrated by the Sierra Club's 1998 provision of fourteen hundred volunteers, who eradicated from 245 acres in two state parks 600,000 alien plants that were threatening 22 endangered species and habitat for half of Kauai's rare plant species (Hattam 1999, 73).

Government agencies also rely heavily on volunteer labor to enhance and conduct their operations. Volunteer labor has been an integral part of the Hakalau Forest National Wildlife Reserve (NWR) habitat restoration for endangered and native birds since its inception in 1987 (see photograph 19.2). In fiscal year 2003, volunteers provided $26,592 worth of labor (valued at $8 per hour for adults, $2 for children) (Dick Wrass, personal communication, April 2003). Currently, the reserve utilizes between forty and six hundred volunteers per year and hosts work groups on forty-two to forty-nine weekends (Dick Wrass, personal communication, April 2003). In the first nine months of 2003, volunteers planted twenty thousand koa seedlings plus seven thousand understory species (Hakalau Forest NWR staff, personal communication, April 2003) (see photograph 19.3). Of the groups that provided volunteers to the reserve in 2003, it is important to note that one-third were affiliated with academic organizations (table 19.1). This illustrates the importance

Photograph 19.2. Volunteer field crew at Hakalau National Forest Reserve, 2004. (Courtesy Christopher Jasparro.)

of educators/academics in Hawaiian grass-roots environmental protection. This middle-class, professional character is also evident in the makeup of crews. For instance, of a ten-member Oahu Sierra Club volunteer team that I studied in spring 2004, six members were

teachers/academics and another three highly educated professionals (medicine, law, etc.).

Another example is provided by the Nu'upi Wildlife Management Area on the Kanehoe Marine Corp Base Hawaii. The Nu'upia Ponds are remnants of ancient Hawaiian fishponds

Photograph 19.3. Lines of koa trees planted by volunteers in Hakalau National Forest Reserve. (Courtesy Christopher Jasparro.)

Hawaiian Biodiversity Protection and Alien-Species Control

Table 19.1. OISC Volunteer Support

Target	Staff P/H	Volunteer Manhours	Major Supporter
Miconia calvescens	1,129	1,504	Sierra Club/EEWF
Pennisetum setaceum	208	110	HIARNG/HDLNR
Rubis discolor	62	15	
Leptospermum spp	156	73	Army Environmental
Eleutherodactyulus spp	2	550	DOA, EEWF
Prosopis juliflora	N.A.	147	DOA, EEWF
Melochia umbella	18	12	Army Environmental

Source: OISC (2002, 28).
Note: EEWF = Emergency Environmental Workforce; HIARNG = Hawaii Army National Guard; HDLNR = Hawaii Department of Land and Natural Resources; DOA = U.S. Department of Agriculture.

that form part of the Makapu Wetland Basin. The area is an ecosystem critical to numerous shore- and seabirds, fish and shellfish, including several endangered species. Nonpoint pollution, such as fertilizers, pesticides, grease/oil, and sediment, plus invasive species, such as mangrove, have had an adverse impact on water quality and indigenous species. Over the past decade, however, the removal of invasive aliens has led to better water quality, increased stilt nesting opportunities, and improved native fish incubation.

Military personnel and civilians from the local community regularly participate in service projects, including mangrove removal, nest-island building, trail repair, bird counts, and water-quality monitoring. Community education and involvement are integral parts of the management plan, and these volunteers not only provide valuable labor but also become stakeholders whose support bolsters the project's and the wetlands' sustainability.

The driving force behind the success of the project and particularly the integration of the local community is Dr. Diane Drigot, the senior natural-resources-management specialist at Kaneohe Marine Corps Base Hawaii. The U.S. Marine Corps manages the area in partnership with the U.S. Navy, the U.S. Fish and Wildlife Service (USFWS), the National Marine Fisheries Service, and the Hawaii Department of Land and Natural Resources (HDLNR). Here again we see a combination of midlevel professional leadership, interagency partnership, and volunteerism.

Participatory-Voluntary Institutional Partnerships

Local biodiversity and invasive-species-control efforts in Hawaii are now characterized by extensive voluntary and participatory partnerships between the government, nongovernment organizations (NGOs), private organizations, academic institutions, and business. Much of the progress made since the early 1990s can be credited to the synergy that has resulted from these partnerships. This was not always the case, however. In the early 1990s, alien-species-monitoring, -control, -quarantine, and -eradication efforts were divided among numerous government (federal, state, and local) agencies, environmental NGOs, and private entities. Comprehensive statewide strategy and planning as well as interagency cooperation were minimal, creating coverage gaps and inefficiencies that were particularly problematic given the resource constraints faced by nearly all involved. Furthermore, this fractionalized set-up made it difficult to build public awareness and high-level political support. Finally, TNCH/NRDC (1992) set out recommendations to address these problems. In 1995, in response to this report, the Alien Species Action Plan was laid out as a cooperative effort among government agencies, environmental NGOs, academic institutions, private organizations, and businesses. The group's oversight committee evolved into the Coordinating Group on Alien Pest Species (CGAPS), which has become the primary body for coordinating interagency action on invasive species and promoting public awareness (Jasparro and Shibuya 2002). The CGAPS is administered by the Hawaii Department of Agriculture (HDA) and includes staff seconded from TNCH (see table 19.2).

The next pivotal development in grassroots alien-species control was the establishment of specific individual-island ISCs. These are voluntary partnerships of private, government, and nonprofit groups and individuals. The ISCs (two of which will be examined in more detail later) engage in education and advocacy (e.g., claims making), eradication and control, and mapping and spatial analysis. They are also important community-outreach mechanisms and growing users of volunteers.

Geographic and Information Technology

Hawaii has also been a center of innovation in the experimentation and use of geographic and information (e.g., Internet, GIS, GPS, cartographic) technologies for alien- and indigenous-species identification, monitoring, and mapping (Jasparro and Shibuya 2002). These technologies also provide a means for linking the previously discussed three characteristics. For example, geographic and information technologies facilitate scientific study and the employment of maps and aerial and satellite imagery. They also provide powerful

Table 19.2. OISC Volunteer Administrative Support, 2001–2002

Working Group Meetings	Volunteer Manhours
All-OISC general	270
Control group	105
Strategy	84
Fountain grass	18

Source: OISC (2002, 28).

visual tools and the authority for bolstering claims as well as an interactive means of public participation in invasive-species reporting. Volunteers often take part in fieldwork that involves mapping, the use of GPS technology, and entering information into GIS databases. GIS technology, in particular, facilitates the sharing of data, maps, and other spatial information as well as the providing of partners with a "common picture."

A key component of Hawaiian biodiversity-protection efforts is the Hawaiian Ecosystems at Risk Project (HEAR), which was created to disseminate information and technology to decisionmakers, resource managers, and the general public. HEAR was established thanks to the efforts of key personnel from several agencies, in particular Lloyd Loope of the USGS and Alan Holt of TNCH (Phillip Thomas, personal communication, August 10, 2001).[3] It is administered by the USGS Haleakala field station on Maui but relies on voluntary partnerships with twenty-five (international, government, nongovernment, academic, and business) agencies and organizations to function.

Two databases lie at the project's heart: the Harmful Non-Indigenous Invasive Species (HINIS) Database and the Hawaiian Natural Resources Monitoring Database. The HINIS is an interactive, online database providing names, sightings, locations, photographs, and information on methods of controlling harmful species that allows scientists and resource managers to submit and exchange data. The public can report invasive-species sightings to the HINIS Database via e-mail or phone messages.

The monitoring database facilitates standardized and comparable data collection and analysis by government and other organizations in order to give what the HEAR Web site (see n. 3 above) calls "a 'big picture' approach to the analysis of this type of scientific data never possible before in Hawaii."[4] Personnel from the East Maui Watershed Partnership and Nature Conservancy developed the concept, but the system is now administered by the Research Corporation of the University of Hawaii. Related software licenses are given free of charge to any qualifying landowner, organization, or individual. Data, current maps of invasive-species locations, and links to other databases, organizations, and listservs can be accessed through the HEAR Web site.

Case Studies: Two ISCs

The MISC was founded in 1997 from the Melastome Action Committee (MAC). The MAC was established in 1991 as a cooperative effort to combat the spread of Miconia (*Miconia calvescens*), glorybush (*Toboucina herbacea*), and Koster's curse (*Clidemia hirta*). The creation of the MAC was due to the leadership of two private sector and government professionals: Randy Bartlett and Ernest Rebello (the project director of the Maui section of the Tri-Isle Resource Conservation and Development Office, U.S. Department of Agriculture [USDA]). The HDA, the HDLNR, TNCH, the National Park Service, the U.S. Forest Service (USFS), the USGS, the University of Hawaii, and the USFWS were the primary institutional partners.

The MAC evolved into the MISC in order to address a wider array of invasive-species threats. The MISC functions first as an ad-

vising and coordinating body that facilitates the planning and implementation of invasive-species-control measures and strategies. Other important missions are funding development, legislative and public education/awareness, and alien-species-identification, -location, -control, and -eradication operations. Since 2001, the MISC has grown rapidly, quadrupling its operations (MISC 2006, 3).

The MISC describes itself as "a grassroots partnership that has the capacity to survey, map, and control incipient invasive pests" (MISC 2002, 16). This description exemplifies how Hawaiian biodiversity protection is characterized by partnerships and the use of geographic techniques. MISC partners include Haleakala National Park, the USGS Biological Resources Division, the USDA Forest Service, the USFWS, the Hawaii Army National Guard, the USDA Tri-Isle Resource Conservation and Development Council, the HDLNR, the HDA, the University of Hawaii, the Maui County Office of Economic Development, the Maui County Board of Water Supply, TNCH, the Maui Land and Pineapple Company, the East Maui Watershed Partnership, and the Community Development Block Grant Program (according to the HEAR Web site and the MISC manager, Teya Penniman [personal communication, 2003]).

The nature of the partnership illustrates the salience of environmental professionals in Hawaiian biodiversity protection. The partners are all professional agencies with environmental responsibilities/obligations, but the actual labor and drive behind the partnership comes mainly from individual dedication: "Really it is the fact that there are real people at each of these agencies or organizations who are dedicated to working on the issue of preventing, controlling, or eradicating invasive species. Of course that commitment arises from and is nurtured by a commitment to Hawaii's incredible biodiversity" (Teya Penniman, personal communication, 2003). The MISC has not relied on volunteer support as much as other organizations and, until recently, was using only a few volunteers, mostly friends and relatives of staff (Teya Penniman, personal communication, 2003).[5] This is partly because the MISC has found it difficult to recruit locally workers who posses the desired skills and training. It is, however, looking to start an internship program in order to develop a base of volunteers as well as to expand its relationships with the wider community. To this end, it has already sought direct public involvement in eradication efforts, public education, the publicizing of success stories, and the involvement of a broad range of other entities such as hotel associations, farm bureaus, chambers of commerce, and visitor organizations.

Even though the importance of volunteers was initially limited, the success of the MISC has partly been due to the importance attached by its founders to the early creation of a wide support base and participation balanced against the need for timely action. On the basis of the MISC experience, Randy Bartlett advises: "Organizers must attempt to include as many stakeholders as early in the initial stages as possible, but should not wait for all possible stakeholders to join before meeting to plan strategies and future actions" (personal communication, August 2001).

Perhaps the most important success of the MISC has been its use as a model for the creation of other ISCs, such as the recently established Oahu ISC (OISC). The OISC's mission statement reads: "The Oahu Invasive Species Committee (OISC) is a voluntary partnership of private, governmental and non-profit organizations and individuals united to prevent new invasive species infestations on the island of Oahu, to eradicate incipient invasive species, and to stop established invasive species from spreading. The group is concerned with all non-native invasive species threatening agriculture, watersheds, native ecosystems, tourism, industry, human health, or the quality of life on Oahu" (OISC 2002, 1).

The OISC's four main activity areas are the following:

- the on-the-ground control, containment, or eradication of invasive alien species;
- the recognition of and preparation for rapid response to control new incipient alien species not yet found on Oahu;
- the education of community members, legislatures, and businesses about the threat of invasive species and what can be done to ameliorate it; and
- the support of statewide efforts by the CGAPS and other ISCs to affect policies related to invasive species (e.g., plant-importation screening, the revision of the state Noxious Weed List) (OISC 2002, 3).

As is evident from its mission statement and primary goals, the OISC is a voluntary partnership that also works in concert with other ISCs and CGAPS. Indeed, more than sixty participants representing a diverse array of agencies and organizations attended the OISC's inaugural meeting. Currently, the OISC has about twenty-five active members, including the USFS, the USFWS, the HDLNR, and the Hawai'i Community Foundation.

Since it began operation in August 2000, the OISC has cleared 1,015 acres of invasive flora and removed over 3,070 miconia plants (OISC 2002, 3). Much of this was accomplished with volunteer assistance. Over the 2001–2002 operating period, volunteers supplied 2,339 person-hours in support of plant-eradication efforts versus the 1,575 put in by full-time OISC professional field crews (see table 19.3). Volunteers also contributed 477 hours of both general and working-group administrative support (OISC 2002, 3) for the 2001–2002 period (see table 19.4).

Although volunteers are important, it is critical to note that the sponsoring organizations are generally government/professional agencies or "elite" NGOs such as the Sierra Club. Again, we see evidence that professional leadership is a defining characteristic of Hawaiian biodiversity protection.

The use of geographic technologies is also an essential aspect of OISC strategies and activities. All professional staff members are trained in the use of GPS technology and various GIS skills. An major objective for eight of the top ten species targeted by OISC control efforts in 2002 was mapping with the application of GPS and/or GIS technology (OISC 2002, 8–23). Field data are also digitized to facilitate the analysis and mapping of target-species distribution and the tracking of the number of species found and removed (OISC 2006, 2). The OISC and other state ISCs are working to standardize database procedures (OISC 2006, 2).

Conclusion

Grassroots biodiversity protection (and, in particular, alien-species control) in Hawaii is characterized by the four key characteristics described in this chapter. Does a movement of this sort have the ability to effect lasting and sustainable change?

Observers and participants alike generally agree that these efforts have led to significant improvements and that Hawaii now has a strong foundation plus powerful tools on which a sustained, comprehensive, and effective biodiversity-protection and alien-species-control regime could emerge (Jasparro and Shibuya 2002, 12). The innovative use of geographic and information technologies has greatly facilitated cooperation and coordination between actors and agencies, scientific research, biodiversity management, alien-species eradication, and public awareness and participation efforts. In terms of biodiversity protection and alien-species control, many of the programs and techniques pioneered in Hawaii are viewed as models by others throughout the world (Loope and Canfield 2000). According to Paula Warren, the principal policy analyst for New Zealand's Department of Conservation, who has examined Hawaii's situation: "The basic elements of the

---

Table 19.3. Volunteer Support of Plant-Eradication Efforts, 2001–2002

Environmental
Molokai Environmental Preservation Organization
Big Island Sierra Club
Oahu Sierra Club
Youth Conservation Corps
Hawaii Nature Center
TREES
Wilderness Volunteers

Educational / Academic
Hawaii Community College Forest Team
   Management
Hawaiian Community College Biology Class
University of Hawaii Biology Class
Environmental Education Enrichment Program
University of Hawaii Alumni
Bishop Museum

Other
Boy Scouts/Cub Scouts (Troops 23, 56, 25/Pack 50)
Hilo Missionary Church Youth Group
Hawaiian Medicine Group
Big Island Pig Hunters Group

Table 19.4. Volunteer/Administrative Support, 2001–2002

State Agencies
Hawaii Department of Agriculture
Department of Land and Natural Resources
Hawaii Farm Bureau
Hawaii Department of Health
Hawaii Department of Transportation
Hawaii Visitors Bureau
The Nature Conservancy of Hawaii

Federal Agencies
U.S. Department of Agriculture
U.S. National Park Service
U.S. Navy
U.S. Customs Service
U.S. Fish and Wildlife Service
U.S. Postal Service
U.S. Postal Inspection Serivce

---

system are sitting there, waiting to be plugged in, and there's a lot of enthusiasm among individuals" (Conrow 2006).

However, the intensity and magnitude of the threat to Hawaii's biodiversity ultimately requires sustained government funding, political will, and comprehensive strategies and organization. Meanwhile, public awareness and participation have not yet reached the critical mass needed either to instigate greater political/government action or to enable both professional and voluntary organizations to keep pace with the threat.[6] The professional/ elite nature of grassroots biodiversity protection has been critical in lending legitimacy and gravity to biodiversity issues as well as a scientifically and technically sound basis for action. Furthermore, volunteer labor has been essential in helping fill gaps created by limited professional and full-time human resource capacity as well as in adding stakeholders to the process. Volunteerism and grassroots actions cannot, however, in this case, fill a role that only government can.

Yet, if the movement remains largely one of environmental professionals and elites, it may not be able to garner the political support necessary for building and sustaining a government effort that, by its very nature, will be (in all practical terms) unending. Thus, it remains an open question as to whether recent grassroots biodiversity-protection efforts will bear even greater fruit and become a launching pad for a wider, broader, and more inclusive political and public effort or just a gallant but desperate struggle by a dedicated but all too small cadre.

Presently, public awareness and funding are increasing, which indicates that a new stage in the evolution of Hawaiian biodiversity protection and invasive-alien-species control may be emerging. Yet it is still not certain whether this is a long-term trend or whether these developments will be bounded by socioeconomic and structural factors.

A 2006 telephone survey of Hawaii residents conducted for CGAPS seems to bear this uncertainty out. According to the survey, public awareness of the invasive-species problem has increased since 2004, with 43 percent residents believing that it is a "very serious" problem, compared to 36 percent in 2004 (CGAPS 2006). Furthermore, 88 percent of respondents said that they would support a law that increases service fees on incoming cargo so that the HDA can conduct adequate inspections.

However, there is still the potential for awareness and support to backslide. For example, in 2004, 50 percent of Hawaii residents surveyed said that they would be "very likely" to report an unknown insect, a drop of 3 percent since 2004 (CGAPS 2006). Eighty-three percent of those surveyed in 2006 said that they would be "very likely" to report a snake if they saw one, a drop from 91 percent in 2004 (CGAPS 2006). There also remain clear socioeconomic divides on the depth and breadth of public concern and awareness. Awareness of invasive-species problems is still most developed among Caucasians and Japanese and in households earning more than $75,000 per year and lower among other income groups and ethnic groups such as Filipinos and native Hawaiians (CGAPS 2006).

Public and political support for funding for alien-species programs has also grown, in large part thanks to the growing level of awareness. In 2004, Governor Lingle proposed that the state spend $5 million on invasive-species control, up from the $628,000 spent in the previous two years. Ultimately, the legislature approved $3 million—a substantial increase, but still not what most experts think necessary (Leone 2004). In 2006, the state legislature increased funding for the HDA's plant-quarantine and plant-pest-control divisions by 60 percent ("Hawaii— Rich in Endemic Biodiversity" 2006).

Furthermore, voluntary partnerships and the goodwill of individuals from multiple agencies and sectors alone cannot produce or sustain a comprehensive and effective "whole of government" response and the efficient use of resources (even if they are growing). Paula Warren says: "Hawaii's system is fragmented and poorly coordinated . . . , [while,] in New Zealand, the momentum is within the system . . . , [but] in Hawaii, it's mostly built by individuals working against the system to get around the problems. So you lose it if that person changes jobs, or retires" (Conrow 2006).

In conclusion, the Hawaiian model for biodiversity-protection and alien-species control has produced a host of innovative approaches that have achieved considerable progress, relative to the speed and scale of the problem and the historically limited resources, the low level of public concern and political will, and the fractured array of actors (often with differing interests and agendas), all of which offers valuable lessons. The question now is whether the foundation can be built on and sustained and more cohesiveness brought to the system—at a pace and scale sufficient to the challenges at hand.

Notes

The ideas and opinions expressed in this chapter are those of the author alone and do not represent those of the U.S. Marine Corps, the U.S. Department of Defense, or the U.S. government.

1. At this point, it is fair to ask why this chapter devotes little attention to native Hawaiians. First, Hawaiians are limited in numbers and political-economic power. Second, although concern over the protection of native lands and species has developed alongside the Hawaiian cultural revival and sovereignty movement (the revitalization of taro production is, perhaps, the best-known example), the goals of native Hawaiians and mainstream conservationists are quite different. Hawaiians have focused primarily on fending off developers, obtaining the return of native lands from the military, and promoting traditional culture. Ecological arguments have been employed in these struggles, but, unlike environmental organizations and government agencies, the Hawaiians

are seeking to create or protect, not conservation areas or capitalist economic interests, but, rather, "the social reoccupation of space according to sustainable principles" (Ross 1994). Also, the driving force behind mainstream efforts has come from conservation biology, which generally approaches problems from etic perspectives, which employ outsider position and privilege "data drawn from empirical observation, operationalized definitions, and replicable situations," as opposed to emic ones, which seek understanding from the "unique and culture-bound definitions" of others (Drew and Henne 2006, 36). Some cooperation and sympathy between the different actors and approaches does occur, but it is not systematic or synergistic on a large scale.

2. In precontact times, Hawaiian land management was the responsibility of nobles organized according to the *ahupua'a* system (land divisions based on altitudinal zonation roughly coincident with watershed boundaries) and enforced by various taboos. The first modern forest-conservation efforts were taken in the late nineteenth century by various Hawaiian monarchs (Woodcock 2003, 624).

3. See www.hear.org.

4. See http://www.hear.org/monitoringdatabase.

5. The MISC now organizes Saturday volunteer miconia and coqui frog volunteer work trips for schools and hiking clubs of approximately twelve people (MISC 2006).

6. In my conversations and interviews with various biodiversity professionals and leaders of environmental organizations, a recurring and common theme that emerged was the difficulty of keeping policymakers' attention on biodiversity/alien-species issues as well as building and maintaining public awareness. Poor relations and mistrust between, on the one hand, government agencies and mainstream environmental groups and, on the other, the native Hawaiian community has also limited cooperation and community participation in some instances.

References

Bisignani, J. D. 1999. *Hawaii Handbook*. Chico, CA: Moon.

Conrow, Joan. 2006. "Snakes on a Plane." *Honolulu Weekly*, August 23. http://honoluluweekly.com/cover/2006/08/snakes-on-a-plane-2.

Coordinating Group on Alien Pest Species (CGAPS).

Christopher Jasparro

1996. *The Silent Invasion.* Honolulu: Coordinating Group on Alien Pest Species.

———. 2006. *Tracking Awareness of and Attitudes toward Invasive Species in Hawaii: 2006 Telephone Survey of Statewide Residents.* http://www.hear.org/presentations/cgaps 20060531omnibusreport/index_files/frame.htm.

Dalton, Russell. 1994. *The Green Rainbow: Environmental Groups in Western Europe.* New Haven, CT: Yale University Press.

Drew, Joshua A., and Adam P. Henne. 2006. "Conservation Biology and Traditional Ecological Knowledge: Integrating Academic Disciplines for Better Conservation Practice." *Ecology and Society* 11, no. 2:34–42. http://www.ecologyandsociety.org/vol11/iss2/art34.

Hattam, Jennifer. 1999. "Alien Species Repelled." *Sierra* 84, no. 3:73.

"Hawaii—Rich in Endemic Biodiversity and the U.S. Region Most Susceptible to Biological Invasions." 2006. http://www.hawaiiconservation.org/_library/documents/hispecialprotectionneeds.pdf.

Jasparro, Christopher, and Eric Shibuya. 2002. "Environmental Security and Ingenuity in the Pacific." *Regional Development Dialogue* 23, no. 1:1–17.

Kahng, Sam. 2005. "A Silent Invasion Threatens to Overrun Pristine Black Coral Beds and Alter Hawaii's Deep Reef Community." NOAA's National Centers for Coastal Ocean Science, December 14. http://www.cop.noaa.gov/ecosystems/coralreefs/features/fs-2005-12-12-cr.html.

LaPierre, Lance. 2001. "Fencing out Threats on Oahu." *Nature Conservancy in Hawaii Newsletter* 24, no. 2:4.

Leone, Diane. 2004. "Alien Species Battle Gets $3 Million Boost." *Honolulu Star-Bulletin,* July 1.

Loope, Lloyd. 2003. "Needs and Possibilities for Better Quarantine to Protect Biodiversity in Hawaii: NZ as a Model." Presentation to Native Hawaiian Plant Society, June 6.

Loope, Lloyd, and Joan Canfield. 2000. "Hawaii: A Model for Addressing Invasive Species." *People, Land, and Water* (U.S. Department of the Interior), July/August, 29.

Maui Invasive Species Committee (MISC). 2002. *MISC 2002 Action Plan.* Maui: Maui Invasive Species Committee.

———. 2006. *Ki'ai Na Moku o Maui Nu* (Maui Invasive Species Committee newsletter). Fall 2006.

Meindel, C., D. Alderman, and P. Waylen. 2002. "On the Importance of Environmental Claims Making: The Role of James O. Wright in Promoting Drainage of Florida's Everglades in the Early 20th Century." *Annals of the American Association of Geographers* 92, no. 4:682–701.

Mlot, Christine. 1995. "In Hawaii, Taking Inventory of a Biological Hot Spot." *Science* 269, no. 5222:322–24.

The Nature Conservancy of Hawaii (TNCH)/National Resources Defense Council (NRDC). 1992. *The Alien Pests Species Invasion in Hawaii: Background Study and Recommendations for Interagency Planning.* Honolulu: Nature Conservancy of Hawaii and National Resources Defense Council.

Oahu Invasive Species Committee (OISC). 2002. *Strategic Action Plan, 2002–2003.* Honolulu: Oahu Invasive Species Committee.

———. 2006. *Strategic Action Plan, 2004–2005.* Honolulu: Oahu Invasive Species Committee.

Ross, Andrew. 1994. "Weather Report (Ecoculture in Hawaii)." *Artforum International Magazine,* March 1. http://www.encyclopedia.com/doc/1G1-15329473.html.

Stone, Richard. 1999. "A Plan to Save Hawaii's Threatened Biodiversity." *Science* 285, no. 5429:817–19.

Woodcock, Deborah. 2003. "To Restore the Watersheds: Early 20th Century Tree Planting in Hawai'i." *Annals of the American Association of Geographers* 93, no. 3:322–24.

Contributors

STANLEY D. BRUNN is professor of geography at the University of Kentucky. He is the author of several books and articles on cyber geography, technology, and urban geography.

DAN CAREY is a hydrologist with the Kentucky Geological Survey, University of Kentucky. His research interests are in geographic information systems and environmental systems, and he has authored land-use planning maps for Kentucky counties.

NATHALIE CAVASIN is a geographer with research interests in high-technology industries in Japan. She is currently a visiting researcher at Waseda University, Tokyo.

RICHARD FORREST is a private environmental consultant in Washington, DC, doing work with organizations including the National Wildlife Federation and the World Bank.

AKIKO IKEGUCHI is a geographer on the faculty of Yokohama National University, Japan. Her research interests are in the environment and development geography, and her current research is centered in Laos.

CHRISTOPHER JASPARRO is associate professor of national security affairs at the U.S. Marine Corps Command and Staff College, Quantico, VA. He has published several articles on environmental geography of Asian countries.

PRADYUMNA P. KARAN is professor of geography/Japan studies at the University of Ken-

tucky. He has published several books on the environment, development, and Japanese landscapes.

SHINJI KAWAI is a social science researcher at Stahlbush Island Farms, Inc., an environmentally friendly farm in Oregon's Willamette Valley. His research interests are in the environment and community-development programs.

JOHN J. METZ is professor of geography at Northern Kentucky University. He has authored several papers on environmental geography and social and political geography.

YASUKO NAKAMURA is a geographer on the faculty of Tokyo Gakugei University, Tokyo, with major research interests in agricultural geography and geographic information systems.

KENICHI NONAKA is on the faculty of Rikkyo University, Tokyo. Formerly, he was a senior research scientist at the Research Institute for Humanities and Nature, Kyoto, Japan. He has published many papers on environment and cultural geography.

KOHEI OKAMOTO is professor of geography at Nagoya University, Nagoya, Japan. His research interest is in economic and urban geography. He is the author of several books and papers on urban behavioral geography.

RACHEL PENROD is a program assistant with the Association of Zoos and Aquariums. She is responsible for administering Conservation Endowment Fund grants, maintaining and updating the Conservation Programs Database, and publishing the monthly *Conservation and Science Report*.

KIM REIMANN is on the political science faculty of Georgia State University, Atlanta. Her research interest is in the environmental politics of Japan. She is the author of several papers on environmental policy.

SATORU SATO is a sociologist on the faculty of Akita Prefectural University.

MIRANDA SCHREURS is the director of the Environment Policy Research Center and a professor of comparative politics at the Free University of Berlin. Formerly, she was a professor of political science at the University of Maryland. She is the author of *Environmental Politics in Japan, Germany, and the United States*.

TODD STRADFORD is professor of geography at the University of Wisconsin, Platteville. His major research interest is in the economic and human geography of Japan.

UNRYU SUGANUMA currently serves as associate professor at J. F. Obirin University and adjunct professor of international affairs at Temple University, Japan Campus, where he specializes in Sino-Japanese relations and the political geography of East Asian and Chinese politics. He is the author of *Sovereign Rights and Territorial Space in Sino-Japanese Relations*.

YOSHIMITSU TANIGUCHI is professor of sociology at Akita Prefectural University. His major interest is in eco-friendly farming systems and sustainable agriculture.

MASAO TAO is professor of economics at the Graduate School of Government, Kyoto Uni-

versity. He is the author of several papers on economics and environmental aspects of local government policy.

JONATHAN TAYLOR is associate professor of geography at California State University, Fullerton. His research interest is on environmental policy in Japan. He is the author of several research papers on environmental issues in Okinawa.

TEIJI WATANABE is associate professor in the Faculty of Environmental Earth Science and Graduate School of Environmental Science at Hokkaido University, Sapporo, Japan.

NORITAKA YAGASAKI is professor of geography at Tokyo Gakugei University, Tokyo. His major interest is in cultural geography, land use, and immigration. He is the author of several books and research papers.

KENJI YAMAZAKI is on the faculty of Iwate University, Morioka, Japan. He is the author of many papers and books on geographies of the environment as well as in broad fields of economic and cultural geography.

TOMOKO YAMAZAKI is professor in the Faculty of Education at Iwate University, Morioka, Japan. She has published many papers on environmental and educational problems.

DAVID ZURICK is a geographer living in the Kentucky foothills near the Bluegrass Army Depot Ordnance. He specializes in studies of the Himalaya and is the author of several books, including *Himalaya: Life on the Edge of the World* (1999, with Pradyumna P. Karan) and *Illustrated Atlas of the Himalaya* (2006, with Julsun Pacheco).

Index

Agricultural Fair (Nerima Ward, Tokyo), 138

Agricultural Land Act (1952), 134, 142

Aichi Prefecture: alternative waste disposal site, 240, 241

Ainu, 18

Air Pollution Control Act (1955), 17

Akita Prefectural Agricultural College, 165, 169, 171

Akita Prefectural Agricultural Experiment Station, 169

Akita Prefectural University, 171

Akita Prefecture, 9

Alliance for Nuclear Accountability, 93

American environmentalism, 5; key schools of thought in, 15; rise of, 14

American Farmland Trust (AFT), 146

American Fisheries Society (1870), 15

American Forestry Association (1875), 15

American Friends Service Committee, 93

American Ornithologists Union (1883), 15

Anniston, Alabama, 9, 120

anti–golf course movement, 170

Anti–Itai-Itai Disease Council of Toyama Prefecture, 31

antinuclear movements: types of in Japan, 65, 67–68

antipollution movement in Japan, 20

architectural heritage, 10

Ashio copper mine, 20–21, 187

Ashio struggle, 22

Asia Wetland Symposium, 238

Association to Protect the Yaeyama-Shiraho Coral Reef (APYSCR), 247–51

Association to Save the Children from Dioxin, 240

Association to Walk in the Nature of Yanbaru, 276

asthma. See *zensoku*

Audubon, John James, 14

Audubon Society, 16

bakufu, 18

Battle of Okinawa, 271

Berea, Kentucky, 118–19

Besshi copper mine, 21

biodiversity: loss of in Hawaii, 282; preservation of tidal flats, 229; use of geographic and information technology in protection of, 287

Biodiversity Network Japan, 46

Bluegrass Army Depot, 9, 111, 114, 116–19, 125

Bluegrass Conservancy, 146, 150, 152

Bluegrass horse farms, 9

Bluegrass landscape, 145, 150, 151

Bluegrass Rails to Trails Foundation, 154–55

Bluegrass region of Kentucky, 145; local groups to save open space in, 151; loss of farmland in, 146, 149; most-endangered cultural site designation, 149; subdivisions around towns, 149; urban planning in, 150

Bluegrass Tomorrow, 150, 151–52

Bluegrass Trust for Historic Preservation, 150, 153

Boone and Crockett Club (1885), 15

Buddhism, 22

Bunkyo Ward, Tokyo, 133

Bureau of Reclamation, 15

Carson, Rachel, 24

Catlin, George, 14

Central Kentucky, founding families of, 121

Chemical Stockpile Disposal Program, 111

Chemical Stockpile Emergency Preparedness Program, 117

Chiba Prefecture, 10, 32, 33, 187, 189, 200, 202
Chiba Prefectural Environmental Council, 200–201
Chisso Corporation, 30
Chiyoda Ward, Tokyo, 133
Chubu Recycle Action Network, 239, 240
Chuo Ward, Tokyo, 133
citizen advisory boards, 94, 102; at DOE facilities, 93; boards that functioned well, 95
citizen involvement, 8, 34, 75, 76, 94, 102; in chemical weapons safe disposal, 124; in nuclear waste cleanup process, 90; in preservation of the Bluegrass region, 151; results in Bluegrass region, 163–64
Citizens' Alliance for Saving the Atmosphere and the Earth, 67
Citizens' Clearinghouse for Hazardous Waste, 123
Citizens for Saving the Kawabe, 211, 217–18
Citizens' Nuclear Information Center, 67–68, 72
Climate Action Network, 34
Colorado Department of Public Health and Environment, 96
Common Ground (1987), 113, 119, 122
Communist Party, 67
Community Garden Promotion Act (1990), 142
community gardens, 140–41
Concerned Citizens of Madison County (1984), 113, 118–19, 122
conservation easement, 161
Consumers Insurance Corporation, 212
Control of Wildlife Conservation Law, 220
Cuyahoga River fire, 24

daimyo, 22
Daisetsuzan National Park, 259
deforestation, 18, 19, 20, 30, 31, 32; in Hawaii, 282; in Kinkai Basin, 18; in national parks, 261, 263; under the Tokugawa bakufu, 18
dekasegi, 207
development and preservation interests, 10; in Kyoto, 178–81
Domoto Akiko, 33, 196, 202

donated easements, 163
Dugong Network Okinawa, 275, 276

Eastern Kentucky University, 118
Edo, 19
Edo period (1603–1867), 19, 20, 187
Ehime Prefecture, 21
Ehrlich, Paul, 24
Emerson, Ralph Waldo, 14
Endangered Species Act (1966), 24, 27
Environment Basic Law, 236
Environmental Agency, 232, 241
Environment Assessment Law, 231
Environmental Defense Fund (1967), 24–25
environmental groups, 3, 5, 9, 16, 23, 24, 25, 26, 27, 28, 33, 34, 35, 39, 46, 56, 90, 92, 93, 120, 151, 164, 275, 276, 279, 281, 284, 292
environmental movements, 23, 45, 118, 120–21; differences between Japan and the United States, 5, 13, 65; influence of sociocultural and political factors on, 14; leadership, 11–12; in Ogata, 169; in Okinawa, 11; to preserve tidal flats, 229; role of National Geographic Society in, 7; shifting tactics in Bluegrass, 123–24; silenced by Peace Preservation Law, 22
environmental politics, 5, 17, 19, 48, 49, 72, 73, 138, 296
environmental risk, 115, 127, 185

Farm and Ranch Lands Protection Program, 102
farmland loss, 9
Fernald Citizen Advisory Board, 85, 90, 101
Fernald Feed Materials Plant (nuclear facility), 84, 98, 100–102, 104; citizen advisory board, 95; contamination of drinking water by, 86
Fernald Residents for Environmental Safety and Health (FRESH), 85, 104
First National Comprehensive Plan, 232
Forum on Environmental Administration Reform, 238, 239
Friends of the Earth: –Japan, 47, 56, 60; –U.S., 25, 34, 53
Fujimae Neighborhood Association, 235, 240
Fujimae Tidal Flat, 10, 33, 202, 229, 232; evaluation of conservation movement

in, 240–41; media attention to, 238; movement to preserve, 234–40
Fukugawa, 189
Fukui Prefecture, 7, 65
Fukuoka City, 205, 212
Fukushima Prefecture, 69
Funabashi, 10, 187, 201
Futenma Air Station, 11, 271; relocation of, 273, 276, 278

gaiatsu, 50
Gandhi, Mohandas Karamchand, 14
Global Network for Anti–Golf Course Action, 46
global warming, 9, 42, 46
grassroots groups, 75, 90, 131, 185, 239, 281; to influence DOE, 92–93; at various nuclear sites, 93
grassroots movements, 4, 118, 120–21; for biodiversity protection in Hawaii, 291; fishermen in, 232; involvement of women in, 170; in Kyoto, 177, 182, 184–85; in national parks, 261–65; in Nerima Ward, Tokyo, 137; in Ogata, 165, 173; to save Sanbanze tidelands, 196–202; for urban farming, 139
Great Hanshin Earthquake of 1995, 69
Greenpeace: -Japan, 34, 47, 67; -USA, 15, 26, 123, 250
Green Movement, 4

Hakata Bay, 238
Hamburg Place Farm, 145, 150
Hamaoka, 68
Hanford, Washington, 77, 84, 96, 102, 104; citizen advisory board, 90, 95; uncontrolled dumping of waste at, 89
Harumi reclaimed land, 189
Hawaii, 11; biodiversity protection in, 281; grassroots organizations in biodiversity protection, 288–90; U.S. Army's role in natural resource management in, 286
Hawaii Sugar Planters Association, 283
Heian (later Kyoto), 18
Henoko, 276, 278–79
Hirosaki, 18
Hiroshima, 82, 211, 212, 215, 232, 271
Hiroshima Forestry School, 211
Hitachi copper mine, 21
Hitoyoshi City, 207, 210–11

Historical Spot, Scenic Beauty, and National Monument Preservation Law (1918), 23
Hokkaido, 18, 169, 207, 259, 260
Hokkaido University, 263, 266, 296
Hosogochi Dam, 33
Hull House, 17

Ibaraki Action Coalition against Nuclear Power, 71
Ibaraki Prefecture, 8, 21, 65, 71
Ichikawa, 10, 187, 201
Idaho National Engineering Lab, 77, 101; citizen advisory board, 95
Ikata nuclear power plant, 71
Ikushima Toru, 246
Imperial Game Law, 22
Institute for Energy and Environmental Research (IEER), 93
International Comprehensive Ban on Chemical Weapons, 111
International Council for Local Environmental Initiatives, 28
International NGOs, 49; links with local and national, 59
International Union for the Conservation of Nature, 249–50
iriaichi, 18
Isahaya Bay, 33, 45, 51, 54–59, 216, 232, 237
Isahaya Bay Land Reclamation Project, 45, 51, 54–59, 232, 237
Ise-Mikawa Bay, 242
Ishigaki Island, 33, 245, 246, 248, 250; airport construction, 251–56
Itai-Itai disease, 31
Itsuki village, 207
Iwadare Sukio, 201
Izumi village, 207

Japan Atomic Power Company, 71
Japan Environment Association, 171
Japan Environmental Lawyers Federation, 211
Japan Federation of Bar Associations, 67
Japan Scientists Association, 67
Japan Wetlands Action Network, 46, 55, 237
Japanese environmental NGOs, 34–35
Japanese movements, 5, 7
JCO Corporation, 68–69

Jinzu River, 29, 31
Johnston Atoll, 9, 116, 120

Kaburagi Tokuji, 21
Kadena Air Base, 271
kami, 22
Kamogawa, 182
Kansas City plant, 85
Kaneyuki Nakane, 211
Kawabe River, 10, 207, 212; white-water
    rafting in, 210
Kawabe River Dam, 211; expert opinions
    on, 215; farmers' opposition to, 209;
    supporters of, 215–16
Kawabe River Study Society, 218
Kawasaki, 188
Kawasaki Reclamation Project, 189
Keihin Industrial Zone, 189
Keiyo industrial complex, 10
Kentucky Department of Environmental
    Protection, 96
Kentucky Environmental Foundation (KEF),
    113–14, 123, 125
Kentucky Horse Park, 149
Kentucky Rails to Trails Council (KRTC),
    154
Kiko Forum, 46
King, Martin Luther, Jr., 14
Kitakyushu Environment Bureau, 29
Kitakyushu, pollution in, 29
kogai, 21, 22, 29
kombinato, 31
Kosaka copper mine, 22
Kumagawa Shiranui Sea Research Group,
    218
Kumagusu Minakata, 23
Kuma River, 207, 211
Kuma River Fishing Cooperative, 209
Kumamoto Prefecture, 29, 207
Kumazawa Banzan, 19
Kunitoku Yasuyo, 211
Kushiro, 56, 237, 238
Kyoto, 10, 18, 177; Business Association,
    184; City Hall, 181, 183; conflict
    between preservation and development,
    178–81; Kyoto Protocol, 28; Station
    Building, 181, 183; Tower Hotel, 183

Lacey Act (1900), 16
Lake Biwa, 32

Lake Hachiro, 170
Land and Nature Trust, 146
Land and Water Conservation Fund Act
    (1964), 24
Land Earth Outdoor Sports Club, 210
Land Evaluation and Site Assessment, 159
landscapes, 3, 5, 7, 9, 10, 13, 29, 41, 129,
    153, 259
Law for the Promotion of Utilization of
    Recycled Resources, 240
Leopold, Aldo, 15, 17
Lexington/Fayette County, Kentucky, 111,
    118, 145, 146, 149, 152, 154; planning
    commission, 164; urban service area
    concept, 155
Liberal Democratic Party, 28, 240, 251
local activism, 3, 59, 65, 119
local environment, 35, 46, 59, 92, 126, 164,
    229, 238, 284
local governments, 3, 11, 29, 31, 142, 154,
    155, 162, 165, 177, 179, 216, 231, 236,
    239, 264, 265
local movements, 3, 7, 35; connections with
    global movements, 47; influence of, 12; in
    Japan, 35, 47; Okinawan, interface with
    national, 277
Los Alamos, New Mexico, 95, 96, 97, 102,
    103, 104
Love Canal, 25

Madison County, Kentucky, 117–18
Mallinckrodt Corporation, 98–99
Manhattan Project, 77, 82, 84
Marsh, George Perkins, 15
Marshall, Robert, 17
Matsumoto Han, 18
Meiji Era (1868–1912), 13, 188, 230
Mie Forestry Union, 283
Mie Prefecture, 10, 221, 224
Mihama nuclear power plant, 65
military activity, 5; direct environmental
    impacts of, 273–75; U.S. Army assistance
    with biodiversity protection, 283
mimaikin, 20
Minamata, 29, 187, 212
Minamata Disease Patients' Families'
    Mutual Aid Society (1956), 30
Minato Ward, Tokyo, 133
Mitsui Real Estate Company, 189
Miyagi Prefecture, 3, 274

Monju reactor, 7, 65
monkeys in Japan, 10, 219; GIS map, 223; integrated network to control damage by, 226; protecting crops from, 222; radio tracking of, 223; SIS display of distribution, 224. See also *saru-mawashi*
Morse, Edward, 19
Mound Plant, Ohio, 85
Mountain Hawk Preservation group, 218
Muir, John, 15, 16
mura hachibu, 208

Nader, Ralph, 24
Nagara River Estuary Dam, 33, 45, 51, 59, 238; link to Narmada Dam campaign, 53; participation of international actors in, 53
Nagasaki, 33, 55, 82, 271
Nago Citizen's Network, 276
Nagoya, 234, 239
Nagoya Port Tidal Flat Association, 235
National Academy of Sciences, 89, 115, 123
National Conservation Commission (1908), 16
National Environmental Policy Act (1969), 25, 90, 122
National Federation of Women's Clubs, 16
National Federation of Workers, 212
National Forestry Commission (1873), 15
National Geographic, 39–43
National Geographic Education Foundation, 43
National Parks Association (1913), 23
national parks in Japan, 23, 259; management style in, 266; problems facing, 261
National Parks Law (1931), 23
National Parks Law (2002), 259
National Parks Law of Japan (1957), 29
National Parks Service (1916), 16
National Resource Council, 122
National Resources Defense Council (NRDC), 93
National Resources Defense Fund (1970), 24
National Toxics Campaign, 123
National Wildlife Federation (1936), 16
Nature Conservancy (1951), 24; Kentucky chapter, 153
Nature Conservancy of Hawaii, 284

Nature Conservancy Society of Japan, 28–29, 47, 56, 200, 277
Nature Restoration Act (2000), 232
Nerima Ward, Tokyo, 133, 135, 136, 137, 139, 144; aging farm population, 143; community gardens in, 140–42; farmland in, 135; local farmers' cooperatives, 140; Urban Farming Division, 138
Nevada Test Site, 83, 99, 101
New England Climate Action Coalition, 28
New Food Act, 168
Newport Army Ammunition Plant, Indiana, 120
New Tokyo International Airport, 32
nihon-zaru [Japanese macaques]. *See* monkeys in Japan
Niigata, 29
Nikko, 23, 29
Nikko National Park, 28
Nonprofit Organization Law (1998), 34
nonviolent protests, 4
no-till rice farming, 169
not-in-my-backyard (NIMBY) movement, 45, 46
Nuclear Non-Proliferation Treaty, 103
nuclear power in Japan, 65
nuclear radiation, 5
Nuclear Waste Management Organization of Japan, 68

Oahu Sierra Club, 285
Oak Ridge Environmental Peace Alliance, 93, 98
Oak Ridge plant, Tennessee, 82, 84, 85, 95, 98, 102; dumping of wastes into rivers at, 89; leaked mercury at, 90
Ogata Environment Creation 21 (OEC21), 165, 169, 171–73
Ogata farmers' protests, 166
Ogata Low-Input Sustainable Agriculture, 169
Ogata Reclamation Office, 166
Ogata Village, 165–73; founded in, 172; organic pesticide use, 168; pesticide residues in drinking water in, 170; reclaimed land in, 165; women's group in, 170
Okinawa, 256, 277, 278, 279; damage to natural environment of, 271; environmental groups in, 275–76;

environmental problems in, 275; military
bases in, 271–72; U.S. land confiscation
policy in, 272–73
Okinawa Protection Fund Committee, 276
Okinawa University, 256, 276
Okinawa-Yaeyama-Shiraho Association for
the Protection of Sea and Life, 275
Olmsted, Frederick Law, 15
Open Forum for Citizens on Environmental
Disruption, 31
organic farming, 9–10, 165. See also *teikei*
Organic Farming Study Group, 170
organic rice, 165
Osborne, Fairfield, 24
oshidashi, 20
Ota Masahide, 251, 273
Owari domain, 18
Oze Marsh Conservation Union, 28

PACE Corporation, 161–62
Paducah, Kentucky, 82, 84, 95, 96, 102
Pantex plant, Texas, 85, 95, 96, 102
PDR. *See* Purchase of Development Rights
Peace Preservation Law (1925), 22
People's Network against Construction
and Strengthening of Military Base
(Okinawa), 275
People's Research Institute on Energy and
Environment, 67
Perry, Commodore Mathew, 20
Physicians for Social Responsibility, 93
Pinochet, Gifford, 15, 16
Pine Bluff Arsenal, 120
Pinellas plant, Florida, 85
political systems in Japan and America, 11;
authoritarian and democratic tradition, 20
Portsmouth, Ohio, 82, 84
preservation tools, 9; for farmland and open
space, 155
Private School for Green and Agriculture,
136
Public Information Disclosure Law (1999) of
Japan, 34
Pueblo Depot, Colorado, 120
Purchase of Development Rights (PDR)
program, 156–59

Ramsar Convention, 56, 57, 58, 200, 237, 238
resource management concepts: influence of
Confucian scholars on, 19

Resources for the Future (1952), 24
Richmond, Kentucky, 118
ringi proposal system, 181–82
river management policy, 10
Rivers Watch East and Southeast Asia,
212–13
Rocky Flats plant, 84, 85, 87–88, 89, 96,
102, 103, 104
Rocky Mountain Peace and Justice Center,
Boulder, 93
Rokkasho, 8
Roosevelt, Theodore, 16
Rural Land Management Board, 150,
152–53, 156
Rural Service Area Land Management Plan,
155–56
Ryukyu University, 247–56

Sagara Village, 207
Sanbanze Forum Urayasu, 200–202
Sanbanze Land Reclamation, 10, 33; impact
on fishing, 192–96; movement to save,
196–202
Sanbanze Study Circle, 200–201
Santa Barbara oil slick, 24
Sardar Sarovar Dam Project, 53; linkage to
Narmada dam campaign, 53
saru-mawashi [monkey performances], 220
Savannah River Site (SRS), 77, 82
Save Isahaya Bay, 58
Save Our Health and Environment! Aichi
Residents' Action, 239
Save the Dugong Foundation, 276
Save the Fujimae Association, 235, 238, 239,
240–41
Save the Ozone Network, 46
Seikatsu Club Consumers' Cooperative
Union, 32
Seki Tenshu, 21
Senior Citizens' Club, 139
shaku-ayu, 209
Shibuya Ward, Tokyo, 133
Shiga Prefecture, 32, 211
Shimojima Tetsuro, 246
Shinjuku Ward, Tokyo, 133
Shinto, 22; state Shinto, 23
Shiraho Coral Reef, 11, 33, 245–46
Shizuoka Prefecture, 68
shufu [homemaker], 69
Sierra Club (1892), 15, 16, 23, 24, 53

Social Democratic Party (Japan), 240
socialist movements, 22
social movements, 4, 65
Society against Nagara River Estuary Dam
    Construction, 52, 53
Society for Monkeys, 223, 226
Society for Zero Emission Nagoya, 240
Soil Conservation Society of America (1944),
    24
Special Farmland Loan Act (1989), 142
Staple Food Control Act, 168
Sumida River, 188
Sumitomo Corporation, 21
Sumitomo Metal Mining, 69

Taisho Era (1912–1926), 22
Takagi Jinzaburo, 72
Tanaka shozo, 19, 20
Tanaka Yuko, 18
teikei food-distribution system, 32
Thoreau, Henry David, 14
tidal flats, 230; Awase, 276; conservation
    groups from, 237; creation of artificial,
    232; decrease in area, 231–32; feeding
    area of migratory birds, 235; Fujimae
    reclamation project, 234–35
Tohoku Electric Corporation, 67
Tokai village, 8, 65, 68, 69, 70, 71
Tokugawa Ieyasu, 23
Tokushima Prefecture, 33
Tokyo: environmental preservation, 144;
    local food production, 144; urban
    livability, 144
Tokyo Bay: fishermen in, 232; large scale
    reclamation of, 189; pollution by Asano
    Cement Corporation, 189; reclaimed
    area, 192
Tokyo Bay International Cheer Group, 200
Tokyo Disneyland, 232
Tokyo Electric Power Company, 65
tomeyama, 18
Tooele Army Depot, Utah, 9, 111, 120,
    122–23
Toshima Ward, Tokyo, 133
Town Branch Trail (Lexington, Kentucky),
    154
Town Planning and Zoning Act (1968), 133,
    134
Toyama Prefecture, 29
Tsukuda, 188

Tsuruga, 7
Tsurumi Reclamation Association, 189

Ui Jun, 31, 276
Umatilla Depot, Oregon, 120
UN Conference on Environment and
    Development (UNCED), 49, 52–53, 58
uranium industry, 83–64
Urayasu, 10, 187, 201, 232
urban farming, 133, 134–39
U.S. Department of Defense (DOD), 89,
    115, 119
U.S. Department of Energy (DOE), 8
U.S. Forest Service (1905), 15
U.S. Public Health Service, 17

Wajiro Tidal Flat, 33
Watarase River, 20
Water Conservation Act (1958), 29
water pollution, 29
Wetlands Convention, 56–57
Wild and Scenic Rivers Act (1968), 24
Wild Bird Society of Japan (1934), 23, 47,
    56, 58, 202, 237
Wilderness Act (1964), 24
wilderness preservation, 15
Wilderness Society (1935), 16–17, 23
World Equestrian Games, 149
World Monument Fund, 149
World Wildlife Fund: –Japan, 34, 47, 58,
    200, 202, 212, 238; –USA, 24

Yamaga Soko, 19
Yamashita Hirofumi, 55
Yellowstone National Park, 16
Yodo River, 18
Yokkaichi, 29, 31, 235
Yukawa Reiko, 201

zensoku [asthma], 31